Strength & Honor:
America's Best in Vietnam

by Terry L. Garlock

Dear Judy,
I hope you learn alot and are happy with the content of so many vets. It was my honor & privellege to serve with great men.

"Strength & Honor: America's Best in Vietnam," by Terry L. Garlock. ISBN 978-1-60264-715-2.

Published 2011 by Virtualbookworm.com Publishing Inc., P.O. Box 9949, College Station, TX 77842, US. ©2011, Terry L. Garlock. All rights reserved. No part of this publication may be reproduced, stored in a retrieval system, or transmitted in any form or by any means, electronic, mechanical, recording or otherwise, without the prior written permission of Terry L. Garlock.

Manufactured in the United States of America.

Table of Contents

Preface 1
Foreword 4

The Veterans

(M)	R.J. Del Vecchio	One Bad Day	11
(N)	J. A. Sagerholm		18
(A)	Norm McDonald	Grunt Melody	19
(M)	Jay Standish		46
(N)	David Wallace	Take Another Little Piece of My Heart	47
(A)	Sig Bloom		54
(A)	Dexter Lehtinen	Honor and Pride	55
(A)	Bob Babcock		70
(A)	B.G. Burkett	Everybody Knows . . .	71
(A)	David Walker		83
(A)	Ron Current	Helicopter Pilot Factory	84
(A)	Jim Oram		102
(A)	Ron Current	So Others May Live	103
(N)	Drew Johnson		118
(A)	Larry Hogan	LRRPs and Small Men	119
(A)	Gary Linderer		136
(A)	Andy Burleigh	Magnet Ass	137
(M)	David Crawley		146
(N)	Alex Nides	Shipmates Forever	147
(A)	Frank Pratt		159
(A)	Terry Garlock	The Snake and the Ox	160
(A)	Dan Britt		184
(A)	Bill Neal	The Burden of Angry Skipper 6	185
(A)	Ted Reid		207
(N)	Larry Bailey	Thank You Mrs. Gussye Roughton	208
(A)	Roger Soiset		213
(A)	Nick Donvito	Balls of Steel	214
(A)	Dave Pearsall		239
(A)	Tony Armstrong	In the Valley of Death	240
(A)	John Galt		255
(AF)	Nick Stevens	Charlie's Elephant Walk	256
(DW)	Laura Armstrong		310
(M)	Donald Johnson	Ghosts of Torment, Ghosts of Love	311
(A)	Michael Leonardos		323
(A)	Donna Rowe	Just One Life	324
(A)	Bill Stanley		335
(A)	Wayne King	Robin Hood & Brotherhood	336
(AF)	William Haynes		350

(M)	James Warner	To Return With Honor	351
(A)	Bob McFadden		371
(N)	John O'Neill	A Story Never Told	372
(N)	Mike Minor		377
(A)	Steve Crimm	Sideways	378
(A)	Gary Noller		384
(A)	Alan Walsh	Going Home	385
(A)	Carl Bell		399
(A)	Claude Newby	Benediction	400

Epilogue ... 413

Appendix

The Good, the Bad and the Ugly ... 417
Glossary ... 446

Table of Contents Legend:
- (A) US Army
- (AF) US Air Force
- (M) US Marine Corps
- (N) US Navy
- (DW) Daughter, Wife of vet

Preface

THERE ARE A FEW things readers should know about this book.

The Vietnam veterans I know are proud of their service, as they should be. They served their country well, and yet America believed the worst about them as the public's patience grew thin over ten years of war while anti-war activists portrayed the war as immoral, our troops as rogues or victims. I promised myself I would teach my children the truth about them.

There was plenty wrong with the war, but the real defects seemed to get lost in the arguments. Many vets, like me, wanted nothing to do with the political struggle, and we put the war behind us to get on with life, keeping our stories to ourselves.

Over the years Hollywood movies fueled misgivings about vets by characterizing our troops in Vietnam as inept or depraved. TV dramas would introduce a Vietnam vet when they needed a deeply troubled character, or an unemployed and homeless man, perhaps suicidal or liable to snap into violence at any moment. Writers followed suit, applying the same dysfunctional vet stereotypes in the period's fiction. A stridently anti-war academia decided what to write in our children's schoolbooks about the war.

Is it any wonder that so much passing for common knowledge about these vets is simply not true? Is it any wonder so many of them kept their service and their story to themselves for a long time to avoid conflict?

A few years ago I responded in anger to a man who said something belittling about the service of Vietnam vets. After I cooled off, I realized the real problem was our story had not been properly told to counter the false stereotype. Even at this late date, if we don't tell it true, who will? It was long past time I kept my promise to myself, and so I started to work on this book.

I decided the best way would be to help a number of vets tell their story to you in their own way, and let you decide for yourself what to believe. I selected vets who were ordinary people, not larger-than-life heroes or generals. I sought the points of view from a variety of wartime jobs representing the four branches of our armed forces. Each chapter is a different vet telling a piece of their story.

I wrote most of the vet chapters from long interviews, piecing together the story from out-of-sequence facts and memories, going back to them again and again to get it right. Not everyone is a writer and I was striving to help them tell their story in their own voice, using their language, their feelings, keeping even their own profanity and derogatory nicknames for our enemy to keep it real. A couple who like to write gave me drafted chapters or bits and pieces as a starting point and I rewrote and edited. Some chapters are long, others quite short, and in between each chapter is a short "postcard" from another vet. Each story is a little different just as the vets themselves are different. Each vet reviewed carefully to keep it correct, whether a long chapter or short postcard.

With this mixed bag, I decided the best way to be faithful to the vets and their tone was keeping the first person tense in these stories even though it is unorthodox.

Helping some fellow vets tell their story now, in the autumn of their life, is an honor for me. Every story came from the heart and I thank them for trusting me with their memories and sharp-edged feelings, some hidden away for a long time but still as fresh as yesterday when unwrapped.

Joe Galloway and B.G. Burkett threw in their lot with me to make the case that Vietnam vets deserved far more respect than they received. Mary Yost, an energetic retired English teacher, reduced my mangling of the English language with her editing; the errors that remain are, of course, my own. Dr. Alex Nides, whose story is in this book, persistently challenged me to find better words to convey an idea. Bill Neal, whose story is also in this book, advised on some matters where my knowledge was thin. My good friend, Tom Ward, read each piece as my sounding board. I am grateful to each one for their indispensable help.

Two more things for readers. First, this is a long book, but each chapter stands alone and you can read them at your own pace. Second, if I could reach out from these pages and forcefully drag you to the very best starting point, I would have you begin by reading two chapters that are a bit of medicine: they may not taste good but are likely good for you.

The first is background about the Vietnam War, which may be the most misunderstood event in American history. This chapter is in the Appendix, titled *The Good, the Bad and the Ugly*, in which I tell you what *I believe* is the real big-picture story of the war, warts and all. Because my purpose in this book is to help vets tell their story to a public that still believes myths and half-truths about them, it is the most important chapter in the book in my judgment. Even so, I placed this long chapter in the Appendix, far out of your way, because readers would rather be entertained than get down to work. I hope you read it now, before starting on the veteran stories.

The second chapter I would have you read is titled *Everybody Knows*, B.G. "Jug" Burkett's tale of America believing the disparaging stereotypes of Vietnam veterans for decades.

I can't make you read those two chapters first, but if you do, you will better understand the stories of the vets and you may more readily recognize the debt of gratitude America still owes to the men and women our country sent to Vietnam.

"It ain't what you don't know that gets you into trouble. It's what you know for sure that just ain't so."

Mark Twain

Foreword
by Joseph L. Galloway

IF YOU HAVE FED from a steady diet of Hollywood movies about Vietnam you probably believe that everyone who wore a uniform in America's long, sad involvement in that war is some sort of a clone of Lt. William Calley, that all three million of them were drug-crazed killers and rapists who rampaged across the pastoral landscape. Those movies always got it wrong, until *We Were Soldiers*, a 2002 Hollywood film that finally got it right. Ask any Vietnam veteran who has seen the movie.

The movie is based on a book that I and my lifelong friend, Lt. Gen. (ret) Hal Moore, wrote together, *We Were Soldiers Once, and Young*. We wrote that book precisely for the same reason Terry Garlock wrote this book, because we believed a false impression of those soldiers had taken root in the country which sent them to war and, in the end, turned its back on both the war and the warriors.

I did four tours in Vietnam as a war correspondent for United Press International: 1965-66, 1971, 1973 and 1975. In the first three of those tours at war I spent most of my time in the field with the troops and I came to know and respect them and even love them, though most folks might find the words "war" and "love" in the same sentence unsettling if not odd. In fact, I am far more comfortable in the company of those once-young soldiers today than with any other group except my own family. They are my comrades-in-arms, the best friends of my life and if ever I were to shout "help!" they would stampede to my aid in a heartbeat. They come from all walks of life. They are black, white, Hispanic, native American and Asian. They are fiercely loyal, dead honest, entirely generous of their time and money. They are my brothers and they did none of the things Oliver Stone or Francis Ford Coppola would have you believe all of them did.

ON THE WORST DAY of my life, in the middle of the worst battle of the Vietnam War, in a place called Landing Zone X-Ray in the Ia Drang Valley of Vietnam, I was walking around snapping some photographs when I caught a movement out of the corner of my eye. It was a tall, lanky GI who jumped out of a mortar pit and ran, zig-zagging under fire, toward me. He dove under the little bush I was crouched behind. "Joe! Joe Galloway! Don't you know me, man? It's Vince Cantu from Refugio, Texas!" Vince Cantu and I had graduated together from Refugio High School, Class of 1959, 55 boys and girls. We embraced warmly. Then he shouted over the din of gunfire: "Joe, you got to get down and stay down. It's dangerous out here. Men are dying all around." Vince told me that he had only ten days left on his tour of duty as a draftee soldier in the 1st Battalion 7th U.S. Cavalry, 1st Cavalry Division (Airmobile). "If I live through this I will be home in Refugio for Christmas." I asked Vince to please visit my mom and dad, but not tell them too much about where we had met and under what circumstances. I still have an old photograph from that Christmas visit. Vince was wearing one of those black satin Vietnam jackets, with his daughter on his knee, sitting with my mom and dad in their living room. Vince Cantu and I are still best friends.

When I walked out to a Huey helicopter leaving Landing Zone X-Ray I left knowing that 80 young Americans had laid down their lives so that I and others might survive. Another 124 had been terribly wounded and were on their way to hospitals in Japan or the United States. I left with both a sense of my place, among them, and an obligation to tell their stories to any who would listen. I knew that I had been among men of honor and decency and courage, and anyone who believes otherwise needs to look in his own heart and weigh himself.

LZ Albany was a separate battle one day after ours only three miles away in which another 155 young Americans died and another 130 were wounded.

HAL MOORE AND I began our research for the book-to-be in 1982. It was a ten-year journey to find and ultimately to bring back together those we could find who fought in the battles of LZ X-Ray and LZ Albany. We had good addresses for perhaps no more than a dozen veterans, but we mailed out a questionnaire to them to begin the process. Late one night a week later my phone rang at home in Los Angeles. On the other end was Sgt. George Nye, retired and living very quietly by choice in his home state of Maine. George began talking and it was almost stream of consciousness. He had held it inside him for so long and now someone wanted to know about it. He described taking his small team of engineer demolition men into X-Ray to blow down some trees and clear a safer landing zone for the helicopters. Then he was talking about PFC Jimmy D. Nakayama, one of those engineer soldiers, and how a misplaced napalm strike engulfed Nakayama in the roaring flames, how he ran out into the fire and screamed at another man to grab Jimmy's feet and help carry him to the aid station. My blood ran cold and the hair stood up on the back of my neck. I had been that man on the other end of Nakayama. I had grabbed his ankles and felt the boots crumble, the skin peel, and those slick bones in my hands. Again I heard Nakayama's screams. By then we were both weeping. I knew Nakayama had died a day or two later in an Army hospital. Nye told me that Jimmy's wife had given birth to a baby girl the week he died. Nye had encouraged Nakayama to apply for a slot at Officer Candidate School, and when he returned to base camp at An Khe he found a letter on his desk approving that application with orders for Nakayama to return immediately to Ft. Benning, Ga., to enter that course.

George Nye is gone now, but I want you to know what he did with the last months of his life. He lived in Bangor, Maine. The year was 1991 and in the fall plane after plane loaded with American soldiers headed home from the Persian Gulf War stopped there to refuel. It was their first sight of home. George and some other local volunteers organized a welcome at that desolate airport. They provided coffee, snacks and the warm "Welcome home, soldier" that no one ever offered George and the millions of other Vietnam veterans.

George had gone out to the airport to decorate a Christmas tree for those soldiers on the day he died.

HONOR AND DECENCY AND uncommon courage were common among the soldiers who served in Vietnam. I think of how they were, on patrol, moving through jungle or rice paddies, nervous, on edge, trying to watch right, left, ahead, behind, all at once. A friend once described it as something like looking at a tree full of owls. They were alert for sign, sound or smell of the enemy, and they watched each other closely. At the first sign of the oppressive heat and exhaustion getting to someone the two or three guys around would relieve him of some or all of the heavy burden that the Infantryman bears: 60 or 70 pounds of stuff. Rifle and magazines. A claymore mine or two. A couple of radio batteries. Cans of C-Rations. Spare socks. Maybe a book. All that rides in the soldier's pack. A soldier's buddies would make it easier for him to keep going. They took care of each other, because in this situation each other was all they had.

When I would pitch up to spend a day or two or three with such an outfit I was, at first, an object of some curiosity. Sooner or later a break would be called and everyone would flop down in the shade, drink some water, break out a C-Ration or a cigarette. The GI next to me would ask: "What you doing out here?" I would explain that I was a reporter. "You mean you are a civilian? You don't *have* to be here?" Yes. "Man, they must pay you loads of money to do this." And I would explain that, no, unfortunately I worked for UPI, the cheapest news agency in the world. "Then you are just plain crazy, man." Once I was pigeonholed, all was right. The grunts understood crazy like no one else I ever met. The welcome was warm, friendly and open. I was probably the only civilian they would ever see in the field; I was a sign that someone, anyone, outside the Big Green Machine cared how they lived and how they died. It didn't take very long before I truly did come to care.

They were, in my view, the best of their entire generation. When their number came up in the draft they didn't run to hide in Canada. They didn't turn up for their physical wearing pantyhose or full of this chemical or that drug which they hoped would fail them. Like their fathers before them they raised their right hand and took the oath to protect and defend the Constitution of the United States. It is not their fault that the war they were sent to fight was not one that the political leadership in Washington had any intention of winning. It is not their fault they had to fight the war with a crazy patchwork of rules that tied their hands. It is not their fault that over 58,000 of them died doing their duty while America tore itself apart contesting whether the Vietnam War was a noble cause or a mistake.

As I have grown older, and so have they, and the book and the movie have come to pass I am often asked: "Doesn't this close the loop for you? Doesn't this mean you can rest easier?" The answer is no, I can't. To my dying day I *will* remember and honor those who died, some in my arms. I *will* remember and honor those who lived and came home carrying memories and scars that only their brothers and sisters can share and understand.

In recent years Hal Moore and I completed a second book, a sequel titled *We Are Soldiers Still: A Journey Back to the Battlefields of Vietnam*. Just as Hal Moore and I felt the need for a second book, Terry Garlock felt compelled to write this book with true stories of Vietnam vets to counter the twisted history still serving as common knowledge. As you read the stories in Terry's book, think about this: they were the best you had, America, and you turned your back on them.

Joe Galloway in 1991, a couple of hours after the ceasefire in the Persian Gulf War, with the 24th Infantry Division (Mech), a few miles outside Basra, Iraq. Photo courtesy of Joe Galloway, all rights reserved

Joseph L. Galloway is retired from his role as author of a weekly syndicated column on military and national security affairs. He spent 22 years as a foreign and war correspondent and bureau chief for United Press International, nearly 20 years as a senior editor and senior writer for U.S. News & World Report magazine, and had been a senior military correspondent of Knight Ridder Newspapers since the fall of 2002. He was a special consultant to General Colin Powell at the State Department in 2001 and 2002. Galloway covered a number of combat operations in different parts of the world in his own way, close to the soldiers doing the fighting. General H. Norman Schwarzkopf has called Galloway ". . . the finest combat correspondent of our generation - a soldier's reporter and a soldier's friend." On May 1, 1998, Galloway was decorated with a Bronze Star Medal with V for rescuing wounded soldiers under fire in the Ia Drang Valley, in November 1965, the only medal of valor the U.S. Army awarded to a civilian for actions during the Vietnam War. Galloway has received a long list of honors in recognition of his work. A native of Refugio, Texas, he lives in Bayside, Texas.

The Veterans

Vietnam was a war that asked everything of a few and nothing of most in America.

Myra MacPherson, 1984

One Bad Day
R.J. Del Vecchio
Raleigh, North Carolina

I JOINED THE MARINES because the other services all wanted to make me a Limited Duty Officer; I had two degrees in chemistry. But my reason for enlisting was because I believed in what JFK said in his inaugural address about stopping the spread of communism and helping the South Vietnamese stay free. I was also joining because my classmates were already there fighting on behalf of my country and our ideals. I wanted to be part of that service and accept my share of the danger.

The odd thing was that because I was a chemist, instead of an infantry job I was assigned to be a photographer. In the end, this was one of the funny breaks life brings you, because it took me to the war in a way very few people experience.

Leaving Hill Ten
Photo courtesy of R.J. Del Vecchio, all rights reserved

IN FEBRUARY OF 1968 I was sent out to a Marine base called Hill Ten to take part in an operation called Pursuit. Elements of the First Battalion, Seventh Marine Regiment, were going out into the bush, onto the mountainous range we called Charley Ridge, to hunt NVA.

We wound our way through the convoluted path between barbed wire lines that protected the approaches to the hill, crossed rice paddies, climbed foothills and then entered the jungle of Indian Country. As we climbed up one ridgeline, where the vegetation was not too thick, a sharp-eyed Marine spotted movement in the plateau a few hundred feet below us.

The column stopped and everyone went low to avoid being seen. Artillery was called in and we watched as explosions blossomed around the tiny stick figures of VC below, some of which flew through the air and never moved again after hitting the ground. As soon as the shelling lifted, two Phantom jets arrived and sprayed rockets into the area where a few people could still be seen running. It was impossible to have a precise count of how many enemy had been down there, and how many of them were injured or killed, but certainly it was more than a few.

That was a success for our side and smiles and murmurs of "Get some!" went up and down the line of Marines who watched, many of them smoking as they sat and watched the show.

Then it was time to go on, and the column snaked its way further into the hills and we went under the jungle canopy. There it was darker, moister, and quiet, and movement became slow as the point team did their slow waltz down the pathway that had been chosen to follow. At times the column would stop, and a different platoon would move up to take its turn in the lead. I was about a third of the way down the column, hot, sweaty, listening and looking, carrying my two cameras and a heavy WWII vintage .45 caliber Greasegun a friend had brought to me from the first days of the battle of Hue, which was just winding down about then.

We'd just climbed up one of the hills and had started down the increasingly steep reverse slope when from ahead and below the sudden racket of heavy machinegun fire broke out, with the crackle of M-16 rifles immediately joining in. Faint shouts and screaming could be heard through the intervening jungle, and the column froze in place, every other man turning right or left to watch the jungle around us. The call for "Guns Up!" drifted back through the sticky wet air, meaning get the M-60 machinegun team to the point to help in the fight. A faint explosion was heard, which sounded like a grenade going off. I knew an ambush had taken place, and fighting was going on, and if I wanted to do the job and get combat pictures, it was time to go to work. So I moved up the column quickly, past the lead element of that platoon and started encountering the end elements of the lead platoon.

The path became very steep, to something like a 45-degree angle. A few feet down was the radioman, and next to him the platoon commander, a fairly new lieutenant. The firing was taking place another 50 feet down the steep slope, where it appeared to level off, and I could see a knot of Marines there, one M-60 firing sporadically into the heavy brush ahead of them where the path turned slightly and went out of sight.

There were bodies of men laid out just off the path just before where it turned, with Corpsmen working on them. Firing continued from around the corner of the path where I couldn't see anything. The jungle was a thick triple canopy here, with almost no sight at all of any sky, and little visibility to either side.

The Lt. was on the radio, but I couldn't make out what was being said. He stopped speaking and was yelling at the men at the bottom of the slope, who yelled back that it was an L-shaped ambush and the gook machineguns had the path zeroed in, shooting anyone trying to reach the men who'd been hit in the first burst of fire. One wounded man could be heard screaming for help. I took a picture looking down the hill, with the body of the first dead Marine I'd ever seen in the foreground.

During a moment's lull in the firing, there was the sound of movement in the jungle to our right, and I turned and opened up with the Greasegun, knowing its slow but heavy .45 cal slugs would penetrate the vegetation better than the M-16 rounds. When the magazine was empty everything was quiet on that side, but the officer told me not to fire again unless he asked me to.

In the meantime, one of the wounded men had been carried up to where we were, and one of the Corpsmen was working on him. He'd been shot through the legs, hit by 2-3 bullets, and was moaning in shock and pain. Since the ground was so angled, the Corpsman was having trouble keeping the man positioned and I went to help. We worked on bandaging the wounds and stopping the bleeding. When I was elevating the one leg to let the Corpsman wind the bandage around it, it was a like holding a long sack with jelly because the bullets had shattered both leg bones so there was no firmness any more. The noise of firing resumed with confusion all around, and there was no time to do anything but shrug off the horror and wipe the thick blood off my hands onto my jungle utilities.

We got the Marine bandaged up and the bleeding stopped, but he continued to slip into shock, his breathing became irregular, his eyes rolled up, and the Corpsman went into a frenzy of yelling at him to hang on, beating on his chest, and finally, doing mouth-to-mouth on him. With the firefight going on below us, I realized with a sense of total helplessness that the man had slipped into death despite every effort to help him. The Corpsman gave up, and for a moment sat by the dead man, with his own head hung low, beating on the ground with one fist.

I looked down the slope again, and saw a slim, blonde Marine burst out of the jungle around the corner of the path, carrying a wounded man. I realized then that he'd been doing this before, going out into the kill zone of the ambush by himself, to drag back the wounded and dying. Without thinking I slid down the hill the rest of the way to join the small group at the bottom, and started taking pictures. The Corpsmen there were working on another wounded man, who had two sucking chest wounds but was hanging on. Someone spoke to the man who'd just brought back another body; he had collapsed on the ground and I realized he was crying. They called him Jones. After a few seconds, he pulled himself up, turned and went back around the corner of the path again. I was stunned at that degree of bravery, the readiness to face death again and again, alone in the firing path of an enemy machinegun, to go get others who are dead or dying.

We heard a chopper above us trying to lower a jungle penetrator through the canopy for the man with the chest wounds. A Corpsman looked up, again and again, hopefully, waiting to see the metal frame break through the heavy vegetation above us.

It never did, and the chopper could be heard leaving. The wounded man's chest stopped moving, and the exhausted men working on him slumped to the ground. I turned, and Jones came around the path once more, with another body. He collapsed again, cried again, and finally lifted his head and said there was one more out there.

Looking for the jungle penetrator from medevac helicopter
Photo courtesy of R.J. Del Vecchio, all rights reserved

He seemed to be having trouble getting himself ready to go again, and I put the camera down, picked up the Greasegun, and went to him. There was no conscious decision to help him, I simply could not let him go out again by himself. When I got up with him, two of the Marines who'd been watching the path with their M-16s leveled looked at me, then at each other, then got up to go with us.

These are not moments of rationality; they are instants of the most intimate experience that you, and the people around you, can have, when your inner being allows you to act without thought. Your pulse is thunderous, time slows down and your body floats along as you take the big ball of fear in your mind and somehow push it to the back of your head and you go on.

The four of us went around the corner and down the path, the smell of gunpowder, blood, and burned flesh filling our nostrils. We stayed in a crouch with guns pointing forward, muzzles moving in sync with our eyes, looking for one motion, one glint of metal, something, anything to fire at. We came to the last man's body, which was in terrible condition. Something had set off the grenades in his belt, and blown one of his legs off at the upper thigh.

There was now an eerie quiet, broken only by the buzzing of large flies that gathered at the blood pools on the ground. Later I concluded that the NVA had decided they'd stayed long enough and done enough, and it was time to leave before the bombs or napalm or artillery came.

Jones and one of the other grunts picked up the man's torso, the remaining grunt took the leg and thigh, and they slowly backed up the path. I became the rear guard, staying down on one knee with the Greasegun fanning back and forth at the jungle. In only a few seconds I was alone, as I backed up tiny step by tiny step to give the others time to get their burdens back to relative safety. The sensation hit me that the entire world was a hidden gun barrel, pointed at me.

R.J. Del Vecchio at China Beach.
Photo courtesy of RJ Del Vecchio, all rights reserved

After a span of time impossible to describe, I found myself back around the corner of the path and with the others, at which time I became aware of breathing again. As I started to relax, the strained faces of the others, five altogether, told me something was wrong. While we were down the path there had been an order to get back up the hill, and the entire platoon had retreated, so we were now cut off from the main body of our people. No one was in sight up the steep path, we couldn't even hear them anymore.

There was unspoken agreement it was time to go; one guy offered to be rear guard, I took the leg and thigh over my shoulder, two others took the body, and we all scrambled up the hill, pulling at roots and branches and slipping on the loose earth. Finally we got up to where the slope was less steep, and found the last two or three Marines waiting for us, along with the body of the man who died from his leg wounds. Everyone was exhausted, but starting to feel less threatened, and we needed to get all the way back up the main part of the hill to where the regrouping was taking place.

The main problem was the body of the man who'd died on the slope from his wounds. Finally I said I'd take him, others would take the parts of the man who died in the ambush, and they helped me get the body across my shoulders in the Fireman's Carry position. Dead men are amazingly heavy, much heavier than they were in life, and the ascent up the hill in the heat and humidity became a nightmare in which my vision got narrower and narrower and the buzzing in my ears louder and louder as I fought step by step, upwards. When the ground leveled and I realized I was finally at the hilltop area, I fell to my knees, others took the body from my shoulders, and I fell forward in a daze, never noticing the impact with the ground.

When I had recovered and could think again, everyone was packing up to move back. There was a lot to carry, and I wound up with two M-16s, one broken M-79, and the thigh & leg, which by then had been wrapped in a poncho. The column made its way back to another hilly area with less vegetation, and choppers came in using hoist lines to pick up the wounded and some of the other materials, so I was relieved of the extra weapons. We moved a mile or two further, and then night came, and we settled into a perimeter, two in each forward hole, one awake at all times. Movement could be heard outside our lines, and we waited for an assault but none came. The next day we made it back to Hill Ten without further incident. That was one day, one incident, one tiny bit of the war.

THAT WAS A LONG time ago, but it's not a memory that fades. More than anything else, I remember Jones, going out again and again by himself into the killing zone to help his fellow Marines and crying in between those insanely brave trips into the mouth of Hell.

When, 34 years later, I started attending the reunions of that unit, more than anyone else I wanted to find Jones, the bravest man I ever saw, and just shake his hand. But he was not there, and finally I looked through the company records and found him. He is on The Wall, having been killed in another battle a month later. I trust he is with God, in comfort, light, and peace. But I wish so much he were still here, having lived a good life.

In war, it seems too often we lose our best. I recall them often, and honor them in my heart and my prayers. They deserved so much more than we could give them, and infinitely more than the disrespect that became part of the antiwar movement.

In my job as a photographer I traveled all over I Corps, able to roam widely as I wished with open travel orders. I served with different Marine units in many situations from rice paddies to jungle to mountains to cities. I talked to people at a number of bases and hospitals, including a few other foreigners like the Germans I met. I saw parts of the command structure, including officers discussing tactics. Most guys assigned to a unit in Vietnam had a range of experience limited to that unit's operations and movements. I had the benefit of an unusually broad exposure of what was going on in that theatre at that time.

I saw the worst and the best of it. I saw real heroes, fine officers and men, and sometimes I saw bad leadership, fear, suffering, bureaucracy, inefficiency, military careerism, burnt-out men, and destruction and waste enough to make anyone wonder and despair. But I never saw a policy of abuse of the natives or the enemy, and I did see plenty of kindness and care for the villagers we met, kindness you would never see in TV news reports about the war in Vietnam.

Photo courtesy of R.J. Del Vecchio, all rights reserved

R. J. Del Vecchio is very active in veteran groups and co-authored a historical booklet on Vietnam, titled "Whitewash/Blackwash: Myths of the Vietnam War." He is a frequent lecturer in high schools and colleges on the war, and Director of a charity for disabled ARVN vets still living in Vietnam.

Postcard to America

I SAW SERVICE IN the Navy from 1946 to 1985, and during the Vietnam War was skipper of a ballistic missile submarine, engaged in making strategic deterrent patrols in the Pacific. Although based in Pearl Harbor where we resided during the period when the alternate crew had the boat on patrol, we operated out of the advanced base at Guam to maximize the actual time of the boat on patrol. Thus, we were relatively sheltered from the news about the war and the events transpiring on the mainland, and were immersed in our own ops. That is, until one day in 1969.

I had just returned from three months at sea and at Guam, and was with my wife on Kapiolani Boulevard in Honolulu, in my summer whites. As we walked down the street, I began to hear the beat of what sounded like a bass drum, accompanied by some kind of chanting that I at first could not understand, but soon I recognized the words: 'Ho, Ho, Ho Chi Minh, Hell yes, he will win!'" being shouted over and over again.

We turned the corner, and there was a scraggly looking band of fifty or so parading down the street, carrying a Viet Cong flag over the upside-down U.S. flag. I was shocked, and then terribly angered, and my wife grabbed my arm as she saw me start to go out into the street to confront them, as did a nearby police officer. When they saw me in my uniform, they started shouting obscenities, and one of them cut the American flag from the pole and set it on fire. They then continued down the street, waving the Viet Cong flag and continuing their chant.

I was shaking with anger, and the officer stayed with us until they had disappeared from sight and hearing, counseling me to stay clear of them, and telling me he was sorry that I had to see such a display.

That was my wake-up call to what was happening in our country. We didn't lose the war in Vietnam, we lost it in the United States. Those of you who fought in Vietnam never were defeated, and you should hold your heads high with the knowledge that you served your country with honor, courage and sacrifice.

J. A. Sagerholm, Timonium, Maryland
Vice Admiral, USN (Ret.)

Grunt Melody
Norm McDonald
Orem, Utah

THE MAN I SHOT moments before had his eyes locked on mine as he died. He was a uniformed NVA soldier, my enemy, a small man about my age of 20. I hit him four or five times in a firefight with my M-60 as he was running on the other side of a stream about 75 meters away. I was a machine gunner, and a very good shot.

The firefight lasted just a couple of minutes and two of my squad members reached the man I knocked down before I could cross the stream, working on him and talking to him, trying to keep him alive. They called medevac on the radio so our helicopter crew could risk landing in a hot area to rush a wounded enemy soldier to a US hospital. I guess that's part of the insanity of war; we do our damndest to kill each other, then, when the enemy is helpless, we're supposed to instantly morph from lethal to paternal. Our enemy was never encumbered with the same scruples. Besides, a live enemy, even if wounded, was a prime source of intel on unit strength, plans and movement. Some of them were conscripted, brutally treated by their officers and eager to talk.

Whether the guy I shot was dedicated to the cause or fighting because he was forced to, I don't know. I never felt guilty about shooting him, but I did feel bad. I guess you'd have to kill someone in combat to know the difference. After all these years, I still wish he had not waited to die until he was looking in my eyes, knowing I was the one who shot him.

My part in the war was about 40 years ago. Even so, some things are as fresh in my mind as this morning and when I reflect on our days and weeks trying to keep each other alive in the jungle I think about that NVA soldier, my buddies, Flash and Joe, and three young women who made their permanent mark in my memory.

OUR HELICOPTER APPROACHED THE side of a hill to insert us since there was no open and level LZ. We threw out our rucks and jumped about seven feet to the ground. This was my first mission and I was thoroughly unprepared. Even though a jump of seven feet is pretty quick I thought to myself on the way down: "How the hell did you get yourself into this mess?" Maybe it was equal parts bad luck and my own fault.

I was a good kid, never in trouble, active in my church and an Eagle Scout. When I graduated from high school in Salt Lake City in 1968, like many other young Mormons I was thrilled to enroll at Brigham Young University in Provo. But I was also a child of my time and I was drawn to the hippie view of the world. I ignored my studies, let my hair grow, dressed like a flower child and broke the rules I had been taught to discover the self-indulgent delights of beer and marijuana. My reaction to images of war on TV was to strum my guitar, wonder why America didn't just refuse to go to war and thank my lucky stars for college draft deferments. However, I spent my time doing the things that made me feel good at the moment, not the things that would keep me in good standing with the draft board.

Not long into my first college year, BYU uncovered my secret: I did not even resemble a student. At about the same time, the US Selective Service devised a new way to select draftees by a lottery system. Since 1942 they had drafted the oldest man first to determine order. The new lottery method was introduced on December 1, 1969, with radio and TV coverage of Congressman Alexander Pirnie, a member of the House Armed Services Committee, reaching into a large glass container to select and open, one by one, blue plastic capsules, each containing a date in the year 1970. As each capsule was drawn, that determined the 1970 sequence of drafting men age 18 to 26 whose birthday fell on that date.

BYU must have notified the draft board promptly that my college draft deferment was dead and buried deep because I received notice in January of 1970 that my draft number was 100. On April 22 I reported for induction into the Army. By 4AM the next morning I was at Ft. Lewis, Washington to start basic training, where, over an eight week period, they squeezed out every visible particle of our individuality, forcing us to conceal in the shadowy corners of our mind who we really were while we grew a thin Army veneer. The Army taught us vital skills like how to buff an ancient linoleum floor to a high shine in the middle of the night before an inspection, pushed our bodies to the limit every day for conditioning, and forced into our heads the things one must know to be a soldier. Well, at the least a beginner and reluctant soldier.

In the last week of basic, the Drill Sergeant called a company formation to hand out orders. The National Guard, Reserve and RA troops, the Regular Army guys who volunteered, received their orders to various training schools and fell out, leaving a dozen of us out of about 50 in the platoon. The Drill Sergeant gave us a sly grin and informed us we were now in the Infantry. He marched us across the parade field to our new and temporary home for AIT – Advanced Infantry Training. That meant we did eight more weeks of the same stuff we did in basic, just more intensely. We spent a week in the woods, learned to shoot the weapons we heard about in basic, learned tactics like how to set up and defend against an ambush, and generally became more conditioned to enduring the discomfort of an infantry soldier.

Despite rumors to the contrary, we all got orders for Vietnam. After a long airplane ride and a few days waiting at the 90th Replacement Company in Long Binh, I was assigned to the 5th of the 7th Air Cavalry, the 5/7, at Fire Base Snuffy, a huge artillery base near the Cambodian border in northeast III Corps in the Song Be area, not far from a part of the border that looks like and was referred to as the *Parrot's Beak*. Fire base Snuffy was used as a staging area for the Cambodian incursion earlier that year. The enemy had been very active in our AO, which was interposed between the Ho Chi Minh Trail, the enemy's supply route from Hanoi just across the Cambodian border, and the free capital city of Saigon, their primary target.

I know what you are thinking. The 7th Cavalry was Custer's unit at the Little Big Horn. You'd think the Army would have retired that unit!

My buddy Stanley grew up with me, got kicked out of BYU with me, was drafted with me, went through basic and AIT with me, and we flew to Vietnam together. When I was assigned to the 5/7 Cav, Stanley went north to the 11th Brigade, 5th Mechanized Division. I was grateful that we didn't go to the same unit because I didn't think I could handle it if something happened to him near me.

Another of my Basic and AIT buddies, Gordon Pitts from eastern Oregon, was assigned to my 5/7 unit and we paired up. We called him "Flash." When we arrived at firebase Snuffy, the platoon we were assigned to was not there, it was on patrol in the jungle, or as we soon learned to say, in the *bush*. While we waited about ten days for their return, Flash and I got the dirty jobs at the firebase, mostly shit-burning detail.

If you were never in that war, shit-burning detail sounds like something polite people shouldn't mention, but we all should know the realities of the daily life of troops sent to war on our behalf. Every American in Vietnam who served outside the comfort of a few large bases with flush toilets knew all about the foul shit-burning detail. Imagine the worst smell possible, then double it and you have a close approximation for the smell of burning our own sewage. Latrines, the Army term for where you go to poop, were built to accommodate a plywood slab with holes sawed in them for sitting, with half a 50-gallon barrel under each of the holes. Some holes were even sanded to remove splinters. Many latrines were built in the open with no privacy whatever, so you could enjoy nature without confinement, even continue your conversation with your buddies while taking a dump. We had *piss tubes* for urinating, many out in the open, artillery shell casings planted at an angle and partially filled with sand.

Flash and I tackled the filthy task of wrestling the putrid half-barrels out from under the hole, pouring in JP-4, the kerosene-gasoline blend burned in jets and helicopters, and lighting it off. The black smoke was considered semi-lethal and we always hoped for a breeze that would engulf the officer's mess at lunch or dinner time but the smoke usually drifted upward in the still heat. Helicopters would divert their course to avoid shit-burning smoke.

As some of it burned off, we had to stir it up, pour in more fuel and light it off again because it burned in layers. We were prepared for this awesome responsibility by the Army's fine-tuned method of training recruits to clean, shine or paint anything that does not move, unless it is shit, then you burn it. The awful smell and filthy job of burning shit is forever imprinted in our memory and, boy, were Flash and I glad to see our platoon arrive from the field! If getting shot at was the price of our ticket out of shit-burning detail, we were ready. At least we thought we were ready.

Our platoon arrived from their jungle patrol on several helicopters, looking ragged, unshaved and grubby from 20-30 days in the bush, now looking forward to 4-5 days of rest in a secure area. They were wound up, ready to party, and uninterested in questions from new guys. So, while they yucked it up, Flash and I gathered our M-16 rifles, ammo, a pile of grenades, and we packed our backpacks that grunts called a *ruck*. We had no idea what we really needed and what we didn't need in the bush so we packed pretty much everything, and it's a good thing we packed when we did because the Lt. promptly returned to tell us to *mount up* to answer a *QRF* call, whatever the hell that meant.

It turned out QRF was a Quick Reaction Force. When our platoon was relaxing at the firebase it was on QRF standby to respond when a unit in contact with the enemy called for help. So we rushed to the helicopter pads, loaded up and took off. Flash and I were pushed to the center of the helicopter while the experienced guys, bitching and moaning about their party being cancelled before it got a good start, sat on the ledge with their feet dangling over the side. I don't know if my eyes were wide with apprehension, but my mind certainly was. Flash and I had no idea what we were doing. It was rainy and cold, and when the helicopter edged close to the side of a hill that's when we jumped and, well, that's how I got into this mess.

WHEN I HIT THE ground from that first helicopter insertion I rolled around and got tangled in a bed of *wait-a-minute vines*, cut up and bleeding. I knew nothing about them but the experienced guys explained they had backward facing little thorns that grabbed your skin and uniform when you went by, and you had to stop, go back and unhook, thus the wait-a-minute name. I soon learned to spot and walk around wait-a-minutes. Here we were on a side of a hill, cold and wet in the fading evening light. I assumed we would try to find the unit that needed help but the Lt. told us to wrap up in our ponchos and sleep for the night.

Flash and I stood out like a neon sign. Our rucks weighed a ton and were unbalanced to the point we could hardly carry them, our fatigue uniforms were brand new while the grunts' fatigues were weathered, beat up and tucked into their boots. They had tight straps at their leg bottoms and calves and arms with sleeves rolled down, shirts buttoned at the top. They ignored us as they set up a perimeter for the night and made their sleeping spot. Oh, well, I figured, we'll ask them for advice in the morning. Flash and I settled in a spot to try our best to sleep on the ground under our poncho.

I didn't sleep much at all that miserable night in the cold rain. I was itching all over and scratching, couldn't get comfortable. In the morning Flash, and I were both covered in blood. We looked under our fatigues and found we had leeches all over us. Resisting the urge to be hysterical and embarrass ourselves even more, we burned the leeches with cigarettes to make them drop off while the grunts explained they used the straps and tucks, and slept in hammocks off the ground, to keep out ground leeches. I guess all those slimy little bastards found us instead of them. They showed us if we stood in one spot for a few minutes we would see leeches slowly come out from under leaves sliming their way toward us, like they had little infrared sensors in what passed for their heads.

Leeches were drawn to warm areas between our legs, our groin, between our butt cheeks, our armpits and behind our ears. They didn't hurt, but they itched and, when we scratched in our half-sleep, we popped the blood-bloated damn things making us look pretty much like a horror movie. It was a horrible experience, filed away in my memory for when I need a really good nightmare.

Flash and I set out on our virgin hump in the mountain jungle. We didn't know it then but this first hump would last nearly a month and would turn us into real grunts. We were way back in the column, struggling to keep up with our overloaded rucks and shiny new fatigues. We learned our mission was to find a LRRP team that had not reported in by radio when expected the previous day and had not been heard from since. I didn't even know what a LRRP team was.

We humped and I thought I would pass out from the heat and humidity. What rescued me from that humiliation was the slow pace and many stops as the point hacked through jungle. Sometime that morning I discovered what *contact* means. When the front end of our column opened up in full automatic gunfire, which we learned to call "rock-n-roll," I freaked out. I fell backwards with this huge heavy pack on my back and couldn't get out of it, stuck like an upside down turtle. Grunts were on their bellies, their rucks off and their rifles out, locked and loaded, safety off and poised to fire. One of them helped me off my back. We were the last squad in line, not close enough to see anything through the thick jungle or to shoot at the enemy.

One little grunt who had been in-country about nine months came over after the firefight was over and started going through our packs, throwing away all the junk we didn't need. He threw a bunch of our grenades into a sack to send back to the fire base, then he said, "Now get in your pack." We did and remarked how much lighter and easily balanced it was. Then he told us to grab the little straps near our armpits and pull them. We did and to our surprise the pack fell off. Flash and I didn't even know about the quick-release we would need when we came in contact with the enemy and they shot green tracers at us while we fired red tracers at them.

Flash and I were probably the perfect example of why grunts tended to stay away from FNGs as each guy showed up on his very own one-year tour schedule. New guys made lots of mistakes that might get themselves and those nearby killed.

We moved on and eventually found the four-man LRRP team, all dead, their weapons gone. On my first day in the bush I got shot at, was incapacitated by my own ruck, got my first sights and smells of death and filled up my nightmare files with images of leeches. Welcome to Vietnam.

I soon learned firefights were the exception, and long, steamy hot humps, boredom and fatigue were the norm.

We went out that first day as absolute cherries, and we stayed in the bush for a month. I was in pretty good shape but that first month in the bush got me conditioned to the suffocating humid heat and the work of hacking through thick brush and tangle uphill over steep mountains covered in triple canopy jungle. When the jungle was thick we could do 300 meters in a good day, and be exhausted. When we had time-pressured orders to cover a klick, or 1,000 meters, we all griped because it would be a day from hell. Our purpose was to cover ground, to patrol, to hunt the enemy while the concealed enemy decided whether they wanted to hunt us or avoid us.

Helicopters that we called *slicks* gave us mobility to get to our patrol area, and more importantly to take us out of the bush and back to the sanctuary of firebase Snuffy. I discovered I liked being on the lead slick going into an LZ, not knowing whether it would be hot or cold. I liked the thrill of the uncertainty, and the excitement of two cobra gunships circling and firing miniguns and rockets while the door gunners on our slicks worked out with their 60s to keep the enemy's head down as the pilot brought it to a near stop just off the ground so we could roll off and run to cover to set up a perimeter, and even before the last grunt was off the pilot would nose it over and haul ass. When the LZ was big enough the entire flight would offload and haul ass within seconds of each other, but most times the LZ was small and we'd have to do it one slick at a time, an open invitation to the enemy to shoot us and our slick. We'd sit on the edge of the helicopter doorway with our legs dangling in the air and no security straps, held in place only by centrifugal force when the helicopter turned and banked, so steep sometimes we were looking straight ahead and straight down at the same time! On my maiden flight I knew nothing and my stomach came up to my eyeballs, but as I became a grunt I got used to it fast.

When the jungle was thick and the going slow, we spaced ourselves in single file with about five meter intervals, trying not to bunch up in case we got hit, making it harder to hit many guys at one time. If the bush was more open our interval spread to ten meters and we staggered left and right, whispering when we needed to talk, heads turning, eyes searching, looking for signs of our prey, trying to spot them before we became the prey.

Something happened on that first hump that would change my experience in Vietnam. The machine gunner went home, and our squad needed someone to take the 26 lb M-60 machine gun. Since I'm a big guy they asked me if I wanted to be the gunner. I eagerly took the 60. I liked the idea of having a powerful punch when the shooting started.

I carried the M-60 locked and loaded, meaning ready to fire, but instead of using the shoulder straps most guys used, I carried the gun in my hands with a hundred rounds draped over my left arm. It was a fine gun and is still in wide military use. It fired belts of 7.62 mm rounds at about 550 per minute, over nine rounds a second if you do the math, with tracers every fifth round. The assistant gunner followed me with more bandoliers of ammo, ready to link to my ammo belt when we made contact. My new role changed how I moved through the jungle, and it changed a lot of other things.

Our patrols were usually platoon size, which should be four squads of eight to ten men each plus the CP, the Command Post composed of the Lt., the radio/telephone man called an RTO who stayed near the Lt., the Medic, Platoon Sgt. and sometimes a Forward Observer to call and adjust artillery strikes, the FO's RTO and maybe a Kit Carson Scout, a former enemy soldier now applying his special jungle fighting skills for the good guys. We were always under strength and had just three squads and the CP. Sometimes we were company strength, three or more platoons hacking through the jungle. I don't know how that many men sneak up on anything with the noise they make no matter how hard they try to keep it quiet, and I don't know why we bothered whispering as we moved but we always did in the jungle.

When we took a break there were always a few grunts who needed to answer nature's call. They would tell a couple buddies what they were doing, and their buddies would keep their weapons ready while the grunt took his rifle and backed off a little ways into the jungle to do his business after digging a small hole, then covered it up so we didn't leave too much of a scent trail. Our body cycle got used to a daytime routine because it was too risky after dark.

When my squad was up front, the point man had his shirt off, hacking through vines with a machete to make our trail. The point man not only hacked a path, he also watched constantly for signs of the enemy like footprints, broken limbs, trampled vegetation, campfire residue, dropped items, trash or even their scent. Point was the first contact when we found them . . . or they found us. If there was a trip wire to a booby trap, the point man was the first to see it, or to trip it. Point was the first to be seen by the enemy, the most likely to be shot, the most at risk from booby traps. He had to be stealthy and observant. He had to be alert. Most guys avoided point because it was dangerous, and because it was a lot of work, so point was often rotated unless one man did the job well and liked it. Flash liked point.

I became a machine gunner and Flash turned into a wild man walking point. He liked being on the edge, seeing, smelling, sensing tiny little signs that warned the enemy was near. I think he also fed on the thrill of being in the middle of it when the shit hit the fan.

Right behind point was the slack man, point security, rifle ready and watching the point man as he hacked a trail, looking ahead as point looked down, keeping point headed in the right direction. Behind slack was the squad leader, the man who would give direction during contact. I followed the squad leader with my M-60 to keep the heavy gun up front where it would be needed fast in a firefight.

When a man, especially point or slack, held up their hand everybody behind them stopped and waited, and when they held up a fist everybody froze, not making a movement or a sound.

Step by step we moved as quietly as we could through the jungle, climbing or descending steep inclines, stepping around the bushes and vines with thorns or noxious secretions, ever-vigilant for pit vipers, small snakes of many varieties with a body heat sensing indentation between their eyes and a deadly venom, indigenous to Asia. We watched for cobras, too.

I got used to the bugs, like foot-long centipedes with a thousand legs. Kit Carson scouts would catch big black scorpions, knock the stinger off their tail with a knife, eat them live and declare, "Numba One!" That means yummy. I learned to ignore the mosquitoes in the dry season even as they covered my sweaty arms by the hundreds, but I hated the ground leeches. If I stood still for a moment, they'd start coming out from under the leaves, a couple dozen of them, then I'd step to another spot. The bug repellant the Army gave us didn't do anything to the mosquitoes and not much for the other bugs either, but it literally melted the ground leeches. When we stopped I would see guys squirting bug juice and knew they were melting leeches; I loved watching those slimy little nightmares just boil away. We didn't have much of a leech problem in the mountain streams because they were usually fast-moving and we quickly crossed. Crossing the slow-moving streams in the lowlands was altogether different.

We moved and climbed steadily on, one foot in front of the other, watching the man in front of us, constantly turning our head to look side to side, now and then stealing a glance behind, moving and listening and waiting for something that seemed out of place, something that would try to kill us.

As the point man hacked a path, sometimes he would find an open area where he suddenly emerged onto a trail, like a tunnel thru the jungle. The enemy had lots of trails, or *trotters*, to transport men and weapons and supplies from the Ho Chi Minh Trail across the border in Cambodia, moving deep into South Vietnam and toward Saigon or other target locations. Sometimes the trail was ten feet wide, big enough for heavy traffic. When we found a trail, we covered our tracks with brush, backed off about 15 feet into the jungle and set up an ambush with claymore mines, clackers to trigger them on command, machine guns on each end and rifles in between. Usually the ambush was a squad size, in a line parallel to the trail, while the rest of the platoon backed off about 100 meters back in the jungle, close enough in case the ambush squad needed help. If the enemy came by on the trail we'd watch them to see if we wanted to hit them or wait, and if we hit them we tried our best to kill them.

Most of our firefights weren't planned, though; most were intense contact surprises that didn't last long. The noise of opposing forces firing weapons on rock-n-roll, violating the jungle semi-quiet, was always startling. I was rarely firing at people I could see; I was usually just shooting in a direction because the jungle was thick, though I did return fire back along the path of the enemy's green tracers coming at us. I would start with the bandolier over my left arm, my assistant gunner would link up his bandolier to mine while I was firing, then guys would pass up 7.62 bandoliers or ammo cans they carried and the assistant gunner would link them up so the M-60 had a continuous supply of 3,000 to 4,000 rounds. We never needed that much. We would normally have a short firefight, try to sort out and report by radio the enemy strength and suspected direction they took, call in medevac for our wounded, call in slicks to take our dead, and then hump some more to do it again, day after day.

THE DAYS WERE LONG, hot and weary. Relief came at night.

When we stopped for the night, we called it *night log,* probably a derivative of "logistics" because log was our slang for resupply. You might assume nights in the jungle would be tense, and sometimes they were but night log was when each man had a little time to himself, relaxing and doing the daily nesting things that might give him a little comfort. The Lt. would select a spot where we could get between the trees, maybe high ground, maybe a small area with one side protected, hopefully level but sometimes we didn't have much to choose from. The squads would be spread around a perimeter, with the three squad machine guns spread evenly facing outwards, rifles in between, the CP in the protected center. Guard duty was split up a few hours for each man. Once we were set up, then each man went through his own nightly routine, his personal daily ritual, a strange bit of privacy in the open for all others to see, his unique way of soothing himself in a miserable existence. For me, my ritual was my unrelenting grip on a small daily piece of normal life. Other than that, of course, we had each other.

When our night log was settled and my machine gun set up properly on the perimeter, I set up my sleeping spot for the night very near my M-60. Some units dug holes at night but we didn't in this steep mountain terrain with very tough ground. There was already plenty of cover for concealment even though leaves don't stop bullets, but digging in the jungle floor would have been futile with all the roots we would have to hack through. The Army issued us air mattresses to sleep on but they were heavy and the bugs would bite holes in them and they leaked. There was always some new guy using the air mattress but most grunts used a hammock. We bought them from the Vietnamese for almost nothing, they folded up real tight and weighed just a few ounces and were very comfortable. In the summer I would hang my hammock very low, just off the ground, and in the rainy season I would use my poncho to make a tent over my hammock to keep in a little warmth and keep out the cold rain. It was comfy.

When I had my hammock prepared, my ritual continued with heating my night meal, the only real meal I ate every day in the jungle. Because it was late in the war we had the good fortune to have prized LRRP rations for daily meals, freeze-dried meals developed for the rigorous demands of LRRP teams, tasty and loaded with calories and nutrition. I used heat tabs and a cook kit cup to boil water about 20 minutes, mix it into the dried LRRP meal, reseal it and let it hydrate with the hot water five or ten minutes. Then I added salt, pepper, Tabasco, or whatever . . . and it wasn't bad. When I didn't want to wait, I'd take a small piece of C4 plastic explosive, roll it in a little ball, light it with my Zippo and it burned real hot, boiling the water in no-time. The brass didn't like it but who cares?

We had boxes of C-rations, too. I liked the pound cake and beef and potatoes in the little cans and the peaches; the other stuff I gave away. After heating my meal and eating I would settle in for the night, always tired. Our sleep had to fit around our two hours of guard duty, but to a grunt in a war zone two hours of guard duty is nothing and I looked forward to my rest in my hammock in the jungle, catching up on sleep. The man coming off guard duty would quietly wake his replacement. Everybody took guard duty very seriously because our lives depended on guards being awake and alert.

One night in the pre-dawn chill of rainy season, one of our guys went to wake his replacement and found him already awake, frozen still, lying on his air mattress with eyes wide open. Lying on his chest was a small cobra about three feet long, apparently having crawled up on his warm chest while he was sleeping. The cobra was curled up and would raise its head now and then to look around. Mr. Warm Chest told us later he had been lying there for hours, terrified and waiting for first light so we could rescue him. Our squad leader, Bill, cleared people out of the way and took a prone firing position with his M-16, took careful aim, waited until the snake lifted its head then shot his head off with either a very good or lucky shot. Since dawn had arrived and everybody was now awake, we didn't need the guy for guard duty any more, which was a good thing because he was a basket case and had to be medevaced back to firebase Snuffy.

There were critters everywhere. When we set up night log, we would often set an automatic ambush on a trail or other path we expected the enemy to take approaching us, with a daisy chain of five to eight claymore mines and high trip wires. We ran the clacker wire back to our night log position so we could manually trigger the mines as well. When it went off with loud bangs that demanded our immediate attention, most often it had been tripped by wind or small animals. Once it was a tiger, a small one, dead with one side turned into hash by the claymores.

Every three days or so we were *logged*, re-supplied by helicopter. The chopper would make several trips to bring what we needed, like one of the cooks and containers of heated food to serve us a hot meal, maybe soda and beer, water and ammo, replacement weapons, other supplies and clean clothes. We just shoved the clean shirts, pants and socks out of the helicopter into one big pile and pawed through the pile to find something that came close to fitting. We didn't wear underwear because it caused crotch rot. In the rainy season, the clothes would get pretty moldy and jungle rot was prevalent especially for light-skinned guys like me. I always had jungle rot spots on my arms and ankles but the medic had ointment for it and the problem was manageable. With clean clothes on our back and hot food in our belly, we were ready to *Charlie Mike* . . . continue the mission.

WE MADE IT BACK to firebase Snuffy every few weeks to rest and party for a few days. The first time I came back to Snuffy I was not even aware of how dirty, smelly and wild I was now, just like the guys who scared the hell out of me on my first day just a few weeks before and, just like them, I had no patience for naïve questions from new guys. We probably wore the thousand-yard stare, too, the timeless look of weary soldiers who would respond to the command "mount up!" robotically, having done it too many times to care any more.

I had become a grunt.

By this time in 1971 the draft had injected a constant stream of unwilling soldiers fresh from the hippie counterculture, soldiers who had a far more refined taste for marijuana than they did for anything remotely military. Even if you rejected the hippie mindset, nothing will make you dive face-first into self-indulgent excess like coming off a month in the mountain jungle where we traded fire with shadows once in a rare while, and another month in the jungle coming up in a few days. The early war years were pretty much drug-free, but as the war wore on year after year and the ratio of draftees increased, there was an increasing demand for pot by guys with money in their pocket. In such a setting, a supply line will somehow develop no matter how illegal. We smoked pot and drank beer when we stood down for a few days. Maybe some of the officers didn't know about our pot, but others looked the other way or even joined us in the infraction. This late in the war delusions of victory were morphing into expectations of our eventual withdrawal, and morale was tenuous, even among some officers.

Norm McDonald's squad relaxing at firebase Snuffy
Photo courtesy of Norm McDonald, all rights reserved

So, two or three days a month while we were on a break back at the firebase, if anyone could find a stash we smoked a few joints. That doesn't mean we were sloppy in doing our job in the bush; we never, ever smoked pot in the bush. I never even heard of anyone who did. Smoking pot in the bush would be irresponsible to our unit and our buddies, never mind suicidal. If I had found a guard on duty smoking pot, I would have been tempted to shoot him myself.

During our stand-down we clowned around and tried to have a little fun. I would show off at the range with my M-60 shooting up old ammo that was wet and dirty. The gas recoil was just enough that I could fire it on rock-n-roll with one hand holding the gun by the pistol grip, letting the barrel *float* as the recoil of automatic fire held it up, and I even hit my targets like a small tree I was trying to cut down. I was smart enough never to play like that in a firefight. While you might be thinking *Rambo,* I was not the type, I was definitely one of those reluctant draftees. Besides, the stupid character in those Rambo movies is a joke to Vietnam veterans because it is so far from reality, which I hasten to add because Rambo carried an M-60 like me. Don't confuse Hollywood bullshit with the real thing.

OUR REST DAYS AT Snuffy were always over too soon and we would pack our rucks, mount up one squad to a helicopter and begin our next three or four-week hump. Most of our AO was a free-fire zone. Some people think a free-fire zone in Vietnam was where we shot civilians with abandon. That's wrong. A free-fire zone was an area with few if any noncombatants and frequented by the enemy, designated as a free-fire zone so we did not have to call on the radio for permission to fire on the enemy, we could decide ourselves in the field. That's what a free-fire zone was, and I resent how that term was used by the anti-war left and the media to portray us as murderers. I particularly resent it because I considered myself to be part of the anti-war left!

In all my time in Vietnam, in all the firefights I was part of, and all the villages we passed through, I never saw an American soldier intentionally harm a non-combatant, the key word being intentionally. Shit happens in war. Here's one example.

A squad from one of our other platoons found a trail in the jungle one day and set up an ambush. When the enemy came along the trail one of their guys triggered the ambush and killed two guys, but it turned out the two guys were not what they appeared; the rifles they had on their shoulders turned out to be sticks and they were civilian Vietnamese men. We never knew what they were doing there or whether they were helping the enemy but the squad felt horrible about the incident and the word spread through the company fast; we all felt like turds about it. And we went on with our humps in the jungle.

I did see mistreatment of one VC prisoner – he was slapped around by his American interrogator, very mild stuff in a combat zone. I assume there were isolated incidents of dark and unmentionable things because it was a war zone and people sometimes cross the line, but I never saw or heard of any war crimes. When I came home and heard the stories spread by the anti-war people about rampant war crimes by American troops in Vietnam, I thought they were incredibly stupid, especially since that story fit our enemy very well.

Our free-fire zones were not completely free of civilians. In the high mountains, way up there, were Montangnards, honest-to-God stone age village dwellers. They were a small, dark, peaceful people who built their communal hooches on stilts, persecuted by the VC and NVA who stole their food and killed them at every opportunity.

In October of 1970 we found a complex of enemy bunkers in the mountain jungle near the Cambodian border. The bunkers were in disrepair and the punji stake pits prepared as booby traps were seriously deteriorated and little threat. This discovery was unremarkable but memorable to me because years later I would recall this insignificant spot in the jungle as I rose in a college class to inform the professor he was wrong to say President Nixon's 1970 incursion into Cambodia was a failure. I told the professor and the class it was about time in 1970 that someone took off their blinders and recognized leaving the enemy's supply line intact on the Ho Chi Minh Trail was costing American lives and Nixon told our forces to cross the border and hit the enemy supply line hard. The American press portrayed the incursion in a negative way, I suppose because America was sick of the war, and college campuses exploded in protests by students who, by the way, didn't want to be drafted like me. Recognizing where self-interests lie sometimes adds a little clarity.

While I was slogging through the mountain jungles, Ohio National Guardsmen, who tried to disperse angry mobs of students with tear gas on the campus of Kent State University in Ohio, made the mistake of loading live ammo and ended up shooting and killing four students. America went berserk over the photo of the girl with outstretched arms, leaning over the body of a friend, a photo shown on TV a few million times. The media forgot to emphasize the students were rioting, breaking up businesses in the town and burning buildings on the campus. Hell, if I had been there I probably would have been right in the middle of the riot, maybe lighting off a building while I smoked a joint. But aside from all that, I also like to keep a firm hold on the truth. Nixon's Cambodian incursion was a huge success. That operation kicked the hell out of the enemy, set them back a long way, and saved American troops' lives. The enemy bunker complex we discovered, reclaimed by the swiftly-growing jungle during the months of its neglect, was just one tiny reminder that the enemy used to own these mountains but he was now licking his wounds elsewhere. By a grunt's rotten luck, I was about to be sent to find him.

IN MARCH 1971, ABOUT halfway through my year in Vietnam, I left the mountain jungles. My 5/7 Cav unit was standing down, pulling out and returning to the world. Since I had more than 90 days left on my tour, I did not go home with them. Instead, I was transferred to the 2/8 Cav based in Bien Hoa, operating in the jungles east of Saigon, toward the ocean in the Xuan Loc area. Delta company of the 2/8 Cav had the nickname *Angry Skipper* for some reason I never knew. Grunts like me didn't care about that stuff anyway; we left it to the officers to waste their time on unit pride, nicknames and such. We just wanted to finish our time and get out of that stinkhole.

Flash and I asked to stay together in the new unit. We got our wish, me in my comfortable role as machine gunner and wild man Flash walking point as he loved to do.

The change for me meant I was climbing hills instead of mountains, and the jungle was not quite so thick. There was another change. The NVA were active in this area, maybe because there were more villages, more people, more potential converts to their cause and more people to control if they prevailed. Two years prior, in 1968, the enemy's Tet Offensive began in III

Corps with an enemy assault on Xuan Loc, the provincial capital in the area and it was a hell of a battle. Now things were quiet but the enemy was definitely here. We still humped and hacked through jungle, constantly hunting for the enemy while the enemy watched us. We found him once in a while though I figure the NVA mostly decided when and where they were ready for a fight. Sometimes it was a surprise to both sides.

One of the times we came in surprise contact, the point man, not Flash, was hacking through the jungle with his shirt off while slack had him covered, and the point man found a small clearing that turned out to be a trail. We reported by radio and higher-ups told us to walk the trail to see what we encountered. Sure, they didn't have to walk it and run into booby traps or enemy ambushes; they were just fishing, using us for bait. We spread out 10 to 12 meters apart, flipped our weapons off safety, watched for trip wires and punji pits as we walked slowly and quietly with eyes wide, almost tiptoeing in our jungle boots, edging around a bend in the trail when point unexpectedly walked up on an NVA soldier. Not 50 meters apart, they opened up on each other in a rock-n-roll panic but both missed. We dove for cover and I cut the jungle apart with my M-60, firing toward the source of green tracers. Then it was suddenly over; they were gone, disappearing shadows turning into the smell of cordite and drifting gunfire smoke.

We met the enemy by surprise on another day near a creek. They shot at us, our squad up front opened up and that's when I shot the guy on the other side of the creek about 75 meters away, then walked over when the firefight was over and watched him die while the other guys worked on him and he looked in my eyes. I didn't want to kill him. Not a single grunt I knew wanted to kill anybody. We even had a grudging respect for our enemy, never mind that we gave them derogatory nicknames, because they were so tough and sometimes fought with so little resource.

THE MOST SCARED I have ever been came suddenly one evening after our company-sized patrol stopped for night log on the top of a hill. The perimeter was set, my gun was in place, and I was going through my nightly ritual when I heard a commotion on the other side of the perimeter. I thought maybe we were in contact so I got down and took my gun off safe, a little irritated that the small comfort of my routine had been interrupted. The shouting got louder and louder, grunts were running into the jungle in various directions and I couldn't figure out what was happening. All of a sudden, a bee stung me and then there were bees everywhere. I grabbed my gun and ammo because grunts were running away in the jungle leaving their weapons. The bees were EVERYWHERE!

About a hundred grunts without weapons had scattered into the jungle just before dark. This was not funny. By the time I got to the perimeter to escape, I had bees down my neck and in my pants, and I don't know how many times I was stung but I was all swollen up. After a while, the bees settled down and the grunts drifted back inside the perimeter. Some of them were getting sick from stings. Bees settled on the trees and leaves in our night log, crawling around, not in any hive. We carefully gathered our stuff to move about a klick away where a medevac landed to take away the worst cases. The bees didn't bother us any more but, if my memory is correct, one of our grunts died from the stings.

My eyes were half swollen shut, but the CO, Angry Skipper 6, had eyes swollen shut so bad he couldn't see. He reluctantly went on the medevac later, replaced by Capt. Bill Neal, who remained as our CO the rest of my time with that unit.

What disturbed the bees? One stupid grunt was throwing a stick at a hive about 20 feet off the ground. He finally hit it and knocked it to the ground where it exploded. The terror of the bees was worse than a firefight, worse than the leeches. Well, maybe not worse than the leeches

AT THE 2/8 IN the jungle I made another friend, a man named Joe Hall, a huge black guy from Little Rock, Arkansas. We talked a little bit in patches but we became real buddies one night back at the firebase during a few days of rest. The party hooch was too full and both Joe and I were tired of the noise and sat on top of the hooch with a couple other guys. Joe and I talked a long time about home, family, friends, girls and life in general back in the world. Joe was amazed that my mom's youngest sister married a black man in Salt Lake City in 1958 - still married more than 50 years later - and that I had black cousins. Joe scratched his head over that but figured I must be OK for a white dude, and I liked him too, so we got to be tight. The humanity of real friendship amidst all the struggling and dying gave me a tiny lift. I also liked the idea of having a guy as big as Joe watching my back in a jungle rumble.

THERE WAS A NEW Army program for leave from Vietnam. In lieu of the normal two week R&R in Hawaii, Australia, Bangkok or Hong Kong, if we had enough leave accumulated and our CO approved, we could apply for a two-week leave in the world, in the continental US, at home!

I had already met my new 2/8 Commanding Officer, Cpt. Bill Neal. I didn't hobnob with officers but his wife had suddenly informed him in a letter that she was converting to become a Mormon, and he nervously needed a Mormon like me to explain part of the mystery to him. I did what I could, and I'm sure that had nothing to do with his decision, but he approved my leave back in the US. I was very excited about seeing my friends and family, and about getting out of the jungle for a while. I cleaned myself up, pulled a uniform out of my duffel bag and hitched a helicopter ride to the air base at Bien Hoa.

The Army provided my charter flight from Vietnam to the LA airport and I had to arrange a connecting flight to Salt Lake City. The charter flight was a new 747, a huge plane full of troops returning from Vietnam. I guess with a target that big, protestors somehow knew we were coming and had gathered.

As we were walking down the LA concourse we could hear shouting and I could see security guys and people behind a rope carrying signs on sticks. As we came closer to them we had to walk right by the rope line and their yelling was louder. I couldn't help but feel conflicted because I was a hippie in my heart and I wished the war would end, too, and if I were not a machine gunner in the jungle, maybe I would be there with a sign and yelling right along with them. With that feeling of distant kinship, I noticed one of them was a very pretty girl with blonde hair parted in the middle, blue eyes and wearing a granny dress, hippie clothes like I wore, fair young skin like mine but hers looked so soft and tender.

Damn, it had been too long away from women and I couldn't help myself staring at her right up on the rope line and as I passed by very close she spit in my face. I tried to turn my head but was too late and, with arms full I walked along with disbelief, spit dripping from my face as the shouts of "Get out of Vietnam!" and "Killer!" and "Murderer" penetrated down to the center of my soul where a little sensor began glowing to tell me I had been betrayed.

Didn't they know I was just like them and doing what I had to do because my country called me? Did they know better than our country's leaders about helping a country resist a communist takeover? As I slogged along, having mastered the art in mountain jungles with a heavy load, I burned with a confusing mix of unwarranted shame and resentment.

My burn slowly diminished on the flight to Salt Lake City where I was met by my parents and brothers. We had a good reunion and after arriving home I contacted my buddies, who arranged a party for me. You must understand, however, my buddies were Mormons, and part of their life was avoiding not only alcohol, but even coffee or tea to keep caffeine out of their body, which they had learned was a sacred temple. At the party they brought beer for me as a gift, an extraordinary concession since their beliefs told them it was wrong. Well, mine did too but I leaped that hurdle a long time ago!

I loved these guys for our high school comradeship, and I was grateful for their generosity at the party, but while they chattered about the same stuff we talked about as kids my mind was drawn to Flash and Joe and the other guys, wondering what they were running into while I sat and listened about who was going on their Mormon mission, who was getting married soon and other things that seemed so insignificant while my other buddies were struggling daily to stay alive. I could not get wrapped around what was important to my high school friends, what used to be important to me. I forced myself to make it through the party but after saying my goodbyes I fled and spent the rest of my leave time driving around in the little GTO I had left behind, soaking up the beauty of the Wasatch mountains. Driving up the various canyons I began to lose myself in what God surely meant us all to see: the monumental rocks thrust up in the sky, streams full from April snowmelt pounding down the canyons in a roar, throwing up cold mist that made rainbows when the light hit them just so, glacial carving of U-shaped canyons and moraine structures with entire communities built on top, so vast that they were recognizable only when driving down the twisting switchback turns of Little Cottonwood Canyon with its panoramic view of the south end of the Salt Lake valley. There is a feeling of insignificance amidst this vastness that is impossible to capture on any canvas and has to be seen with the naked eye. For a few minutes at a time, I was restored but something was bothering me.

People in the world continued their lives as if the war didn't matter. Shouldn't they be worrying every day about the young Americans like me sent against their will, on their behalf, to fight and maybe die? In my head that seemed crazy, but in my gut it seemed true.

I reconnected to my old source and scored some joints so I could get high and pass the time, and I struggled with the feeling that I was disconnected from my family and friends because they had no clue what was important to me now, no understanding of life in the jungle on edge every step, trigger finger itching to shoot someone before they shot me. While I was supposed to be enjoying my leave, there was a barrier between me and my family and friends, and I was missing my guys back in Vietnam. How crazy was that?

With each passing day on leave I became more anxious for it to be over. When the door opened on the charter flight after landing back in Vietnam and the hot, humid air settled on us like a blanket, I had the embarrassing thought, "I'm home." I was anxious to get back to Flash and Joe and the guys, the only ones on this planet who understood me now.

WHEN I REPORTED TO the 2/8 Angry Skipper company headquarters in Bien Hoa to wait for a flight out to my unit in the field, something was wrong. Usually the only ones here were the First Sgt. and the mail clerk, but here were grunts sitting on the porch and scattered around, many with bandages and slings, and I found out my Angry Skipper unit had a fierce firefight while I was gone.

Any normal person, meaning those who were never in this war, would privately thank their lucky stars they were gone when something bad happened, but I think I found out what soldiers have always discovered to their surprise, that once you know combat, you feel like you let your buddies down if you are not there for them when the shit hits the fan. It was worse for me when they told me Joe Hall was dead. Maybe I was only 20 years old but I knew if I had been there maybe Joe could have made it because I knew some tricks of staying alive now, I knew how to tear the enemy up with my M-60, and maybe I could have helped Joe keep his big ass alive. But I was safe at home trying to be interested in who was getting married.

Goddammit!

The Angry Skipper grunts had changed. They probably didn't realize they had changed while I was gone but I could see it in their eyes, their movement. Something in their spirit took a hit when they ran into a tough NVA unit in the jungle and they found themselves in a firefight for their lives, not just a brief skirmish like most contacts, but a real fight while I was gawking at the Wasatch mountains like a tourist.

They gave Joe Hall the Silver Star posthumously. I was not surprised to learn Joe was a hero. He was a big, gentle and selfless young man, which is why I liked him. I was told that in the firefight Joe had been way back in the column when SSG Dillon was hit several times, and Joe just scrambled past everyone else up to the front where the lead was flying to help Dillon. Joe Hall died doing God's work, of that I am certain. I should have been by his side with my 60 instead of getting spit on in LA by some twit who thinks she has the answers. The very thought of morons like those protestors slinging insults at people like Joe Hall still makes my blood boil. Screw them!

I went back to the jungle for more of the same. The days and nights blend together, patrolling, searching, on edge, waiting, wondering when we would be in contact with NVA units we knew were in this area. To complicate matters, there were too many civilians to be a free-fire zone, so, unless the enemy shot at us first, we had to have radio clearance to fire on them.

We encountered a trotter one day while hacking through the jungle and backed away, set up a log, not a night log but just a defensive perimeter. Then the strangest thing happened. I don't know how he got through, but a single NVA soldier walked up to the perimeter on the other side from me and sprayed about 30 rounds in on us, then he skedaddled. Two grunts were casualties, one shot through the arm. The other casualty was very straight-laced, a seriously religious guy from West Virginia, who was not hit but just freaked out, went hysterical for some reason. We medevaced both of them. Thereafter, for unexplained reasons, instead of avoiding party animals like me when we stood down, the formerly straight-laced grunt joined us in our drinking, carousing and smoking joints. Go figure.

AS TIME PASSED, AND I found myself getting short, I was more nervous about combat and tried to get a job in the rear where something wasn't always trying to kill me. With 60 days left in country, I got my grunt's dream job.

There was a huge rock formation out in the jungle with steep sides, flat on top where there was a US communications depot with lots of communications equipment and antennae. The only way up there was by helicopter unless you were a Ninja. My dream grunt job was straight guard duty. Six or seven of us short timer grunts pulled about four hours of guard duty a night in one of the five bunkers, and the rest of the time was our own. This was like dying and going to grunt heaven. This was not a safe area, but we were relatively safe on top of the rock except for exposure to mortar fire.

Someone had a guitar and it took me two minutes to refresh the feel of my fingers flying across the strings, making a little music to pass the time and wallow in being dry, clean, away from bugs and snakes and leeches and something always trying to kill me. I could pull two months of this guard duty so easy . . . and of course when a grunt like me was safe and comfy for more than an hour, one of the geniuses in the long chain above me would feel moved to put one of their stupid ideas into action, and when those geniuses moved symbols around on a map at headquarters it usually meant a load of shit was about to drop on grunts somewhere.

And that is exactly what happened.

Some genius decided to move the communications depot to the side of a higher mountain closer to the ocean and we had to go check out the desired locale and help the engineers blast trees to make an LZ. We were flown to a firebase at the base of the mountain and humped our way up, passing through a village where I could see the hate in the villager's eyes. Uh-oh! This was not only deep in the boonies, but the enemy owned this turf and apparently had the hearts and minds of these villagers as well. Not good. Not good at all, but we blasted the trees and made the LZ so the helicopters could make a million sorties bringing in the men and materiel to build a new communications site.

Where there are grunts and a lot of work to do, somehow the grunts get the lousy jobs. We started moving a couple platoons of infantry up from the firebase to the new LZ and they assigned me, the other grunts, the medic we called "Doc" and the communications guy to fill sandbags and build the hooches. It was a lot of work and it totally ruined my sweet deal for the last weeks of my tour. I decided to make the best of it. While we worked our butts off getting the job done the thing that was never far from my mind was the sweet security of the rock we left behind. This place was just begging to be overrun. I kept working and ignored my edgy nerves for a few days.

One night when the wind was strong, a trip flare went off outside the new circles of concertina wire. The grunts on guard duty opened up in the direction of the trip flares and I woke up long enough to decide the wind had probably blown branches against the trip wire, and I went back to sleep. The next night a trip flare went up again in the still, windless calm. Grunts opened up again toward the trip and I was guessing it might be an animal, as it was so often in the jungle. When green tracers streaked toward us, I knew we were in trouble. The contact didn't last long and I fired just a few shots as I was out of position for my fire to do any good. I had a deep sense of foreboding. These guys were not going to last on this hill. I knew it.

One grunt was killed that night. I was helping out down at the helicopter pad when they brought him on a stretcher face-up. He had been lying down in his rack, not even part of the short firefight, but a stray bullet somehow put a tiny little hole right between his open eyes. I didn't know his name, but he had fair skin like mine. But for the bullet hole between his eyes, he looked almost angelic in the surreal glow of the moonlight. His face is still just behind my eyelids.

The next day we grunts did what we do best, we humped on patrol, trying to flush or find signs of the enemy. We put out listening posts at night and heard enemy movement but could not see them. We had brief firefights whenever they decided to fire the first shot. We brought our patrols back inside the perimeter and called in gunships and artillery on spots we guessed they might be. Later we went back to building bunkers and pulling guard duty and I was thinking this was crazy, the NVA were in major strength in the area, they had the locals on their side this time, and they simply were not going to let this communication depot stay here on the side of the mountain. We were far enough in the boonies that when we got in contact and called for gunship support they were at least a half hour away. Their triumphant arrival is no good if you are already dead.

That night on guard duty was my worst night in Vietnam because for the first time I thought I was not going to live through my last few days in-country. I could feel deep within me that, despite all the times I had escaped with my life and body intact, I might not live through this one.

We made it through the night. As the sun was coming up the enemy hit us again. We started firing mortar rounds out into the jungle around us, and we heard their mortar rounds coming in on us exploding shrapnel everywhere. Doc and I dove into our unfinished bunker to ride out the mortar attack. Mortar shrapnel is deadly, traveling too fast to see when it goes through you. In this case a big sharp-edged piece of shrapnel, two to three inches long and one inch thick, flew through the bunker doorway – we had not yet built the door - and sliced through the instep of my right foot, through the boot and stuck in my foot.

Son-of-a-bitch! They finally got me, but if that was the worst they could do I would live. When the firefight subsided I was one of the wounded medevaced down to the firebase below, and I asked Doc to bring my ruck for me when he came because, as I told him, "You guys aren't going to be here long, they're going to take back this mountain. Bring my ruck when you come." He promised he would.

They flew me down to the firebase to see the Battalion Surgeon. He gave me a local anesthetic shot in my foot, cut the boot off and pulled the shrapnel out with pliers. The shrapnel carved a hole in my foot between the bones. I got lucky. It didn't break any bones, but it did cut tendons.

This relatively minor injury would turn out to have a major impact on my health in future years, and it would lead me to two young women who will always be at the center of my memories of Vietnam.

Doc cleaned out the hole in my foot, packed it with clean gauze and gave me antibiotics and crutches. I hobbled out to the helicopter pad since I had nothing else to do and I watched the helicopters bringing in load after load off that mountain, bringing the communications equipment, water, supplies, whatever was up there, evacuating that mountainside depot just as I had guessed. After dozens of round trips they started bringing the men down and I saw Doc step off a helicopter. I yelled at him to ask if he brought my ruck and he said, "I knew I forgot something. Sorry!" I could see as I looked up the mountain that nobody was left up there and the enemy must have been overrunning the place because, as I was watching, jets dropped 1,000 pounders on the compound. Watching the huge explosions and hearing the distant "Wham!" was a small thrill, knowing it fried a bunch of the enemy, but my ruck and the personal stuff in it were toast. All I had were the clothes on my back, a pair of crutches and a sore foot.

That night Doc and I slept in a little bunker on the firebase. During the night I started getting sick and feverish, my foot hurting like the devil. Whatever was wrong was beyond Doc's skills as a medic. I waited for daylight to go back to the aid station to see what was wrong and ask for more pain medication. By the time it was light enough to see, I could see my foot was badly swollen, the ankle dark like bruising. I hobbled down to the aid station. The Battalion Surgeon removed the bandage, took one look, diagnosed *gas gangrene* and called for medevac. In half an hour I was in the 24th Evacuation

Hospital operating room in Long Binh, where they cared for patients until they were stable enough to send home.

MANY YEARS LATER IN my studies of medical terms I would learn that gas gangrene can be fatal if not quickly treated, and commonly develops overnight with spiking fever and delirium as mine did. I also learned what they did in the OR is called *debridement*, the process of removing dead or infected tissue and foreign material from a wound. Now the wound was far worse than when the doctor cut my boot off at the firebase, and they were talking about taking off my leg. Holy shit!

They cut the infected tissue away from the bone so I didn't get a bone infection, and with daily cleaning the wound improved and I was able to keep my foot and my leg. I was very lucky, and it is hard to put in words the gratitude I have always felt for the doctors, nurses, and staff who gave much of themselves to care for us, and they did care. Working on a constant stream of wounded must have been a strain, and that strain had to show itself now and then.

They kept me there in the hospital for a couple weeks, cleaning the wound twice a day, trying to get it to heal from the inside out. Most patients had wounds far worse than my foot. As the days went by I got to know the nurses and orderlies, and they knew me from the music I sometimes played on the guitar they had lying around.

One of the nurses was drop-dead gorgeous. I know I had been away from women too long, but this young woman was very striking: nice figure, dark hair, flawless pale skin, sharply pretty face and piercing blue eyes, the kind of girl that made every male head turn when she walked into the room because that's just the way we are wired. Unlike so many beautiful girls, she was not full of herself. She was sweet and friendly and spread cheer throughout the ward.

One day it was her job to clean my wound and change the bandage, which of course made me happy. She unwrapped the bandage and gently pulled the antibiotic string packing to remove it so she could clean the wound, repack it with antiseptic string, and bandage it again. This time she stopped pulling when the end of the string was stuck, like it had adhered to a scab or something. I told her go ahead and pull it because it happened before. She asked "Are you sure it won't hurt?" I told her to go ahead, so she pulled it loose.

Apparently the packing string had adhered to a small artery because every time my heart beat about a three inch geyser of blood shot up out of that hole in my foot, and she freaked! She pressed her hand down on it but that didn't stop anything. Blood was bubbling up through her fingers, she was frantic about what to do and I was losing blood. Blood quickly soaked the bed and patients next to me were getting wary. She yelled for help and I started getting dizzy, either worried or from losing blood. The surgeon ran down to take over. He gave me a local pain shot inside the wound, which hurt like hell, because he needed to sew the artery back together. I was yelling in pain and the nurse was frenzied, feeling she was at fault. The surgeon finally got the bleeding stopped, repacked the wound and wrapped it in a new bandage. The staff cleaned up the mess, put fresh sheets on the bed and started an IV to give me a bag of blood. My special nurse came back, balling her eyes out, sat down beside my bed and hugged me for a little while, an unspoken apology she did not owe me. I thought to myself, "Well, that was worth it!"

There was a guitar there on the ward and I played tunes like *For What It's Worth* by Buffalo Springfield and *The Cruel War*, by Peter, Paul, and Mary for myself and the hospital staff, so they knew me if not by name. There was one tall nurse, from Ogden, Utah, my very back yard, but I don't remember her name. She was a quiet lady, on the night shift, and we talked about home and other things.

One night I woke up in the wee hours hearing noise down the way from me. There was a young guy who had been in the OR all day long where they worked on his wounds from a friendly fire incident. I was later told he had been hit on a firebase in a *mad minute* exercise when all the artillery batteries, and all the troops on the firebase, fired over the wire into the jungle, making hellacious noise and intimidating any enemy in the area. He was hit by fire from within the firebase. A team of doctors and nurses were gathered at his bed, having a rough time working on him, trying to get him stabilized. I fell back to sleep, and when I woke up later, still in the dark, quiet pre-dawn hours, he was gone. I knew he had died.

The nurses' station was near my bed, and in the quiet of the dark ward I could hear sobbing, so I pushed myself down there in my wheelchair. My tall nurse friend was there alone, embarrassed because I caught her crying, upset that after doing everything they knew how to do they lost the guy. We didn't talk but I could see she was tired and discouraged. I had never done anything like this before, but I wheeled over next to her and put my arm around her. She leaned over on me in my wheelchair and I just held her for about ten minutes while she cried. Then she straightened up in her chair and, when she stopped crying, I picked up the guitar that was laying around and played some soft stuff for her, just jamming, playing some rifts and chords and pieces of melodies, while she started doing a little paperwork. I played for her for about a half hour then she said she had to get going. We never spoke about it.

When my foot had healed well enough for me to travel, even though I was not ambulatory, they gave me my magic papers to go home. After a two-day stopover at a hospital in Japan, I made an unabashedly miserable long flight on a cot in an Air Force C141, outfitted to haul patients with a crew of nurses, straight through to Chicago, then on to my destination in Denver. From the Denver airport I was transferred to Fitzsimons Army Hospital in Aurora CO, where I spent my last days in the Army, healing my foot and getting skin grafts. With less than 120 days left to serve, I was released from the Army and eagerly made my way home, a free man at last.

WHEN I WAS TRAPPED in a place and situation against my wishes, like the Army and Vietnam, fantasies of what life would be like back home when it is over trickled involuntarily through my head like a delusional survival mechanism. But when I finally arrived home, everything had changed. I had changed.

When I left for the Army I was shy and timid. I came home anything but shy. I was self-confident, knowing after the jungle survival game I played I could endure just about anything. Still, I had a lot of growing up to do.

My adjustment to adult civilian life took a long and rocky path, but I can't blame it on Vietnam. I spent about ten years getting high on something every day. I might have done much the same thing had I not been in the Army at all. Who can explain addictive behavior?

From the time I returned, I always had a job because one thing I got drilled into me by my father was a sound work ethic. But in the 1970s I surely took every drug there was on the black market. I became addicted to cocaine with a $300 to $400 a day habit, trafficked in cocaine to pay for it and got in big trouble. But my reliable drug of choice, my convenient friend and constant companion, was booze. I had seven convictions for DUI in the 1970s. Of course that leaves out the countless times I wasn't caught.

I can't honestly blame my alcoholism on Vietnam, but I think what lurked in my mind from the jungles *enhanced* my drinking, making me more self-destructive than I already was, making me even more dangerous because of my willingness to ignore risk. I would have been a drunk anyway without Vietnam, just a different kind of drunk.

I know some will quickly blame my behavior on Post-Traumatic Stress Disorder, but you should be wary of popular, easy, knee-jerk answers substituted for the more difficult truth. That's not to say I brought no baggage from the war, because I certainly had baggage.

Soon after I got out of the Army and was living at my parents' house, my dad came downstairs to my basement room to wake me up for work. When he touched me I was startled and came out of the bed fighting. I was devastated when I realized what had happened but Dad told me to relax. He understood. Other times Dad would come down to wake me and find me in a corner on the floor, back to the wall maybe for safety, arms around my knees like a fetal position, asleep.

I was hyper-vigilant for a while. One day, probably around the 4th of July, 1972, I was ripping down the road in the middle of the afternoon in the 1968 VW campmobile I had traded for my hot GTO. Someone lit a chain of firecrackers and I instinctively hit the dirt; the problem was I was driving down the road at 30mph. The van bounced through a ditch, across a lawn, and hit a house. I told the cop what happened, but he called me a lying, hippie bastard and wrote me up for reckless driving. I guess my long hair and hip duds spoke louder than my words. Over time, these flashback behaviors ended as my alcoholism became chronic.

I was kind of a *player* when I first got home from the war, the bad boy moms warned their girls about . . . and so I had a lot of dates with very popular girls who had a thing for bad boys. Some of these were the same girls who wouldn't give me a second glance a couple years before in high school. I really liked one girl and took her to a drive-in show to see *Soldier Blue.* When the soldiers rode through the Indian camp shooting the people, somehow I went into a panic, fell between the seats convulsing and hyperventilating, and scared the hell out of her. I never saw her again after that night. I rented that same movie in the late 1990s to test my reaction. I didn't have the same reaction but I did become nervous at that critical scene, maybe because I remembered how I panicked 20 years before.

My last drunk was September 30, 1981, 10 years after my return from Vietnam. That night I was busted for many things. I went into a local VFW club after drinking beer in another bar all day, sat down to have one last beer before going home, and saw a lady next to me with a margarita in a pretty long stem glass, salt around the rim and a fresh curl on the slush. I remember saying "I want one of those!" That's it. When I came to I was in the Provo City Jail drunk tank. I had thrown up on my chest and had urinated on myself from my waistband down to my knees. In a moment of clarity, I saw what I was at that moment, knowing what I could be, and I slid down into utter despair, what Alcoholics Anonymous calls "incomprehensible demoralization." I love that phrase.

The railroad where I worked was a pioneer in programs for employee assistance. I called the hotline and asked for help. The guy I talked to was local, a Marine vet who had been in Vietnam a couple years before me. He took me to AA in Provo, and I made a commitment that night to them and to myself to attend meetings. I've been sober since then.

I went to college using the GI Bill for financial assistance, and, a few years after getting sober, I received counseling at the Vet Center on dealing with some of my Vietnam memories. I went further and worked at the Vet Center on a work-study program, helping other vets come to grips with their own issues.

Conventional wisdom says that Vietnam produced an unusual number of vets with PTSD. Like so much of the conventional wisdom on the Vietnam War, it is mostly wrong. PTSD is a an umbrella term encompassing a number of behaviors that are believed to be caused by traumatic experiences which are suppressed at the time of the event, only to emerge in certain behaviors as the person gets older, like hyper-vigilance, a heightened startle response, isolation, recurring dreams, flashbacks, self-medication with drugs or alcohol, strained relationships and others. When trauma is suppressed, like soldiers who shrug it off as they have always had to do, sometimes the pressures of keeping that inside emerges in these behaviors later in life, perhaps decades later, usually with the person unaware or in denial, the behaviors varying wildly between individuals. Some never show any outward signs, some are severely affected. The experts concluded these behaviors in response to suppressed trauma have been with us since man sharpened spears around campfires to prepare for battle, but for our prior wars like WWI and II and Korea, we didn't recognize those behaviors in veterans, we didn't talk about them even though they have always been there. As a matter of convenience, we call them PTSD.

That's what I have learned in coming to terms with my own life, trying to honestly see what I was doing to myself, and what part of it might be coming from my combat experience. There are countless Americans who assume Vietnam veterans are particularly prone to PTSD, one notion among many that seems to have been born out of the political struggle over the war. I chalk it up to ignorance, but there actually is one thing that I believe made it harder for my brothers to adjust when they came home, and that is the reception they received. The indifference to their sacrifice, or the frequent assumption they had been part of something bad, weighed heavy on so many of them. Anyway, that's what I think.

I NEVER REALLY FELT any hostility about having served in Vietnam other than the girl who spit on me in the LA airport. Over the years nobody talked about Vietnam or wanted to hear about it. For a while I thought nobody cared, but looking back I think now maybe friends and family were nervous to broach the subject, having been *trained* by the media to wonder if we were brooding and morose, or that we might react in an unpredictable way *so don't talk about it*.

In 1985 the local Vet Center asked a few of us in a vet's group if we would like to walk in the Provo, Utah Freedom Festival Parade escorting a POW/MIA display. Ours is the fourth largest 4th of July celebration in the country with 80,000 to 100,000 lining the parade route. I'm more comfortable under the radar and I didn't plan to walk in the parade until my buddy, big Tom, called me the night before and dared me to show up and walk with him.

Tom had lost a leg in Vietnam, was self conscious and kept his artificial leg hidden under long pants. He joked and said, "If you come I'll show up in my Hawaiian shirt and short pants wearing my Purple Heart!" I said, "Fine, I'll be there with *my* Purple Heart!"

We showed up early where the parade was to begin on the 4th. There were about 25 or so vets milling about, apparently nervous and quietly speaking to each other as we waited for the parade to begin.

We started down 9th East, right out front leading the parade, about a mile down to Center St. Along the way, folks were pushing their reluctant husbands and fathers and brothers out of the crowd to the street and now there were about 50 vets walking.

The crowd were all on their feet, clapping and yelling things like "Welcome home!" I nearly lost it when two young soldiers in full uniform on the side lines snapped to attention and saluted us. We went down Center St. to University St., about a half mile, and by then there were 200 to 300 vets spread out about half a block. As we turned up University for the last mile of the parade, a girl maybe in her mid-twenties ran up and kissed me on the cheek, gave me a carnation and said "Welcome home!"

On the way down University St. to the end more vets joined us from the crowd and there must have been over 400 of us walking with pride while the crowd's roar was deafening. It was almost as if all the people who had been silent about Vietnam were saying with their cheers, "We don't understand what you had to endure, but we know it was hard and that it was for us and we thank you."

That is the day I came home from Vietnam.

OVER THE YEARS MY foot problem has led to limping, a fallen arch, nerve problems, knee and hip problems because the footbone is connected to . . . well, you know. In those same years I *found myself* in my work and my family. My first marriage was short, ending before I found sober life, but my second marriage in 1981 has made me a rich man: my wife, Maggie, a Navy veteran, my children, all grown now, and seven grandchildren to bounce on my good knee. My son Thad is an Iraqi Freedom veteran. He was there when we took Bagdad by storm in 2003 and I am quite proud of him.

After graduating with a BS in Medical Laboratory Science in 1992 from the University of Utah, I went to work as a Medical Technologist in Transfusion Services, and am now Director of those same services in a multi-hospital network.

Looking back over my life, there are three years that I cherish but would never want to repeat because the accomplishments in each of those years came at a high price. The first was my year as a grunt in Vietnam. The second was my first year of sobriety, when I was emotionally an infant and learning a new way of life at 30 years old. The third was my year-long university study in organic chemistry. I just hated that course of study. It was very tough and tedious with the complication of brutal competition from medical students. I had become accustomed to earning As in this second college adventure, but I never worked so hard in my life and was grateful for my C in organic chemistry. It is the hard things we go though, I think, that make us who we are, and I would not trade away these difficult years.

Joe Hall would be about 60 now if he had lived. He is forever frozen in my memory at 20 years old, as is the blonde girl who spit in my face. The two nurses from the 24th Evac Hospital in Vietnam are etched permanently on my heart like a fine and delicate tattoo. That leaves Gordon Pitts, my buddy Flash, the wild man walking point.

I lost touch with Flash when we left Vietnam and found him again just a few years ago. So far our catching up has been on the telephone. Flash and I used to make private jokes about the lifers in the Army when all we wanted to do was finish our two grunt years and get the hell out of the asylum. The funny thing is, after a short time out of the Army, Flash walked into an Army recruiting office in Washington state to re-enlist. He spent the next twenty years as a Drill Instructor turning maggot recruits into soldiers. Go figure.

Flash might be coming to visit soon. Maybe we'll talk about the war in a far different way than we did when we were in the middle of it. When I was young, I was angry and disillusioned about the whole system: the war, the state of the country, the lousy treatment we faced coming home. It made me an anti-war vet, though I wasn't a *John Kerry* anti-war vet; I always knew he was a liar.

Over the years I mellowed and studied and thought a lot about it. From the distance of time and a cool head, I can now see the geo-politics of the collision between freedom and communism, and that Vietnam was just one hot spot in the whole 50 years or so of the Cold War. I see myself as doing my small part in bringing down the Iron Curtain and I am proud to have served in that capacity. I don't think that is rationalizing, but who knows? Maybe I'll ask Flash, and no matter how much we differ I will treat his view seriously because he and I humped through hell together and watched each other's back.

I want my family to meet Flash. My grandchildren can watch us talking about old times, thinking of us as a couple of old coots telling irrelevant and exaggerated stories, refusing to let go of the ancient past. I hope they never realize the jungles of the Vietnam War were far worse than what we say, and I'll pray they never know in their heart what it means to fight for their life, or to shoot a man and watch him die.

Photo courtesy of Norm McDonald, all rights reserved

Postcard to America

BEFORE I LEFT FOR Vietnam in November 1967, while on leave after completing Marine Corps OCS I took my college girlfriend to see the performance of the play, The Fantastiks, in New York's Greenwich Village. I proudly wore my dress blues and we walked in a little late to the small theater with perhaps 100 full seats. Our seats were on the front row, and after waiting for a little break in the action, the usher took us tiptoeing down to our seats. Seeing my uniform, quite a few in the audience started booing and at least a third of the audience booed us loudly until we were fully seated. My date was mortified and seemed to be embarrassed to be with me even though we had been dating for two years. Some of the actors, as the stage was very close to us, seemed to be uncomfortable at the audience response. She was a lovely girl, but after her reaction I felt she had a problem with my military career. We never dated again.

Seven years later, in 1974 after my service in Vietnam, our family visited a high school friend of my wife, Sue, for Thanksgiving dinner. Sue's friend and her husband, a PhD in Philosophy and a professor at Sacramento State, lived in Sacramento, 400 miles away. At Thanksgiving dinner in their home, the husband discovered from the conversation I had volunteered for the Marine Corps and served in Vietnam. He went ballistic, rushed to a bookcase and brought back a large photo album with news clippings about all the supposed atrocities committed by American soldiers in the war, clippings from bogus sources like John Kerry. He actually yelled to Sue and our young kids about my having killed babies and old people. He was completely over the edge, probably upset to have me in his house. We got up from the dinner without so much as a word, packed our stuff and left. We ate our Thanksgiving dinner at a coffee shop and I drove all night to get home.

The incredible thing to me was how absolutely certain these people were of what happened in Vietnam with no interest at all in comments from vets who were there. Those are the people, in academia, who wrote the schoolbooks telling the history of the Vietnam War to our children.

Jay Standish, Mission Viejo, California
2nd Bn, 9th Marines, 3rd Marine Division, Vietnam, 1968-69

Take Another Little Piece of My Heart
David B. Wallace
Sandy Springs, Georgia

IN MID-DECEMBER, 1969, in the Bassac River just north of Cua Lao Dung, known as "Dung Island" in the Delta region of Vietnam, I learned what it means to hate. I was a 24-year-old US Navy LTjg, Officer-In-Charge of PCF-32, a SWIFT boat on patrol on the Bassac, the southernmost major river in the Mekong River complex.

Our boat call sign was *Elbow Golf 32*. We were part of a three-boat patrol, but the other two boats were far away on this night, one in the patrol area nearest the South China Sea and the other was the furthest inland, reaching as far north as Binh Thuy.

Swift boat (PCF, patrol craft fast) 32 on patrol
(shown before modifications made for David Wallace's crew)
Public domain photo, no rights claimed

Our job was dangerous. A 50-foot-long SWIFT boat is a relatively large target with no place to hide on a river or canal, especially if the canal is narrow, thereby limiting maneuverability. This was an area enemy units used to hide and rest, and to infiltrate troops and weapons on the rivers and canals at night. We were based out of Cat Lo, just inside of the bay called Vung Tau and to the southeast of Saigon. Transit to the Bassac from Vung Tau was three to four hours depending on weather and seas. We could go up the river to Binh Thuy to fuel up and re-arm as required. On a typical four to five day patrol we would plan up to three operations per day and at night would usually set a Water Borne Guard Post (WBGP, pronounced "Wib-Gip"), an ambush, or a modified WBGP, meaning that we would patrol, usually on one engine going north and drift going south with the current. Or, like the night in question, we might anchor in the middle of the two-mile-wide river and let the crew catch up on their rest for a few hours, with a two-or-three-man watch with rounds chambered in all weapons and the NOD - Night Observation Device or Starlight Scope – manned at all times.

The crew later insisted that as OinC I stop standing watches as I would be wakened five or more times a night for messages or decisions and I was starting to show some serious signs of fatigue. But on that night I was to stand the second watch from 0300 to 0700. We were on the third day of a scheduled five-day Operation Sea Lords patrol, aggressively pushing the enemy to expose their positions so we could kill them. Enemy contact in close quarters was always exciting, all in a day's work of the impossible task of keeping bad guys out of a free country.

I remember being wakened from the top rack in the main cabin by the Leading Petty Officer (LPO) at about 2330, just before midnight. This had become, by custom, the OinC's rack. There was some flash traffic, an urgent coded radio message, and he woke me while our radioman was decoding it so I would be somewhat coherent when the message was ready for me to read.

The message ordered us to proceed immediately to a small island in the operations area to "lend assistance" and "suppress enemy fire" in support of local militia in contact with the enemy. The local militia were called Regional Force/Popular Force or RF/PF, our slang was "Ruff-Puffs." Instead of the name of the village or the island, we got the location in the form of the universal grid system, eight numbers and letters that could instantly place you anywhere on the map. The coordinates were on a village we knew, a village we visited just two days prior on a PSYOPS mission, Psychological Operations designed to win the hearts and minds of the people.

The island in question was small compared to Cua Lao Dung but large enough to have its own canal leading into the village. The entry canal was perhaps a quarter klick or 250 meters in length with the village on the north or right side of it. The canal was 20 to 50 feet wide, very tight for our boat. There could not have been more than 30 inhabitants in this village; they had all come out to see us two days before to get medicine, candy, Polaroid pictures, and food packages.

Since we were not far away, maybe 15 klicks, it didn't take long to set General Quarters and get there at the boat's maximum speed of 25 knots. Within 15 minutes of receiving the message we were ready to enter that confining canal at night, with no backup boat or air cover. As we approached, we could see evidence of a firefight in the area and dense smoke was rising steadily, backlit by the fires that had been started. Of course, we could not hear anything but the noise of our own two 12V71 supercharged marine diesels, but we were used to this condition. We *were* able to hear the sound of AK-47s (pop-pop-pop) and the return volleys of M-16s (crack-crack-crack). We had to withhold our fire due to the concentration of friendlies in the area.

This was not a good situation but I had confidence in my crew. We had all been in-country several months and had earned our bones in a number of firefights. I knew what they could do and they had learned to trust my lead. Two of the crew, the LPO and Engineman, were over thirty, ancient in those days, and brought to the party the wisdom of a little age. We were not green youngsters on their first combat run. In a small boat crew, you become a little, well, tight. Like a lot of the more experienced boat crews, we had become a bit jaded by the US propaganda machine. Earlier that day we heard on AFVN radio that there were no longer any combat units left in the Delta. Why were we risking our neck in the Delta if our government was saying we weren't even here?

So we were a boat crew who had shot up some bad guys, done a pretty good job, and were starting to suffer from the inevitable *other face* of combat, the excruciating hours of boredom that come between the seconds and minutes of stark terror. We had begun to talk about what we were doing here in the Delta and what were we supposed to accomplish and why couldn't the Vietnamese do it themselves? The thought was just around the corner in a not-too-distant conversation about whether we should be in Vietnam at all.

LT(jg) David Wallace at the helm of PCF-32
Photo courtesy of David Wallace, all rights reserved

As this crew entered the canal, I think we were all concerned about an operation that left us sitting ducks with no room to maneuver and no backup within an hour or more. We could see there was something very bad going on in this place and the pucker factor was rising. Our radio was playing its usual trick of not being able to raise either of the other patrol boats to request they come join us for the fun that was about to start. Likewise, we could not reach any of the Army helicopter gunships, Navy Seawolves, or Navy Black Ponies - OV-10 fixed wing aircraft specially armed for close air support. We would have no air cover. And . . . we sort of *knew* this village. It was a bit surreal, and it was time to improvise.

We popped a half dozen 81mm flares as high as they would go so we could have some chance of seeing Chuck - the bad guys - and started into the canal. My thinking was, if all hell broke loose nobody would have time to pop flares and we wanted to see as much as we could as fast as we could. An 81mm flare at highest elevation will burn for about two to three minutes. We also wanted Chuck to know that some heavily-armed folks were joining in and that we really wouldn't mind terribly if he and his friends quickly vacated the area before we mowed down everything in sight.

Sound heartless? A Swift boat is 50 feet long, has a lot of firepower, a vulnerable 3/8-inch thick aluminum hull, and no armor. Our only defense was to absolutely saturate an area with gunfire before they could stand up and shoot a rocket-propelled grenade or recoilless round at us. If we kept Chuck's head down, he couldn't aim at us very well. We also had permission from our area commander to release weapons if in danger. If being in a narrow canal at zero-dark-thirty with hostile fire on both sides doesn't constitute being "in danger," I was ready to take the court-martial.

Chuck decided to vacate the village just after we illuminated the area and started into the canal. That was the end of the good news.

Chuck had come into this village on a mission of terror as payback for the village chief's audacity in receiving us two days earlier. They had butchered the village chief in an incredibly brutal manner along with his wife. Several other villagers were killed or wounded, hacked apart with machetes or cane knives. As we were taking in the gruesome scene, an old woman approached the boat. We had run the bow up onto the shore and were holding it there with the engines so we could exit quickly. I was in the pilothouse with the LPO, who was driving, and was still trying to raise support on the radio.

Like a ghost in an apparition, this little old woman came walking out of the smoke, through a bluish haze, carrying a little girl toward our boat. It was the village chief's daughter, a girl who was maybe 12 years old. Somehow this little woman thought we could help. After all, we told them just two days before, "Just call us if you ever need anything!"

At first I couldn't see what was wrong with the little girl; she looked as though she was asleep and very limp. I was in full combat gear carrying a 12-gauge sawed-off shotgun and extremely nervous. As I leaned down to see what this was all about, the Mamma-san started jabbering about the girl and I could see a horrendous amount of blood on both of them. Then she tried to hand the girl to me and at first I couldn't take her because of the shotgun and the fact that I was still carrying the mike for the radio, stretched to its limit. I handed the shotgun to the bow gunner, who was manning the bow M-60 machine gun and I just dropped the radio mike so I could take the little girl.

Someone had chopped at the back of this little girl's head with a machete and had taken a five-inch chunk of her skull off. Miraculously, she was still alive and moaning softly; she even blinked, probably in pain, as the Momma-san handed her to me. But I could feel a pulse and she was alive!

Everything I had ever learned about first aid came rushing into my consciousness in one blinding flash. The rest of the crew sensed that something important was happening and came to an even more heightened state of awareness. Out of somewhere, I was handed a moistened combat dressing and a bed of flak jackets appeared on the stern just forward of the mortar box. A blanket appeared and somehow I was sitting on the flak jackets holding this wounded little child in my arms and trying to keep her brains inside her skull as though I was a surgeon.

I remember talking to her, telling her that everything was going to be "A-OK, in just a little "tee-tee," the Vietnamese slang we used for "a little while, very soon." Our LPO got some other wounded civilians on board and we headed for Binh Thuy, the nearest US medical facilities, but still about 30 klicks to the north, about a 60-minute ride.

I remember sitting on the fantail for what seemed like hours, talking to that little girl and to God. I alternated between asking Him to spare her life and demanding that He cause hideous torture to be inflicted on the animals who had done this to an innocent child. This was one of the most emotional experiences of my life. The little trooper was somehow hanging on and staying alive - I could feel her breath or hear her sigh or feel her twitch and I knew, I just knew that this kid was going to make it. By now I was sobbing like a baby, screaming at the crew to go faster, and alternating between comforting the little girl and my conversation with God.

The kid was hanging on. The crew was throwing ammo boxes, full and empty, over the side so we could go faster. The snipe (Engineman) pumped the forward fuel tank into the Bassac River to lighten us up, and the crew asked if we should start tossing the bedding, radios, guns, and food.

As if in cooperation to make us go as fast as possible, the Bassac River assumed a calmness that, under other circumstances would have been eerie. The surface of the river was like glass and we were fairly flying across it. Even though not normally capable of it, the 32 boat was on plane, riding on the very top of the water with barely screws, skegs, and rudders touching the surface. The engines were shrieking at full speed, but there was an eerie silence shrouding me as if I was in a special zone.

Our little passenger was hanging on. I know it, I know it . . .

Finally we saw the lights from the fuel pier at Binh Thuy. The LPO, who was driving, worked the radio and finally raised the watch at the naval station there, telling them we were arriving with wounded and would make a rapid approach to the piers, keep the medical personnel back, but be sure they are there. We toyed with the idea of just running the boat up onto the beach, but decided against it to minimize the shocks and jolts to our little passenger.

The crew brought the boat to a bone-jarring stop and I was over the side onto the pier with the little girl before the first line was thrown.

Something was wrong.

There were no US medical personnel who could treat these Vietnamese civilians. I told the Navy Chief who relayed this wonderful bit of information to me that he better get an Admiral or a Captain down here quick or our boat would start getting attention the only way we've been trained. He started to get tough with me and then thought better of it. I stood there with a wounded child and what must have been a very wild look in my eyes; he was undoubtedly right to try to avoid trouble.

The Chief was quickly replaced by a Lieutenant who seemed preoccupied with the fact that he outranked me and who was intent on explaining to me why the US doctors can't help a Vietnamese girl. I muscled past him with a hearty "Screw Off!" and he snarled about court-martial for insubordination by junior grade Lieutenants. At this point the Lieutenant and a court-martial were of little concern to me.

A Vietnamese doctor approached me gingerly, having observed what happened since we arrived. He quickly looked at the little girl and pronounced her "finee," or dead, and started to walk away. The Gunners Mate, a tall Mormon from Nevada who saw what was going on, came up behind me ready to help. I quickly handed him our little charge and grabbed the Vietnamese doctor, spinning him around while I jerked my .45 from its holster. The doctor, who said he didn't speak English, suddenly started reciting Shakespeare when I jammed my .45 into his mouth and explained what I wanted him to do. The Binh Thuy Lieutenant and Chief were warily approaching me again with the idea that they could talk me out of blowing this guy's head all over their fuel piers. They promised the Mayo Clinic if that would get this crazed OinC to cease and desist. The Vietnamese doctor examined the little girl and muttered in Vietnamese. I re-holstered my pistol.

A Vietnamese nurse who spoke some English told us they were taking "our little girl" to the Vietnamese hospital right away and she would receive the very best of care available. I think she somehow knew how important this was to all of us. They took the little girl away on a padded stretcher. I don't know if she lived or not. I don't even know if she was alive when we got her there. None of us ever found out.

THAT NIGHT I LEARNED two things that you may believe are completely contradictory. I learned what warriors have known forever, how to hate an enemy without mercy, to be totally unforgiving, to be brutal without remorse because the enemy in Vietnam was doing unspeakable things to their own innocent people to advance their political goals. I also learned for sure that there is a God.

If there isn't a God, I don't think we could have flown across the waters so fast that night. If there isn't a God, I don't think I could have lived after holding that little girl in my lap for what could have been the last minutes of her young life. If there isn't a God, we would have surely killed innocents that night in the crossfire, because, without air cover or backup, conditions were ripe for us to open up with our considerable firepower, the crew was anxious, and it was nearly miraculous that nothing happened to make us pull the trigger.

It was a little bit of magic amidst too much misery. I guess that's war.

Someone once told me that during the US Civil War, a new recruit asked an old soldier after the battle of Fredericksburg, "What is war like?" The old soldier thought for a minute and said, "Can you describe an elephant to a man who has never seen one? I have seen the elephant."

Thanks to Lyndon Johnson and Robert McNamara, the bright minds that committed us to a war then screwed it up with their micromanagement, I saw the elephant and I became something I never wanted to be: a warrior. I was pretty good at it, too, and I'll always resent them for that. I learned to live with the elephant that terrified me.

The elephant brought heartbreak when good people died, and heartache while we all did our job day after day and struggled to keep each other alive while we wished we were home. I learned from the elephant to accept many things, but I will never forgive the ones who chopped on that little girl's head. I regret that I didn't have the chance to kill them myself that night, and still to this day I hope they rot in hell.

On PCF-32, the crew liked to think the little girl lived. Maybe she did. I would be grateful for that, just as I am grateful for the hand of God that surely must have been with us that night.

David B. Wallace
Photo courtesy of David Wallace, all rights reserved

David B. Wallace has worked in the paperboard packaging industry as a manager in the industry or supplier to the industry since early 1973. He is director of the Swift Boat Sailors Association. He is descended from the line of the great Scottish patriot Sir William Wallace. The first born son for a couple dozen generations has the middle name "Bruce" after another Scottish patriot named Robert the Bruce. David continued the tradition with his first born son, who is a graduate of the US Naval Academy and a veteran of the submarine service, and who also continued the tradition with David's grandson. David lives in the Atlanta area with his wife.

Postcard to America

DURING THE INCURSION INTO Cambodia, in May of 1970 we flew a hot lunch to a company of the 8th Cavalry, but it seemed they weren't hungry. That was very odd since grunts got only an occasional hot meal, a treat in a jungle war. I asked why and a Lt showed me a sight which haunts me to this day. The grunts had found 20-25 Cambodian men who had been impressed as carriers for the NVA on the Ho Chi Minh Trail, slaves. Most were in leg irons, all were severely malnourished, and several had a limb black with gangrene. All had thousand-yard stares, and were left behind simply because they were too weak to keep up with the Reds retreating as US forces advanced. We were amazed that the brutal NVA did not kill them.

One of the medics had given a couple of them morphine to ease their pain and they simply went to sleep and died. He blamed himself for the deaths of the men he tried to help. His buddies stayed with him and talked to him. Little was eaten by the grunts; they did try to feed the Cambodians, but with little success.

When we flew back to the firebase we told several reporters there was an atrocity and agreed to take them to the village since they were eager for an atrocity story. When we landed and they learned the details, they were disappointed it was not US troops who had done bad things and had no interest in the story.

I was pretty pissed off about their attitudes. I asked if they were going to write about this and they said no. I asked why and they said this was a just a cultural thing and they didn't see any real story. I asked what they would have done if the bad guys had been US troops, and they indicated they could really get excited about that.

When the reporters wandered away I took off and left them there in the boonies with the grunts. I know it was a crummy thing to do - to the grunts - but I did not want anything to do with the reporters after that and have suspected reporters since. No story to my knowledge was ever published about this incident. All the reporters had to do was tell the truth. That is the seed of how the history of the Vietnam War was twisted, how the truth was lost.

My son came home from Iraq in 2006 and tells me the same thing was happening there. The beat goes on.

Sig Bloom, Jonesboro, Georgia
B Co, 229th, 1st Air Cav, Vietnam, 1970

Honor and Pride
Dexter Lehtinen
Miami, Florida

LEADING A GROUND UNIT in combat sometimes requires a bit of role-play. When I was tired and scared I couldn't let it show because a good leader is the last to rest and sets an example of self-control. When I thought a mission was a waste of time and effort, not to mention men, I could complain to my superior but had to give orders to my men that seemed full of confidence, as if knowing the risks we took had a meaningful purpose. So, whether I believed in our mission of the day or not, if I did my job well my men would never sense my doubts.

I was a US Army Ranger in Vietnam in 1970-71, a Lt. assigned as Platoon Leader of Second Platoon, Troop A, Fourth Squadron, 12th Cavalry, First Brigade, Fifth Mechanized Infantry Division. We were Alpha Troop, the *Red Devils*.

Our AO included the hills and the jungles around the Rockpile, a jungle-covered granite ridge near the DMZ. A French-colonial road named QL-9 wound its way through the area, the only route to Khe Sanh and on to Laos, a key highway to keep clear of enemy activity. We spent most of our time searching for an enemy who usually decided when and where we would fight. They hid in caves and tunnels and spider-holes, or just cooled their heels in the jungle, while we tried to find them, or at least keep them busy, to keep the road open for friendly convoys.

When our position was on the Rockpile, or visible on a hilltop, I always wanted to be on the move because our enemy would use the time while we relaxed to organize an attack on our position. So we moved a lot.

My troops were young men who had recently been boys, and those who lived through their year were aged far more than that period of time. This was late in the war, and a sense of mission was a bit rare among the grunts. The relentless anti-war movement at home had a cumulative detrimental effect on morale of draftees, and though leaders like me hoped for self-motivated troops, sometimes motivation came from the heel of our boot. My role as the platoon leader, or "El Tee" as my men called me, suited me well, and I was prepared to deliver that motivation as the need arose. Most often I was preoccupied with things like a good defensive position for the night, where we were going and whether there was a less risky way to get there, how my men were deployed, where the weak spots might be and other things we hoped might keep us alive.

As an officer, I had responsibilities and could not say out loud the cynical thoughts that sometimes bubbled in my head. There came a time, though, when I was completely relaxed, when nobody weighed my words or looked to me for decisions, when the fate of my men did not depend on my orders, when powerful medication eased my pain and let loose my thoughts and dreams to run free . . .

POSSIBILITY AND CERTAINTY.

If you went down that faint trail, through the brush and crossed the stream, you might get shot. But that was just a possibility. It was the certainty of what happened when you crossed a stream that made it so bad, made your skin crawl. That's why nobody wanted to cross streams.

Getting shot was a real possibility, but not likely on any one crossing. If you crossed streams ten times, it started looking worse.

Booby traps, trip wires, maybe even real mines were a bigger danger. They could scare the shit out of you when you made the mistake of thinking about them, or the mistake of remembering the last guy who tripped one or stepped on one.

This was especially so if he didn't die, or at least didn't die right away, because if he didn't die quickly he was screaming and you had to do something about it, like find the parts of his body that were blown off. If he was dead, you tried to stick loose body parts with the body because you thought it was the thing to do, but there was no urgency. You could be a little more careful about other booby traps or trip wires or mines as you looked for the rest of the guy, and you could keep the rest of the patrol away and a bit more organized because the thing was over, there was nothing they could do.

And if the guy died then you had a little more time to think about what you were going to do, about whether you were going to go on. I mean, you were supposed to go on, of course, one KIA isn't supposed to stop a war. One booby trap shouldn't stop an army; you knew that. I mean, what the fuck kind of army would it be if one booby trap or one KIA stopped it?

But, if the guy was KIA, you at least had some time to consider some options. Maybe you could go around, or maybe you had come far enough already, or maybe, after the chopper took the body away, you could kind of look at where you were supposed to go long enough with binoculars or something that made you feel like you had kind of accomplished your mission and then just say you went there and didn't see anything. After all, you did look at it quite a while and didn't see anything, or maybe the Troop Commander would just decide on his own that you should come back. There were always options for carrying out what seemed like clear orders - if the guy was dead.

But if the guy who was blown apart was still alive, then the shit really hit the fan. The medic was frantic, you were trying a get a medevac helicopter on the radio while your other guys were having real trouble with the screaming. And you think you've got to find his arms or legs right away, as if the medic could reattach them right there, right then, or at least do something useful with them, because he's still alive. So you have to find the separate parts quickly, forget the booby traps and trip wires and mines, because it seemed like the separate parts had to be kept alive too. Forget the biology, those parts were alive. Why else would we want so bad to find them and save them too? If the medic was the guy blown up, that turned a real mess into a real fucking mess.

So, if the guy was alive, you were trying to do everything at once and everything took forever. Helicopters to take out the dead came right away, but medevac helicopters to take out the wounded took forever. When a guy is dying right there in front of you and you can't do anything about it, a minute is a lifetime. You figure you go through a shitload of lifetimes before the medevac arrives.

And if the guy's still alive when the medevac gets there, they come right in to pick him up because you're yelling at them to come in right now, even though everybody knows this could be an ambush to shoot down the medevac and everybody knows that they ought to be a little careful to recon the area or check it out somehow so they don't get killed themselves. But to hell with what everybody knows. You want them in right away, maybe the guy can live, maybe they can put a leg back on or plaster the guy back together somehow, and the medevac comes in right away.

They shouldn't, I guess, without first making a quick run around the area, especially if the guy was shot because that means the bad guys who shot him could still be there to shoot again, but that's only a possibility, not a certainty, and because our army has a shitload more firepower than his army, he usually strikes and then gets the hell out of Dodge in a VC minute, so it's really only a possibility that he'll shoot again and not a certainty. But the guy's alive so the medevac crew just comes in like they're immune from the war. I mean, even though they're picking up a guy just blown up in a war, they act like there's no war at all.

Once in a while you think maybe you ought to warn them that there's a war on and maybe they should be careful to avoid getting shot, but you need them and you really want them acting recklessly for your own interest, so you seldom warn them off even if you should. You figure they know what they're doing anyway, and if you knew for sure that they were going to get shot, you would warn them.

But you're rarely in that situation, where you're pinned down by overwhelming force and you know they've got the RPGs ready to kill a bird. That's the big war; it happens somewhere, I guess. It's on TV but it's not this war. You're always in the situation where one or just a few fucking ass VC shot at you from you don't know where with you don't know what, or he dropped some mortar rounds on you from God knows where or maybe an RPG, or you tripped a wire set by some shit-faced VC three years ago who's now eating rice some fifty miles away while you struggle to get the consequences of the wire medevaced while he's still alive.

Of course, your problems are about twice as bad if there's no clear ground around for an LZ. Tall grass, as high as a man is OK, its trees and shit that stop helicopters. Then you've either got to find an LZ, which means dragging the guy somewhere else, or clear one, which means chop one out with machetes or blow one out with C4 or try to walk in some artillery without getting more guys cut up or blown up by your own or your guy's fuckup in the rush.

All of this assumes you weren't the guy who tripped the wire or stepped on the mine or took the mortar frags or AK round. If you were, well, that's a different story. Either you've checked out or you're the guy on the ground screaming. Either way, the consequences of the stream would no longer be a certainty. That's the only way you'd beat the certainty of what happens in the stream - take a hit.

Assuming you weren't the guy who took the frag or the round or the explosion, then once the medevac was gone, if the guy had been alive, is the first chance to consider your options.

You were supposed to cross the stream. Getting shot or tripping a wire or booby trap or stepping on a mine or taking incoming at the obvious candidate for prearranged targeted fire, was only a possibility. Whether it happened or not, it's always only a possibility in the future. But what happens in the stream is a certainty. If you go down that kind-of trail, through the brush, and across the stream, there's something bad going to happen just as sure as you're wearing Army green.

You look at it, or really the brush next to it, and you figure it's thigh to waist deep. Maybe stream is too clean a term if it implies flowing, clear water. Maybe ditch would do better but ditch means someone dug it, and these weren't dug, they seem to be natural with brush or bushes on the sides that you have to push away or chop through if there isn't a trail, a trail that you both want and don't want, because it means you don't have to push or chop to get through but it means the other guy knows you might come along there, too, taking advantage of the superhighway that trails amount to in the bush, with the occasional toll booth that an ambush is in the bush, the price for using the highway. The water is dark brown, never clear, so you can't see anything, and it's probably flowing downstream a little but you can't really tell.

So you call it a stream because there's nothing else to call it even though it won't ever make it on a postcard like a nice stream. And you call it a stream because the map says it's a stream. It's a long line, for water, between a bunch of other parallel long lines, meaning you're going down when you go toward the line for water and up when you're going away from the line for water, and just once in a while the line for water crosses a line which means the stream is going either up or down, but since a stream's not supposed to go up its got to be going down, even if the water doesn't look like it's going anywhere when you actually get there.

So the map tells you the slope down, the slope up, where it goes, and where the other guys might be sitting to watch you when you cross. But it doesn't tell you what's in it, and what's in it is what's for certain. No matter what you do, it'll happen. No matter how tight you tie your pant legs. No matter how you seal your boots. No matter how you tighten your belt. No matter even how you tape your trousers and jacket. You sit there and make it impossible to happen and it happens. Those bastards are good, really good at what they do. And they're fucking ass bad.

That's why nobody wants to do it. It's not the danger on the other side. The other guy's pretty smart, or smart enough, and he knows you'll stop after you cross. He knows you'll be a good target and he can preset an ambush with direct fire from his guys hiding out across the stream or he can just mortar and RPG you using his preset ranges and azimuths. But we're pretty smart too, or smart enough anyway, so we know he knows and we try to counter the obvious threat, the possible threat from the VC. But you can't counter the certain threat of the stream.

You counter the danger on the other side, or you supposedly counter the danger, by first sending two guys across. They're supposed to clear the far side by walking two boxes, forward and then to the left and right, seeing if anybody's hiding in the grass. This doesn't do much for the indirect fire ambush from mortars on the pre-set target or the long-range direct fire from RPGs and machine guns from far off higher ground, but it's the standard method for crossing a danger area, so you do it. Never mind that almost every area is a danger area; these are, I guess, more-dangerous-than-normal danger areas. You pick guys who take the VC seriously, but even they will be distracted after they cross. And they've heard the advice too: what happens in the stream won't be a long term problem if you act right away. So they should ignore that advice while they check 40 meter boxes, moving slowly and watching for trouble?

That counters the enemy, as best you can, anyway, and that's all you can do. Unless, of course, you called for artillery fire at every danger point just to blow away anything that might be there. Which, once in a while you do, when for some reason your CO and the arty guys are willing, like maybe there was an ambush there before. But usually that's ridiculous; even America doesn't have that much ammunition and that many spare gun tubes. And if you want to get anywhere at all you can't take a milk run with stops for blowing up the trail in front of you at every interesting terrain feature, which are about every hundred meters. And even if you do it just once or twice, you've pretty much told the whole neighborhood what you're doing. So much for the element of surprise. So you pretty much use artillery the way you're supposed to, after you know, at least you think you know, where the fucking guy is, roughly speaking. And that usually means after he fired at you. So if he's good, that means after you've taken a casualty or two.

But he's not always good. He's like us, sometimes good and sometimes he screws up. Knows better, like us, but fucks up anyway. So when you go out, you wonder, is he going to be good today or is he going to fuck up? Is this one of his good days, or one of his bad days? And for us - good day or bad day? Will we know what we're doing, or will we just fuck everything up, even without his help?

So you get ready by lighting cigarettes. Even if you don't smoke. Cigarettes have a lot of purposes, and I always figured they had to keep them in the C rations because of the streams, at least Vietnam streams.

That's why there's one rule about streams at night. Don't cross them. You've got cigarettes lit and they're like goddamned neon lights in the dark. Lighters. Matches, that you don't light 'til you're across, amount to the same thing. Or worse, in a hurry the flash is more likely to be seen than on the other side when you lit the cigarette. So light up on the other side and cup the cigarette, if you absolutely, for some reason, had to violate the first rule of streams at night - which is, don't cross.

When you're in the stream, you think you can feel them but you can't. You just know they're there because you've done this before. Once across and out, you try to get everyone to go in pairs roughly to positions in the defensive arc that's been discussed and practiced and ordered so many times. And to see that the even or odd numbered guys, whichever you picked just before crossing and sent back along the column, stay in defensive alert, watching outward with weapons ready while the other guy in the pair strips down to remove the leeches. From his legs. Inside boots. Crotch. Buttocks. Between his buttocks. Everywhere. They've already got a lot of blood in them. You can't pull. You've got to burn them off while the other guy is getting pissed that it's taking so long.

Boots off. Trousers off. Underwear off if you wear any. Sticking each one with a cigarette, ten at least, probably more. And the other guy has to look to see what's between the first guy's buttocks. All while you're supposedly at war, in a danger area. Not regular all-the-time danger area, but what the manual calls a real danger area, a "more-dangerous-than-normal" danger area. Pants down, searching between buttocks, in the middle of a patrol in the middle of a war.

Leeches look like they're free-lancers. They're so bad they couldn't be on anybody's side. The tactical impact or strategic role of the leech could be a subject for an intelligence analysis, which would undoubtedly conclude that it depends on who needs to cross the stream the most. In other words, it depends on the disposition of forces, nature of the battle, and that kind of stuff. Like other intel assessments, it would fucking "depend." The leech would be bad, that would be clear, but conclusions would depend on so much other shit that all you'd know in the end is that you want to stay away from leeches.

Kind of like VC. You can smoke them out and kill them, but when you come back others that look just like them are always in the same goddamn place that you killed the first ones. No one or group of leeches, or VC, can survive if you have the time to pay attention to them. They're history, after they've bloodied you. After they've stuck to you and sucked out some of your blood. You always get them, but only after they've gotten some of you. The blood suckers know the terms of trade. They get some of you and then you kill them. But they always get some of you.

And when you come back, there's more because leeches and VC think alike: "These guys will get tired of giving up their blood, even if they kill the leeches and VC who do it," they think, they know. It's not the national interest, or the rights or wrongs of the war aims. It's the facts, the reality of the leeches. Because the leeches always get the first bite, they always get some blood, before you get them. Because there will always be more leeches, just like there will always be more VC.

NOW AND THEN, FROM just below the surface of being awake, I could hear noises and voices. I wondered where I was and what happened to me, why I couldn't see, but then I would slip down deeper . . .

REASONS.

You could break out at any time onto an open area or a trail or a kind of trail-tunnel under the jungle canopy. Or so you tell yourself and you tell everybody else so you've got a reason for doing this. I mean, without a reason, you'd just be stupid, this would be senseless, so you've got to have a reason. It'd be better if the reason made sense, but if you can't find that kind of reason lying around then one that doesn't make a whole lot of sense is still better than nothing. And, of course, it'd be better if the reason was a real reason. Not "the" real reason, because you learn right away that there's never a "the" reason; instead, there are shades of reasons, categories of reasons, different reasons for different people, even for the different people who give the orders or pass them down. But if you have to you can settle for anything that will fill in the block labeled "reason" in your mind.

I'm not exactly sure why we need reasons at all. Maybe we should just settle for "We're chopping through this bush for no reason." Why shouldn't that work? You could still swing the machete the same. The vines and brush and the other green and brown stuff you're chopping through would still be the same. They'd give in or fight back just the same. How would the fucking vines know what your reason was? Your reason for chopping through might mean something to you, but would the vines really give a shit?

We're smart, we think things out, we know what we're doing and why. When we fight these vines, which is the particular war we're in at the moment, and when we fight the VC, which is the war we're in once in a while when the VC decide, we know why we're fighting. A vine, on the one hand, is dumber than a fucking rock. It doesn't know why it's here; it doesn't have any goddamned idea why it's fighting. What does it hope to accomplish anyway? I can cut any one of them, even the thickest. I can blow an LZ out of the thickest bush if I have to, with enough C4 and I'll always have enough C4 because the stuff is light and they can drop as much as necessary from a helicopter if it's important enough. So the fucking vines can't win. So what the fuck is this dumber-than-a-rock vine fighting for?

Of course, there's not just one vine. That's where it gets confusing. There are other vines, in fact, a whole shitload of vines, pretty much the definition of infinity, so even us thinking guys, us guys who think we know what we're doing and why, can't seem to chop them all down. The vine looks at you and says: "You're the one who's dumber than a fucking rock. I grow back! You don't know shit about vines! Maybe you can multiply and read and write but you don't know shit about vines! And you don't even know why you're here!"

Even if the vines make some sense in this dialogue, you chop through them anyway. That's your job. That's why they sent you to Vietnam, to chop down vines. They're too many goddamned vines in Vietnam and American national security is at stake. We know what we're doing. We have our reasons.

Even in a vine war, the sun goes down. So we go back. Back along the trail we had cut all day. Or, to be precise, the front of the column goes back. We had chopped through the jungle all right, it wasn't impenetrable, nothing was truly impenetrable. We could penetrate it if we had our machetes and we had our reasons. We worked for about six hours, hacking for about five if you subtract the breaks you had to take because you sweat so much you couldn't see, even if you weren't chopping; we weren't stupid, we could think, so we rotated the front four guys every so often so as to not wear out the one chopper at the front, but you couldn't rotate too easily because this was a very thick jungle and you had pretty much cut just a one-lane tunnel and you couldn't spend forever cutting passing lanes cause you were supposed to go forward, not sideways, and if you went sideways too much the fucking vines would win, so you just crawled over one another any way possible, which made you more tired as you rotated so you didn't get too tired. We made amazing progress, when you think about it, considering the infinity of vines and the total lack of vine consideration for vine life which meant that the front vines were continually sacrificed by the vine army without any chance against our machetes. Even against this senseless sacrifice, we hacked almost ten yards an hour, five hours, fifty yards. Fifty hard-earned, well-engineered, motherfucking yards.

So the front of the column came back. You see, there were eleven guys in the patrol, or, at least, eleven guys assigned to the patrol, prepared to move out with the patrol; I don't know if you could say that the last couple were really "in" the patrol. They were about five yards apart, the separation you want between guys so one trip wire or one booby trap or one mine or one mortar round doesn't get more than one guy at a time. Now you might say, "How the hell would Charlie have been able to set trip wires or booby traps or mines in the middle of the vine army that was so thick you had to hack for an hour to go ten yards?" The answer is I don't know how he does it but sometimes he does, and anyway you've got to keep the habit every damned waking minute cause this is goddamned Vietnam, and anyway Charlie can always drop mortar rounds on you through the jungle canopy. You gotta treat mortars like oxygen, they're everywhere all the time. Even if you bunch some at the front, which you do as sure as you run out of water, it just fucking happens because you could put steel I-beams between the front guys and bring along a firehose and they'd still bunch up and you'd still run out of water.

The last guy was supposed to drop back some as rear security, so it was 50 yards for eleven guys any way you look at it. And that's what the patrol went, 50 yards. I mean, that's what the front of the patrol went. The middle guys, about 25 yards. And the last guy, the LFG, the last fucking guy, well, he didn't go anywhere. When the front of the column finally trudged back to the start, there he was, sitting with his back against the track of the APC. It wasn't that he didn't do his job, he was looking out from the platoon position with his rifle between his legs so he looked as ready as one could expect or demand if the occasional VC sniping or mortaring started, and he even had a machete on his LBE belt for the inevitable and not occasional fight with the vine army. But he just sat there all day, as one guy after another inched down the trail tunnel we were cutting, one new guy going about five yards every half hour. The tenth guy made five yards at least, but the LFG didn't move a goddamned Vietnam inch - he never got his moment of glory in the war against the vine army.

But he apparently did his job as rear security, he and the twenty or so other guys in the platoon that stayed in the platoon APCs that were circled in a temporary defensive position which was the base from which the patrol attacked the vines. I don't know if you could say that he was on or in the patrol. That's a question of metaphysics or some shit like that.

When the front of the column got back to the platoon APCs, the patrol was over, the short, honorable, didn't-accomplish-a-goddamned-thing-except-kill-a-bunch-of-vines patrol, was over and even that was in doubt because the vines were infinite and any fixed number taken away from infinity still leaves infinity, at least it does in vine math, so we were in the same shit-ass place we were when we started. To make matters worse the vines had said we were the fucking dumb rocks because the vines come back to life anyway, so you don't really kill the bastards and you couldn't really ever win against the vine army. Maybe that's why they seem to sacrifice their front line vines without any regard for vine life; maybe they do know what they're doing after all, maybe they have their reasons, vine reasons but reasons nonetheless, that our western culture just doesn't understand.

And when the front of the column got back to the APCs, the day was over too. "You know, El Tee," Johnson said. "I don't think we're gonna win this war. There's too many of 'em. There's just too many goddamned vines."

Another day. Another fifty yards in a long war. Fifty fucking Vietnam yards.

I WAS VERY CLOSE to awake for a moment. I know I'm in a bed, I know by sounds and smells I'm in a hospital, I must have been hit. Why can't I see? Why can't I open my mouth. My face hurts like hell . . .

BEING SURE.

There was only one way to be sure. I mean, to really know for sure, because memories are tricky things.

I remember the horse-drawn milk wagon coming to the house when I was a kid. Clear as a bell. Yet my mother said there never was a horse-drawn milk wagon. There was a milk truck delivering milk bottles to our rural house the old fashioned way, but it didn't have horses, she said. There were horses across the road in a stable, and there was a hay wagon over there. We rode once on the hay bales for Thanksgiving, pulled by a horse. And a runaway horse even broke into our back yard one time, injuring the rider on our clothes line. I remember that clearly, too, and Mom said I was right about that. But not about the milk wagon. She always figured I just somehow mixed up the milk truck with the hay wagon to get a milk wagon. And I guess she was right. How else would something that didn't exist in reality end up existing in my mind? Make no mistake, I believe Mom, but I still remember the milk wagon.

And I remember lying on the ground on my back with my eyes closed. At least I thought my eyes were closed, since I could normally see so they must have been closed because I couldn't see anything.

Why was I lying on my back? Why was the platoon sergeant saying, "You'll be OK, El Tee. You'll be OK, El Tee." I knew his voice. And why was somebody grabbing at the pouch at my right front side, where my first aid packet was? I always carried my basic equipment, even if I went just ten feet. Rifle. Equipment belt with compass, canteen, first aid kit, a smoke and a frag grenade, some ammo, and so forth. Helmet. My father said to always wear my helmet, and, you know, I almost always did. But this time I had been carrying it as I walked across the platoon defensive position to talk to the platoon sergeant. Why wasn't I walking? Why was I lying on the ground? Why couldn't I see?

I heard the wop-wop-wop-wop of helicopter rotor blades. Why was I on a helicopter? Why was someone holding me? Why couldn't I see?

But that's the point. Was I really on that ground? Was I really on that helicopter with someone holding me? Did the platoon sergeant really say I'd be OK? I remember it as clear and plain as yesterday.

But I remember the horse-drawn milk wagon, too. Everybody has dreams, often including slices of reality, even if distorted reality. But you see the point. The logic. If that wasn't real, or was real only in my mind as a collage of other realities, then how should I know whether lying on my back on the ground was real? Actually, I lay on my back on the ground in Vietnam . . . a lot. And how should I know that being held on a helicopter was real? I was on helicopters in Vietnam . . . a lot. And I grabbed first aid packets and sent out wounded, too. So all the parts are there to be pieced together into whatever collage the mind wanders into.

Then there's the woman. She says, "It's your birthday. You're going to be OK." OK from what? Of course I'm OK. Nothing's happened to me. But why was I on my back? And why couldn't I see? And, you know, why couldn't I talk, either?

She says, "You came in three days ago, hit in the head and shoulder, but you'll be OK. There's a thing in your throat so you can breathe, so don't mess with it." Or mess with any of the other stuff she's going to mention. "Your eyes are swollen shut and your jaws are wired shut and some of your left cheek and cheekbone are missing and stuffed with gauze but all this is fixable so you're going to be OK." She seemed really nice. All that stuff she said didn't seem like anything because she said it was all fixable and I'd be OK, and she seemed really nice so I believed her.

But could that have been real? It seems like a fairy tale. I mean, you're shot up or mortared or whatever, obviously pretty bad, and a guy who doesn't know anything more about biology or medicine than you do, which isn't a damn thing, says you're going to be OK and you believe him. And then some woman who you know absolutely nothing about says you've been unconscious for three days and some other pretty amazing stuff but says you're going to be OK and you believe her.

And all of this when you can't see and everything obviously isn't really OK, but you believe them anyway because you want to believe them. The fairy tale is, everything is OK because you believe. If you believe, then you can fly. You don't want reality, you want the fairy tale. You want everything to be OK so they tell you everything will be OK and you believe everything will be OK and then everything is OK because you believed.

But there's only one way to know for sure. I just want to know for sure about the ground and the platoon sergeant and the helicopter and the woman. Nobody will see me checking. Nobody has to know that I'm not sure, at least not sure right at this moment as I think about it. If they thought I had to check, now and again, to be sure, they might get the wrong impression, might not understand, and, well, that would not be a good thing. Anyway, it didn't happen to them, so they don't know why you'd have to check to be sure if it happened, or might have happened, to you. If it had happened to them, or might have, then they'd know there are times when you're just not really sure without checking.

So I think about doing it, about reaching up to check. I should do it. It's not dumb . . . at least I hope it's not dumb. All you're going to do is check.

I REACHED UP SLOWLY and put my left hand on my left cheek, to see if it's all there. I reached silently, and it was not all there. Some bone was missing. The skin was tight, drawn in, numb. It's fine, but it's not all there. A reality like this is not a problem. A dream like this would be a nightmare. Right now, and for a while anyway, I know for sure.

FINALLY I CAME OUT of my daze, trading dreams for sharp-edged reality. Waking was slow and sluggish, like a scuba diver having been deep under water for a long time, slowly coming to the surface, decompressing, trading the soft comfortable dark for the harsh brilliance of light above the surface. I found myself in a hospital ward, on a Navy hospital ship off the coast of Vietnam. I still couldn't see, but the nurse told me that should be temporary, at least my right eye should be uninjured. My jaws were wired shut, and my left cheekbone was missing, a gaping hole in its place.

I really had been unconscious for a few days. I really had taken shrapnel that destroyed most of my left cheekbone, left me mostly blind in my left eye and tore up my shoulder, too. Is this what they meant by being OK? It must be. They must see a lot of guys in worse condition. Maybe I would be OK.

It took a while to penetrate the fog and piece together what had happened.

My replacement had arrived that day and I was headed to the Rangers, where I wanted to be. We had returned from a short patrol, got our mail and I was sitting on the APC's lowered ramp, reading a letter when someone heard the telltale swishing sound and yelled "Incoming!"

I grabbed the radio and stood on top of the APC to try to see where the enemy was firing from so I could call an artillery strike on them. Either a rocket or mortar hit the explosives box on top of the APC, it spewed shrapnel everywhere when it exploded and that's apparently when I was hit.

My men helped me, held me, put out the fire where my hair burned off and bandaged my face until medevac arrived. Somehow I ended up on this hospital ship near Da Nang.

My thoughts drifted to how my Platoon Sgt., Earl Brooks, was doing with the new Lt.? I wondered about Copper and Hawn and Barnes and Bradley, Cowboy, Pillsbury Dough Boy, Ducky, Supersport, our Kit Carson scout, Boy Song and all the others. I thought maybe it wasn't so bad after all that they wore love beads and peace symbols, and there were worse things in the world than giving their war machines, Sheridans and APCs, hippie names like "Aquarius" and "Blood, Sweat and Tears." I longed to be out of the bed and back with them.

But for me the war was over. I was sent to St. Albans Naval Hospital in Queens, New York. I didn't know it then but that hospital would be my home for a year and a half while I recovered and had corrective surgeries. One of my earliest recollections in that hospital bed, while still blind and with my jaw still wired shut, was hearing on TV the testimony of a young man before the Senate Foreign Relations Committee.

He said American soldiers in Vietnam " . . . raped, cut off ears, cut off heads, cut off limbs, randomly shot at civilians, razed villages in a fashion reminiscent of Genghis Khan."

Where did that happen, I wondered? Don't tell me we have another My Lai! That idiot William Calley smeared us all with his war crimes at My Lai, and now I wondered where another atrocity had occurred that would tarnish our reputation as soldiers even more.

But as I lay there and listened, it dawned on me this man was saying something far worse than another atrocity. He was talking about all of us. In disbelief I heard him tell the world that I served in an Army reminiscent of Genghis Khan's, that officers like me routinely let their men plunder villages and rape civilians at will, that war crimes committed in Vietnam by my fellow soldiers ". . . *were not isolated incidents but crimes committed on a day-to-day basis with the full awareness of officers at all levels of command.*" He said America's *policy* was to commit atrocities in Vietnam.

Who is this guy? I couldn't see the TV, so I had to ask someone to tell me his name. They said his name was John Kerry, an ex-Navy LTjg who had served in Vietnam. How could he get away with telling a load of lies like that on TV, to Congress no less!

He was talking about me and my men. Me and my men!

WHEN I WAS WITH my men in the bush, just like them I looked forward every day to leaving this miserable war and going home. Now that the war was over for me, I missed them, the ones I lost and the ones still alive. I never saw one of them mistreat a civilian or a prisoner, and they all knew I would not tolerate it. I think they would not tolerate it from one another, either, for they were like you and me, just regular guys trying to do their duty so they could go home. They were like your neighbor's freckle-faced son, like the paper boy, like your high school buddy, and damn, how I missed them. Why did I want to be back with them in that miserable place?

Maybe it was because we lived and trained and fought together, and sometimes when I lost one we gritted our teeth and grieved together. We checked between each other's butt-cheeks for leeches, we learned together the many things we must help each other do in hopes of surviving the insanity of Vietnam. We would always be able to share that experience among ourselves with a wink, a gesture, a few words that conveyed the understanding nobody else could ever hope to grasp.

MAYBE OUR WOUNDED IN all wars have these conflicting feelings, like you're in another world, on an unreal planet. The people around you now, the doctors, nurses, and perhaps your family, are doing all they can to help. You know they are, and it's not that they don't care, it's pretty obvious that they all seem to care quite a lot.

But the gulf between you and them is that *they don't know*. It's not their fault; they can't know, because *they weren't there*. And the ones who were there with you are still there, but you are not. In some ways, you're still there; your body may be here, but you, your mind and your identity, the essence of what you have been, are still there.

So the people who care for you say well-meaning things which they offer as support, but which are often not what you want to hear. At least, not what you want to hear while you're still bed-ridden, on medication, and wondering exactly what the US military is going to do with you. Things like, "Good news! You'll never have to go back." Or, "Don't worry, they'll let you out of the service." They might see such projections as good news, but it's not good news for someone who expected to leave the war zone or the service on your own terms, when you were supposed to, after finishing your job. Indeed, the idea that you'd be happy to be out of either the war or the service kind of demeans your commitment in the first place, demeans the importance that you attached to what you were doing. Whatever the attitudes of society at large may be, most soldiers have a much more developed concept of commitment to their duty. Maybe it's hard for the civilian world to understand, but many wounded soldiers *want* to go back, to finish the job. They're not helped by reassuring predictions that, "It's over for you."

Sometimes you can't help feeling like a failure. Of course, nobody else thinks you were a failure just because you were badly wounded. They even give you a Purple Heart. But what other people think is not the point; you *know* that being wounded just burdens everyone, that *accomplishing the mission* was your duty, and that being wounded does not accomplish the mission. Whatever other people think, you think you left your unit and you let your fellow soldiers down.

All this is complicated by being helpless and dependent on others, having lost options and control over choices. You are now under the complete control of others and even if you were to regain some control, many options are likely to have evaporated. You feel that others will make, and are making, decisions for and about you, *without really knowing* what you did, where you've been, or where you wanted to go.

When a Physical Evaluation Board convenes, they're going to say in more delicate words "You're no good anymore, we can't use you." You're about to be discarded like a broken machine part or last week's spoiled leftovers, like a bar fight in a grade B western where they throw some money on the table for the broken furniture before they walk out, only this time you are the broken furniture left behind.

When a young man, like me, is processed out of the military because of wounds received, there is much compassion and care provided along the way, but the key to the best result for the wounded man does not lie in the medical care provided or the disability compensation to be awarded, as important as those things may be. It lies instead in the wounded warrior's mind, his heart, his self-worth and pride. Or hers. Family members love you enough to chatter in anticipation of your separation from the armed forces, eager for your freedom, and you wish you could tell them they are making you miserable with the irrational feeling you didn't quite make it as a man. The feelings of isolation, failure, and helplessness seep in no matter how hard your head works to resist. If you are a fortunate wounded warrior, eventually your heart loses that sinking feeling and understands that, no matter what, nobody can take away from you what you were and what you did. No one can take away from you the honor and pride of having done your duty. What it means to anyone else doesn't matter.

I CAME TO THINK of myself and my men with those two words: honor and pride. The war was crazy, the vines were infinite just like the VC and the leeches, but we did our duty and we did so honorably.

That's what cuts so deep about John Kerry. He was supposed to be one of us. He could have found plenty to criticize about the war without telling outlandish lies that painted all of us - my men and even me – as war criminals. My men were good people who struggled under sometimes impossible circumstances just to do their duty for their country and for one another. War criminals?

What is even more insulting is the press bought it and served it up to the American public, which for decades treated Vietnam veterans like 3rd class citizens.

I later was told John Kerry went to Paris to meet with enemy officials, while our soldiers continued to fight and die in the field. Can that be true? If he did that, how the heck did America let him get away with it?

The pain and disbelief I felt listening to Kerry's words went deeper than the pain I felt from the enemy fire that decimated the left side of my face. After eighteen months in hospitals, much of it spent wishing I were back with my unit, I was discharged, the enemy-inflicted wounds mostly healed. But more than four decades later, the wounds inflicted by John Kerry remain raw for me and many other Vietnam veterans. Those wounds, from the bearing of false witness against a generation of courageous young Americans who fought and died in Vietnam, are far more serious than any wound warranting a Purple Heart. Those wounds go to the heart and soul. Those wounds never heal.

I'll never know why John Kerry chose to use egregious lies to oppose the war, why he couldn't use the truth to make his anti-war case because there was plenty to criticize about the war.

But I do know this. I wouldn't trade one of my men for a hundred John Kerrys.

Photo courtesy of Dexter Lehtinen, all rights reserved

Dexter Lehtinen was medically retired for wounds received in Vietnam, but later served in the US Army Reserves as a Detachment Commander, Special Forces. During his ongoing career as a litigation attorney, he was elected to serve in the Florida House of Representatives as well as the Florida Senate, and was appointed as US Attorney for the Southern District of Florida. He is active in academics as an adjunct professor in four south Florida Universities and is nationally known for his environmental work, such as preservation of the Everglades. He is often a featured speaker, as shown above as he speaks to a gathering of Vietnam veterans. Dexter's wife, Ileana Ros-Lehtinen, is a member of the US Congress. They divide their time between Washington, DC and Miami.

Postcard to America

EVERY GI IN VIETNAM *was on his own one-year schedule, crossing off calendar days and starting short-timer jokes when the halfway point was passed. The shorter we got, the more we quietly worried about getting hit.*

Another thing we did quietly was daydream about how things would be when we got home. Whether I was in the boonies, straining my eyes to spot any movement in the dark beyond our perimeter, or lying on my cot back at the base, I would remember looking at old copies of Life, Colliers, Look, and Saturday Evening Post magazines and seeing the welcome home scenes from Times Square, from the ticker tape parades in New York, and the troop ships coming past the Statue of Liberty at the end of World War II. I thought about John "Patch" Patton, a local high school football hero who was a released POW at the end of the Korean War. Our whole town turned out to welcome him home. As a 5th grader, I was dressed as Uncle Sam and rode on a float with a girl dressed as the Statue of Liberty. Everyone was singing and the band was playing, "When Johnny Comes Marching Home Again, Hurrah, Hurrah."

It didn't turn out that way for us. My first contact with civilians was at the San Francisco airport and on the plane flying to Kansas City. I was not spit on, heckled, called a baby killer or dope addict, or ridiculed. This was mid-1967 - all of that would begin later. Instead, I was ignored.

I returned to Heavener, Oklahoma, the small town I had lived in all my life, where I knew many of the people. As I walked down the streets and saw someone I knew, the conversation usually went something like this, "I haven't seen you in a long time, Bob. Where have you been?" When I responded I had just come back from Vietnam, I would get an "Oh," an awkward pause, then the subject was changed. Indifference was a disappointment, not at all like the welcome home I expected.

Bob Babcock, Marietta, Georgia
22nd Infantry Regiment, 4th Infantry Division, Pleiku area, Vietnam, 1966-67

Everybody Knows...
B.G Burkett
Plano, Texas

EVERYBODY KNOWS THAT VIETNAM was a bad war, an immoral war, a mistake, and that the GIs we sent to that war were forced to do things that we, as American citizens, find repugnant. Most of us are sympathetic to those veterans, because we know they played an unwilling role in that sad chapter of our history.

Everybody knows that WWII was a good war, a necessary war, a war in which our men fought with honor and courage against an evil enemy and without the lapses typical of Vietnam. We regret that Vietnam did not meet this high standard of how war should be conducted.

Actually, neither of these notions is true, but the American public has been conditioned to think that way. Those ideas are part of our mindset, our preconceived notions. I know you doubt me. I know you don't believe there is a difference between the truth and what America believes about Vietnam veterans. In fact, there is a substantial difference.

Some years ago I was motivated to write a book titled *Stolen Valor* with a co-author, Glenna Whitley, a professional writer, exposing myths, lies and frauds about the Vietnam War and its veterans. The stories in this book Terry Garlock has written are important, and to help you understand why, I will tell you what compelled me to write *Stolen Valor*.

WHEN SAIGON FELL IN April of 1975, like many other Vietnam veterans I felt a deep sadness, an overwhelming sense of grief as if I had lost a loved one. So much work and hope was invested in Vietnam by over three million Americans who risked their lives. More than 58,000 paid the ultimate price and still the communists prevailed because our nation at home lost its will.

Countless South Vietnamese had worked with us and were well-respected and staunch allies of the Americans. Now these same people were at risk to be executed, sent to torturous re-education camps, and if they survived, to be discriminated against for the rest of their lives. We Americans turned our back on them. To our everlasting national shame, we abandoned them to their fate, allowing them to die miserable deaths by the tens of thousands.

After my Vietnam years I did not forget, but I did put those memories on a very special mental shelf and moved on with life. I went back to school on the GI Bill and earned my MBA from the University of Tennessee. In Dallas, Texas, I began a satisfying career as a stockbroker.

In 1986 Paul Russell, a friend from my college days at Vanderbilt University, called to tell me about a fledgling effort to build a Texas Vietnam Memorial honoring the 3,427 Texans who died in Vietnam. Paul, a very successful mechanical contractor, had been commissioned through ROTC at Vanderbilt, gone through Ranger school and had served two tours in Vietnam.

The thought of helping to build a Texas memorial was intriguing. Paul and I had been stationed at Ft. Hood together where one of my three roommates in a rental house was Texan Harry Horton. Harry was later killed in Vietnam while serving as a rifle platoon leader in the 25th Infantry. He was posthumously awarded the Silver Star. Getting involved in this civic project would be my way to honor Harry and all the other GI's who lost their lives in Vietnam.

A third Vietnam veteran, Art Ruff, would join our effort as my co-chairman. Paul would serve as president. Each of us focused on different aspects of the project. I was the chief fund-raiser since I worked in the financial world. Paul worked with the city and state planners, the labor unions, the material suppliers and the legislators. Art, our wealthy and connected member, became the rainmaker, often willing to throw in his own money when we hit a financial dry spell.

All of us were naïve. We thought we could raise the $2.5 million we needed in six months and build the memorial, start to finish in two to three months. Wrong! We were not aware of or prepared to deal with the negative perception the average American had of Vietnam veterans. I knew where to find sources of funding but I didn't yet realize the inherent objections we would encounter again and again. When I approached a city council, a foundation, a corporation or wealthy individual, the words varied but the meaning of the response was always the same about Vietnam veterans: *Why should we give money to those bums?*

Some were vicious, some polite, some diplomatic, but the result was always the same: no money, no thanks!

I mentioned the memorial project to a client one day and he asked, "Why the hell are you involved with something like that?"

"What do you mean?" I said.

"Why would you be involved with people like Vietnam veterans?" he asked.

"Well, I'm a Vietnam veteran," I told him.

"You're kidding me," he said, looking at me as if I just confessed I had syphilis.

Another client I told said, "I've dealt with you for years, and I never figured you were a Vietnam veteran." I wondered how it was supposed to show.

I began to understand our memorial project didn't have a financing problem so much as a public relations problem. The Vietnam veteran stereotype was unemployed or unemployable, criminally inclined, prone to substance abuse and wracked with guilt over the horrible things they had done or seen in the war. And yet, of all the Vietnam vets I knew, not a single one fit that description.

I THOUGHT BACK TO my own introduction to the negative stereotype during my trip home from Vietnam in uniform. As I waited for my flight in an airport coffee shop, a waitress ignored me while waiting on other customers around me, even those who came in after me. She ignored me even when I tried to get her attention and a younger waitress came over. "Oh, don't mind her," she said. "She's got this anti-war thing. She won't serve anybody in uniform." She told me in a matter-of-fact way, as if it was perfectly normal. But my introduction had just begun.

My flight was fairly full and I was flying standby, one of the last to board. When I took my seat the man next to me asked if I was stationed at Ft. Dix. I told him no, I was on my way home from Vietnam. Across the aisle another man, who obviously had been on the flight's previous leg and well into his little bottles of Jack Daniels, overheard me and announced, "Oh, a big war hero! Hey folks, we've been sitting here on the runway waiting for a big goddamn war hero!" I kept my silence, but he continued. "Hey, bucko, you spent a year killing women and children, make you feel like a big man, did it?" He kept on needling me the whole flight while I wished either he or I could just disappear. The most disappointing thing was that nobody else said anything to him. The stewardess ignored him. None of the other passengers defended me. I had to keep silent to avoid trouble, feeling like a pariah. If I had been a soldier returning from WWII, someone would have slugged the guy.

I was puzzled at the man's boorish behavior, but I would soon realize he was just acting on what he had learned about Vietnam vets, America's *baby-killers*.

Now years later in 1986, I was trying to raise money for the benefit of Vietnam veterans, and America's perception of them was still decidedly negative, even though that perception was badly flawed. To raise money for the memorial, I would have to deal with this public relations nightmare.

One of the hottest movies that same year was *Platoon*, Oliver Stone's cynical version of Vietnam depicting soldiers doing drugs on patrol, a soldier shooting a Vietnamese woman in the head, another threatening to kill a little girl and a soldier killing his sergeant. The movie got rave reviews and news reports showed vets crying with their support group after seeing the absurd film. I would learn later to suspect the crying vets were likely frauds; I certainly didn't know any vets like that.

Most Hollywood versions of Vietnam vets were losers, misfits, nutcases. The 1976 film *Taxi Driver* starring Robert De Niro portrayed a crazed, pill-popping Vietnam vet obsessed with a teenaged prostitute. In 1978, De Niro starred again as a Vietnam vet in *The Deer Hunter*, portraying vets as alienated and suicidal. Francis Ford Coppala's *Apocalypse Now* in 1979 was an implausible violent and psychedelic trip to some uncivilized, unnamed jungle planet, but it wasn't Vietnam. *First Blood* in 1983 was Sylvester Stallone's first film as Rambo, the pumped-up, traumatized Green Beret on a rampage. A few years later in 1986, *Casualties of War* would be based on a real story about the kidnapping, rape and murder of a Vietnamese girl, but twisted in the film to make it seem the way American GIs behaved in that war as opposed to the rare and isolated crime that it was.

While the big screen was filling American heads with ridiculous notions about the war and our own troops, TV was doing its best to keep up. When a TV drama or even sit-com included a Vietnam veteran, you could be sure the character was some type of psychopathic nut-case, dysfunctional or homeless, mentally fragile, suffering from flashbacks of combat or maybe prone to unpredictable violence.

Literature followed; short stories, novels, any medium that conveyed a message about Vietnam veterans picked up the loser theme that seeped into every crevice of our culture. In the press, stories about successful businessmen rarely made reference to their status of Vietnam veteran, but when a street bum shot wildly at pedestrians the headline would be another "Vietnam Veteran Goes Berserk!" Any mention of *Vietnam veteran* in a group would bring knowing glances and quiet nods of understanding because . . . *everybody knows.*

Curious, I began to clip newspaper stories and read books about Vietnam veterans. I began to see them myself: the squalid man near a traffic light with a poorly scrawled sign saying "Vietnam vet – will work for food" and collecting coins from those feeling sufficiently sorry for him; unkempt men in tattered remnants of fatigues at Memorial Day events or hanging around the Vietnam Memorial; men in pony tails, needing a shave, wearing jungle boots with blue jeans and jackets bedecked with military decorations and patches; men drinking from bottles tucked in paper bags, bragging loudly about their exploits in the Vietnam War or bitterly discussing Agent Orange or their nightmares. I suddenly realized that I had always seen these guys, but not the way the public did. These men, the ones likely not to be real vets at all, were what people thought of when they heard *Vietnam veteran*, not me in my suit.

If I was going to raise the money needed for the Texas memorial, I was going to have to counter the misinformation that people had absorbed over many years from multiple sources. I needed facts.

I STARTED BY CONTACTING every agency and archive that might have primary data on the subject: Department of Defense, Selective Service, Veterans Administration, the National Archives, the Labor Department, etc. As I gathered and reviewed the data, slowly a pattern began to emerge, a pattern of misconceptions about Vietnam *and* WWII, large and small things that *everybody knows,* but are simply not true.

In WWII everyone pulled together. Vietnam was a class war in which young, poor, uneducated minority men were chopped to pieces while white college boys avoided the war with deferments. Brave American soldiers in WWII bested the evil of Hitler and Hirohito. In Vietnam, confused, drug-addicted soldiers killed women and children. WWII veterans came home to stirring parades, eager to sire the baby boom and forge a super nation. Vietnam veterans slouched back one at a time in dishonor, fighting drug habits and inner demons. It didn't matter that these things were untrue; perception creates its own reality.

Some of my findings gave me some discomfort because my Dad, my hero, served in WWII and I will always admire veterans of that war. I certainly didn't want to create any competition for virtue between the generations, or the wars, since that seemed ridiculous. But facts are stubborn things, and I continued my research where the trail of facts led me.

The WWII desertion rate was 55% higher than for GIs in the Vietnam War.

No American unit in Vietnam ever surrendered, while in WWII tens of thousands of GIs surrendered, including entire regiments that did not fire a shot.

In the 11 months of the European campaign, 1,300 GIs committed suicide.

Hundreds of thousands of youths committed draft violations in WWII, and tens of thousands went to prison. In the Vietnam era about 5,000 went to prison for evading the draft. During the Vietnam War nine million served, only two million of those had to be drafted. In WWII, 16 million served, but only five million voluntarily enlisted, making it necessary to draft the other 11 million out of a draft age population of 50 million men. For the Vietnam War 2/3 were volunteers whereas for WWII 2/3 had to be drafted. Of the combat deaths in Vietnam, 77% were volunteers.

Sometimes the combat experience of Vietnam veterans is dismissed as irrelevant, not part of a *real war* like WWII. But in WWII troops were moved slowly in large groups with a lot of waiting between spots of fighting. The average infantryman in the South Pacific during World War II saw about 40 days of combat. The average infantryman in Vietnam saw about 240 days in the bush, where contact could happen at any moment, thanks to the mobility of the helicopter.

The average age of the American trooper in Vietnam was commonly stated to be 19 years old, while the more mature WWII soldier was 26. In reality, the average age of the Vietnam troop was 23 at the mid-point of his one year tour. The average of 26 years old for WWII was at discharge, after several years of war. A contributing factor was in WWII we drafted men as old as 35, for Vietnam the cutoff age was 26.

The American public has been led to believe thousands of 18 year old draftees died in Vietnam, especially black 18 year olds. The correct number is 101. Just seven of those 101 were black. There was some racial chatter about Vietnam being a war that unfairly burdened blacks, but my research showed that blacks actually served and suffered casualties at a slightly lower rate than their proportion to the population.

The draft certainly allowed college student deferrals, giving birth to many perpetual students. That favored more wealthy families, but many college graduates were drafted or enlisted after graduation. Most would serve as officers and likely be platoon leaders or pilots, both high-risk jobs in the war.

Conventional wisdom says hundreds of thousands of Vietnam veterans have been to prison. At the time of my research slightly over one million males were in America's prisons. About 55,000 of those fell within the age range to be Vietnam veterans. Of these 55% were black, while just 10.5% of Vietnam veterans are black. Over 80% of those incarcerated had a felony conviction as a youth offender, which would have precluded their service in the US military. Only 20% of those incarcerated had a high school diploma while nearly 90% of Vietnam veterans had a high school diploma or college degree. In the end I estimated that about 2,500 of the one million incarcerated were likely to be Vietnam veterans. Further data indicates Vietnam veterans, contrary to their image, have one of the lowest incarceration rates of any group in America. There may be more former police officers and politicians in prison than Vietnam veterans.

I often heard of the vast numbers of homeless or unemployed Vietnam veterans, so I gathered statistics from the US Department of Labor. It turns out that, at a time when the unemployment rate of US males was 6%, the rate for Vietnam veterans was far lower at 3.9%, the lowest rate for any group in America. And that did not count the hundreds of thousands who stayed in the military after the war. Vietnam veterans also had the highest rate of home ownership, the highest per capita income and the highest educational achievement; 71% of Vietnam veterans used the GI Bill to support some level of additional education.

Another bit of conventional wisdom I wanted to check out was that more Vietnam veterans had killed themselves since the war than died in Vietnam. That statistic turned out to be ludicrous. According to the Center for Disease Control and the Veterans Administration, the suicide rate for Vietnam veterans is lower than for those of our generation who did not serve in Vietnam, and is one of the lowest suicide rates in the country.

We tend to believe what we want to believe, and we're suckers for those eager to feed it to us.

EVERYBODY KNOWS THE VIETNAM War was unusually brutal. Wasn't it?

Since the Civil War Americans have been isolated from the horrors of war, separated by distance and censorship that cleaned it up for public consumption. But the reality is every war is a brutal business of killing in large numbers, and every war has atrocities on all sides, even the American side, however unaware are the families at home waiting for their soldier to return.

What about the facts? What was the truth about war crimes in the bad war and the good war? 3.4 million American troops served in the Vietnam theater, 2.6 million within the borders of South Vietnam, and from them there were about 600 crimes against civilians reported, many of those proving to be unfounded, during the 10 years of the war. Any city filled with half a million heavily armed and virile young men would envy such a crime rate.

There were many trials and convictions for war crimes, but none were given the death penalty.

In 11 months of the WWII European campaign, nearly 1,000 American GIs were charged with capital crimes, mostly rape and murder. Almost all were convicted and 443 were sentenced to death. Of those, at least 96 were executed. In all theaters of WWII nearly 300 GIs were executed for their crimes.

As in all wars, the criminal actions of a few bad apples should not stain the vast majority who served with honor. That is no different than civilian life. A terrible murder in your home town does not make other residents into murderers.

In Vietnam, while the US military training and was oriented to *prevent* war crimes, and to prosecute offenders when necessary, our enemy actually *trained* their soldiers to use atrocities and terror with maximum effectiveness as a tool in the war. Their atrocities were systematic and institutionalized while ours were isolated lapses. The media, the reporters whose duty is to open your eyes to the truth, failed to bring that news to you.

You might wonder why the Vietnam War is falsely perceived to be a war of American war crimes. Propaganda from the radical left planted the seeds of suspicion that our soldiers and this war were uniquely brutal. Reporters roamed freely in the Vietnam War, and were eager to report American war crimes, even embellish them if the reporter had an agenda. They were free to write their stories as they saw fit, and many stories reflected a preconceived notion that US policy in Vietnam was wrong. Reports on major battles were twisted, such as reports on the enemy's Tet Offensive of 1968 as a US military failure, when in fact it had been an overwhelming US victory. Our own media seemed not to care they were giving aid and comfort to our enemy.

Something else was different in Vietnam. Until then, wars were remote. Vietnam brought the war into our living rooms on TV, showing us things we didn't always want to see at the dinner hour. Since man sharpened sticks around campfires, war has always been an ugly, nasty, brutal and gory business with civilians often suffering the worst of war. If you want glory in war, you have to turn to the fiction of Hollywood. Seeing on their own TV the hideous ways human beings kill each other in war, and the tragic suffering of civilians that results, Americans were prone to think Vietnam was different and evil compared to other wars, but it was not. The difference was Americans were no longer insulated from the blood and gore and misery of war.

In WWII we had censorship to prevent stories that might give aid and comfort to the enemy, and to prevent stories revealing too much of the battlefield horrors that might panic the public at home. But the battlefields of WWII were far worse than depicted in *Life* magazine. Years after the war, it was reported that General Eisenhower wept as he crossed a battlefield in Europe where one could hardly step without touching some kind of body parts. Furthermore, our reporters in WWII considered themselves loyal Americans who had a stake in our victory, and they censored themselves to be certain they did not give our enemy any propaganda advantage.

THERE IS NO CRITICISM of WWII or its veterans implied here. The men who went to war against the Nazis, the Fascists and Imperialist Japan, and the women who supported them at home, did no less than save the world from tyranny. They are our heroes, our fathers and mothers, and their determination and strong character inspired many of us to also serve our country during the Vietnam War when serving was not popular.

But the conventional wisdom that WWII was good and the Vietnam War was bad is seriously flawed. It also serves as an example that public perception can be manipulated, as it was in both wars.

There is, of course, an ultimate difference between the two wars. We were victorious in WWII. In Vietnam, we pulled our military out in 1973, we continued to fund our South Vietnamese allies and hoped with our indirect support they could hold off the communists. In one of its most despicable actions in our history, our US Congress cut off all funding for the Vietnam War in August of 1974 following the Watergate scandal which crippled political supporters. In one vote, Congress abandoned our ally and paved the way for the communists to take control by force. The North Vietnamese army did just that in April the following year.

The American public, with perceptions twisted by a media with an anti-war agenda, seems even today to attribute the loss in Vietnam to *our inept military*. The truth is our military had never lost a significant battle in Vietnam and had been ordered home two years before South Vietnam fell.

ARMED WITH THESE FACTS from my research, I was able to sway some donors and the money for our memorial project ever so slowly began to flow. Still, many sources turned us down, even corporations that had prospered as suppliers of military hardware during the war quickly said "no." Brown & Root made billions as a contractor during the war and they turned me down cold. The same came from Rockwell International and Varo Corp, manufacturer of night vision equipment and missile launcher systems. Bell Helicopter declined to make a direct contribution, but allowed us to approach their unions. Of all these and more companies who prospered from the war, only General Dynamics came through, donating $50,000.

Other corporations seemed to be logical prospects so I mailed requests to the 500 largest public and the 500 largest private corporations in Texas. The answers were no and hell no.

I pitched a mail campaign to the state chapters of the VFW, thinking surely other veterans would be supportive, but even they were a tough sell. I discovered many WWII and Korea vets had little respect for Vietnam veterans, the *inept troops who lost their little semi-war*. After our request caused internal battles, the VFW agreed to send out donation envelopes with a newsletter describing our memorial project.

A week after the newsletter mailing our mailbox was stuffed with the self-addressed, postage-paid envelopes we had supplied, but it was a cruel prank. Apparently some people at the VFW posts had taken the envelopes out of the mailing, sealed them and dropped them in the mail so we would have to pay postage on empty envelopes. Some that trickled in later contained Christian pamphlets or notes like "baby killer," "pot smoker," "Why don't you bums go to work and quit playing GI Joe?" or "Scum, crybabies. WWII vets are real men, you are drug-using wimps." A few had small donations.

Appeals to the American Legion and other veteran organizations didn't work any better. Incredibly, even Vietnam-era vets, the ones serving during the war but who did not serve in Vietnam, were a tough sell, too. Even they believed the worst about what happened in that war. They read the newspaper stories, they saw the movies, they saw the TV programs, they understood the underlying themes on TV news, they read the novels just like everyone else. They weren't there in Vietnam to see what happened with their own eyes, and they bought the bull, too.

While I struggled with the perception problem, an event occurred that added a whole new dimension to our fundraising. A mentally ill vagrant in

Dallas managed to grab a police officer's pistol while he was writing a ticket for a motorist. The vagrant shot and killed the police officer and then just wandered off into a parking lot where he was confronted by two off-duty police officers who shot and killed him. A local Dallas newspaper headline read "Vietnam Veteran Goes Berserk."

I was already in contact with the National Archives about the Texans killed in action in the Vietnam War. I brought up to my contact the story of the cop killing, lamenting how negative it would be for my fund raising. He asked "Why don't you get the killer's military record?" I didn't know military records were releasable. Through some legwork I obtained a death certificate for the dead vagrant and sent a Freedom of Information Act (FOIA) request for his military records.

When the records arrived I was shocked. The vagrant had never served in Vietnam, and had been kicked out during basic training as "psychologically unfit." How could the newspaper say he was a Vietnam veteran?

I contacted the newspaper about their error, sent them the records and naively expected a correction. None came. When I contacted the editor to ask why they didn't publish a prominent correction, he told me, "We have proof enough he was a Vietnam veteran." Of course they had none at all.

Thus began my scrutiny of published statements about Vietnam veterans, and over the years I have witnessed newspapers virtually make up stories and refuse to correct them when proven wrong. This episode also energized me to check every news story with a Vietnam veteran involved. I have checked thousands of such stories over the years in major newspapers – New York Times, LA Times, etc.; stories on national TV – CBS, NBC, ABC, documentaries, news programs etc.; biographies published by major publishing houses; and biographies of presidents of major veterans groups. Of the stories I have checked, approximately 75% of them have major falsehoods. Of those run in major newspapers, the majority of the editors refused to correct a false story printed about a Vietnam veteran.

Here is just one example. In 1988, Dan Rather presented to TV viewers a documentary, *The Wall Within*. In the promotion aired before the program, Rather claimed ". . . a number of these men (Vietnam veterans), haunted by their deeds, became seriously ill." The program went on to explain that the killing and the war crimes his subjects had committed had caused what amounted to PTSD.

This program was preposterous from the start. The first veteran claimed to have been a Navy SEAL, out of the training program just one year, whose primary duty was assassination, which raised my suspicion since junior men would not likely be assigned such radical black ops measures. The program went downhill from there.

I obtained the military records of all six who appeared in Rather's program. Five had just made up their stories. The sixth one did serve in Vietnam, but told some highly questionable tales on the program. I took this information to CBS but they refused to listen.

I gave my information to ABC's 20/20. They did an exposé of CBS' journalistic fraud, winning a CINE award in the process.

As I researched this documentary, I also obtained Dan Rather's military record. He had claimed to be a "two-tour Marine." The facts are a bit different. He did join the Marine Corps, but he was unable to complete boot camp. He had been released for being unable to do the physical activity and designated as "medically unfit."

WE STRUGGLED TO RAISE $2.5 million for the Texas memorial over three long years and wrestled construction problems on the last leg of that journey. Finally, the memorial was complete and the dedication in Fair Park in Dallas was scheduled for Veterans Day in November, 1989. I had never spoken to a group of more than 100 people before, but we expected a big crowd with the President of the United States, George H.W. Bush, as our keynote speaker for the ceremony, making ours the only Vietnam War memorial in the country dedicated by a U.S. President. I wondered if I would be able to squeak out the words I had written when I stepped to the podium.

As we made our way to the platform that day and were stopped by Secret Service agents, I saw a small rag-tag cluster, wearing ratty uniform remnants and unkempt like the stereotype of "Nam vets," and was glad to see they had been kept behind the security fence a short distance away. The crowd was estimated to be 10,000.

When my turn to speak came, I thanked President and Mrs. Bush and all those who had been involved in the memorial effort, then went on. "Every man who ever served in wartime remembers the names of fallen comrades," I said. "Just as our president remembers the names Ted White, Jack Delaney, Jim Wykes and Tom Waters, my generation of veterans remembers the names engraved here." The president turned to look at me, obviously surprised that I had named a handful of men close to him who fought and died in World War II. I thought about my roommate at Ft Hood, Harry Horton, whose name was engraved on the memorial.

"We remember names like Gregg Hartness and Charles Bryan and Russell Steindam. Some of the names represent individuals decorated with Navy Crosses or Medals of Honor, but death makes no such distinction. Each one of these 3,427 young men was an American doing his duty. Each was someone's father, someone's brother, someone's husband, or someone's son."

"Remember them, ladies and gentlemen, remember them all," I said in closing, "and especially remember the pain and suffering of their families, for it is because of their sacrifice and the sacrifice of past generations of veterans that this great nation remains free."

Somehow, I finished without choking up, introduced Texas Governor Bill Clements and sat down, exhausted. All I had to do now was listen to a couple more speeches.

As the speeches ended we looked up into an impossibly blue autumn sky to see four Navy jets screaming overhead, one peeling off to make the *missing man* formation. Four Marine jets followed to do the same. A monstrous Air Force B-52, followed by a KC-135 Stratotanker, rumbled overhead just a few hundred feet off the ground. Finally, the pounding, throbbing sound every Vietnam vet knows announced four Army National Guard Huey helicopters swooping low over Fair Park.

The tune of taps began, a bugler on the ground echoed by another on the roof of the baroque Dallas Music Hall, a sound so lonely and lovely I was near tears.

The handshaking signified it was done at last.

I thought about the long struggle to raise the money and realized success had brought little satisfaction. I was desperately tired and realized little had changed. We had built this stone edifice but a true memorial is what we feel in our hearts. America did not feel in its heart that those who fought in Vietnam were worthy. At best, we had given a handful of Texas families a massive tombstone.

As if to drive the point home, reporters at the event seeking on-camera comments ignored Vietnam veterans like me and the other successful businessmen in suits who worked for three years to make the memorial happen. Instead, they followed a beeline to the rag-tag scraggly guys behind the fence, those most unlikely to be real Vietnam vets. That evening all three local network TV news affiliates showed coverage of the event with emphasis on the rag-tag group. The image viewers took away was Vietnam veterans as bums, derelicts, and losers. I wanted to scream.

INSTEAD OF SCREAMING I became active in researching the background of questionable people claiming to be Vietnam veterans. I found to my surprise that documented cases of imposters claiming to be combat veterans goes back to the Civil War, and if records were available I am confident the same happened in the revolutionary war. I have been privileged to expose quite a number of these frauds, and have worked with the FBI, the Justice Department, the Veterans Administration and others at the Pentagon as an investigator and expert witness in exposing and prosecuting them.

Glenna Whitley joined me in an effort to document how the honor earned by Vietnam veterans was stolen from them. The result of our joint effort and a decade of research is the 1998 book *Stolen Valor: How the Vietnam Generation Was Robbed of its Heroes and its History*.

Maybe now you can understand why vets kept their story to themselves when they came home from Vietnam. The country seemed to have lost its mind, and everybody seemed confident they knew all the answers about Vietnam, even though they were mostly wrong. Who would believe a vet's story? Who would be interested enough to listen? Who would care?

Even now, after four decades, these stories are important to the vets, to their families and to the country. As you read the stories Terry has gathered and written, decide for yourself whether America owes a debt to these people, whether at the very least we should correct the false history.

For young people, there is something more. Study what happened and how reality was distorted in our media and our culture. Learn to question what *everybody knows*. Look at current events, talk to people close to the action and discover for yourself when you are being led astray. Read a couple of good newspapers regularly, but question what you read rather than accepting all that is written. Learn to think for yourself. Don't let the truth be stolen from you.

Photo courtesy of B.G. Burkett, all rights reserved

In 1998 B.G Burkett joined with Glenna Whitley to write the book "Stolen Valor: How the Vietnam Generation Was Robbed of Its Heroes and Its History," a book that won the William E. Colby Award for Outstanding Military Book. Mr. Burkett received the Distinguished Civilian Service Award, the highest award the Army gives a civilian, presented by Former President George H.W. Bush. For his work on veteran issues, Mr. Burkett was inducted into the U.S. Army Ranger Hall of Fame at Fort Benning, Georgia.
Mr. Burkett has exposed and helped prosecute a number of frauds posing as Viet Nam veterans and even active duty personnel claiming honors and awards not earned.

Postcard to America

WHEN PERFORMING A LRRP *mission, we lived in Victor Charles's backyard. Being a lightly armed covert team of six, we were ever in mortal risk of annihilation should our presence become known. I wrote this poem in Vietnam in 1971, while serving with Team Hotel, N/75 Ranger, 173d Airborne Brigade:*

LONG RANGE PATROL

Lurking amidst the jungle gloom, knowing not whether the next sunrise shall greet us with its warm, reassuring rays, a sense of foreboding prevails amongst We Six.
The snap of a twig; an abrupt silence; a disturbance in the triple-canopy rain forest enveloping us;
All can, and have, spelled instantaneous oblivion for others of our unforgiving and deadly profession.
Vigilance, proficiency, and indeed, luck, are our keys to survival in this all-encompassing, hostile environment.
For there is no quarter in this game.
The atmosphere is electrified with His presence.
Our eyes and ears are as one; every sense keenly attuned;
Searching for; listening to; analyzing;
That which is not of ours, or Nature's.
For He is out there; by no timetable must He abide;
Watching for; awaiting; anticipating; our fatal err.
Stealth, discipline, and caution, have, on this mission, proven to no avail.
Our hearts cease; minds race; as we simultaneously detect the firing device's telltale action.
A millesecond's blinding flash of light is accompanied by an ear-shattering thunder.
And then there is nothing.
We Six have embarked upon our journey through the infinite expanse of time...."

David P. "Varmint" Walker
F/58 LRP, N/75 RGR, & E 3/21 Recon Plt.
Phuoc Vinh, LZ English, & Da Nang, RVN 1967-'68-'70-'71-'72

The Helicopter Pilot Factory
Ron Current
Mableton, Georgia

M<small>Y BUDDY</small> T<small>ERRY</small> G<small>ARLOCK</small> gave me two chapters in this book, back-to-back. You might believe our shared history as roommates in flight school led to that generosity, but his purpose was purely practical. This chapter tells part of the story of the Army producing thousands of helicopter pilots for the war. The next chapter tells you my story as a Dustoff pilot in Vietnam, and by breaking these chapters apart, those of you bored with the story of learning to fly a helicopter can easily skip ahead without missing too much.

L<small>EARNING TO FLY A</small> helicopter is a bit like patting your head while rubbing your stomach in a circle at the same time you are hopping on one leg and reciting your ABCs . . . backwards. Difficulties notwithstanding, the armed forces would ultimately send over 12,000 helicopters to Vietnam to gain an air mobility advantage over the enemy on foot in a jungle war, and somebody had to fly them. Most pilots flew Hueys, called *slicks* in Vietnam. Slicks were used to move troops into LZs and out of PZs in missions called *combat assaults*, mostly in formation, as well as *ash-and-trash* missions, meaning hauling people or equipment for one reason or another. Gunships were either Hueys modified with weapons like rockets and miniguns, or Cobras, the new faster and more heavily armed helicopter designed as a gunship. Small scout OH-6 helicopters were used to zip around close to the ground snooping on the enemy, drawing their fire to expose their position. Large Chinook CH-47 tandem rotor helicopters were used for heavy loads, like jeeps or trucks or howitzers or even lifting broken Hueys in a slingload underneath, or a large load of troops inside. Huge CH-54 flying crane helicopters carried the heaviest loads of all, including a self-contained medical operating room pod that could be attached to its underside. The Air Force, Navy and Marines had other aircraft, like the HH-3E Jolly Green Giant often used on search and rescue missions when jet pilots were shot down. Other pilots like me flew Dustoff or Medevac to evacuate the wounded from the field of battle, air ambulance to hospitals.

In all, around 40,000 pilots were trained to fly helicopters for the war. Where did all those pilots come from?

T<small>HE</small> US A<small>RMY CREATED</small> two tracks for developing helicopter pilots needed for the Vietnam War. One track was for commissioned officers, those who had successfully completed the rigorous training, and pressure to screen out the weak, to become a brand new 2nd Lt. Some of these officers attended the Army's ORWAC or Officer Rotary Wing Aviation Course. The officers who were trained to fly helicopters included a lot of lieutenants as well as captains, majors and higher ranking officers. Some would have a flying job in the war, others would become commanders of ground or aviation units and as a trained pilot would have improved perspective of the possibilities and limitations of the pilots and aircraft under his command.

But the Army needed *lots* of helicopter pilots dedicated to flying, and to that purpose developed the WORWAC program, the Warrant Officer Rotary Wing Aviation Course. The course combined the training focused on learning to fly with the pressure and development needed to turn young men into Warrant Officers. The highest ranking WO would be technically outranked by the most junior Lt., but many of us would find our niche as a WO in the Army, comfortably sandwiched in between the highest enlisted rank and the lowest officer rank.

How does the Army take the undisciplined, raw material of 18 to 22 year olds and mold them into a disciplined helicopter pilot to fly in a war? I'll tell you how it happened with me.

YOUNG PEOPLE HAVE LONG marked the completion of high school as the beginning of their adult freedom, and so it was with my generation but the Vietnam War lurked on the horizon, a threatening cloud hanging over the head of every American male over 18 years old who had not voluntarily joined the military. There was a student draft deferment while attending college, a deferment widely used and often abused. In January of 1968 I thought my student deferment was on track. Twenty days after starting the semester I received a Selective Service letter saying since I stayed out of college in the fall, my student classification would not be renewed after the current semester and that I would immediately become classified as 1-A for the draft. That meant I would surely be drafted. I was stunned!

I knew I didn't want to carry a rifle in a jungle war like most draftees. A former Navy Recruiter in one of my college classes mentioned the Warrant Officer flight training program offered by the Army to train helicopter pilots. This was the only military pilot training program that did not require at least a bachelor's degree, and this Iowa farm boy wanted to fly! I contacted the local Army recruiter and with his help I scheduled the battery of tests to screen applicants to the helicopter pilot program. I drove to Des Moines for the tests in the spring of 1968, passed all of the tests, took and passed a physical, and with the Army's commitment to send me to flight school I signed up under the Delayed Enlistment Program and joined the Army on July 8, 1968.

I had married young, and our parting was tearful as I kissed Beth goodbye and took an airplane ride to become a soldier and a pilot. Dallas was my first stop before flying with a load of other recruits in a TTA DC-3, an airline nicknamed Tree-Top Airways but really named Trans-Texas Airways, to Ft. Polk, Louisiana, for US Army Basic Training. I came to appreciate the joke that if the world needed an enema, Ft. Polk might be the best place for insertion!

We arrived late at night and were trucked to the staging area. Each of us was assigned a numbered spot on the asphalt, where we remained the rest of the night, not allowed to sleep. As the sun was rising we were moved to a mess hall to eat, then to pick up our new uniforms, and then we gaggled – we didn't yet know how to march – to the barber shop to receive our new haircuts. It is humbling to learn how ugly we all are without hair. Newly bald and looking a little bit alike, we were indoctrinated into the things we were about to learn by a soft spoken drill sergeant. He answered all the questions we had and gave us a few pointers. He seemed fair enough!

That afternoon we loaded into cattle trucks for transport to North Fort Polk to our company area. When they opened the back doors on the trucks to let us out, all hell broke loose. There seemed to be a hundred drill sergeants all screaming at everyone! "Get out of that truck you Maggots!" "What are you waiting on, assholes?" "Move it, move it, move it!" I didn't think there was any need for all the yelling.

After the sergeants herded this motley group of new recruits into formation and standing at what we thought of as attention, I noticed there were only five of them. We were two hundred new recruits, screamed at if we looked to the side or if our feet were not together or for whatever real or trumped-up reason. The Commanding Officer was a Captain, hard as nails, overseeing his last cycle of new recruits before he returned to Vietnam for his second tour in the infantry.

Like every new guy in the Army I was changed in eight weeks. We marched and drilled and ran in boots farther and faster than I would have believed possible, and as we learned to act as a unit the drill sergeants didn't scream or swear or insult us quite so much. Our individuality was strained out of us by force as we were taught by repetition and threats and punishment to do things as a team, and to execute orders immediately without question. We didn't know it then, but this transformation is essential to a combat setting. We were pushed physically beyond our limits every day until it became routine, we learned how to shoot, how to break down and clean our rifle and even how to take down and kill a man with our bare hands. Theoretically.

Every week we prepared for the PT (Physical Training) test we would take at the end of basic, running a mile in boots, low-crawling through sand, running an obstacle course, etc. Toward the end we spent a few days in the woods, low-crawled under live machinegun fire at night, felt the sting of gas with and without a gas mask and a few other new treats.

For about a month we force-marched - halfway between a walk and a run - exhausting - to the rifle range every day, about five miles in either direction. One day, just before we were to depart for our force-march to the rifle range, the Drill Sergeant asked the company of men if anyone had a current driver's license. I hesitated because I was suspicious, but a couple of guys eager to drive to get out of marching raised their hands. Each of them was assigned to push a loaded wheelbarrow to the rifle range, in addition to carrying their rifle and other gear. The Drill Sergeant got his stuff transported as he wished, and he announced to us all we better pay attention because that was our lesson to never volunteer for anything in the US Army because you will get screwed.

At the end of eight weeks, we were harder, more confident, less individual and more like a team member. We even had a little pride in having become a soldier.

Thirty of us were going on to helicopter flight school. Our CO told us as we boarded the bus to transfer to flight school, "You guys think this was tough? Wait till you get to Ft. Wolters!"

Flight School

FT. WOLTERS, TEXAS, IS in Mineral Wells, about 50 miles west of Ft. Worth. The bus ride was hours long, plenty of time to daydream about the warm reception waiting for us at flight school, the object of our dreams every night since signing papers to join the Army. We knew that we would spend five months in primary helicopter flight training at Ft. Wolters, followed by four months advanced training at Ft. Rucker, Alabama, where we would graduate as brand new Warrant Officers with shiny new pilot wings, products of WORWAC. We were more than eager to join the brotherhood of student pilots, relieved to leave behind the juvenile grind and yelling in basic training. Some pleasures in life are simple, like the anticipation of being treated like an adult as we prepared to become pilots and Warrant Officers.

The bus wheels rolled and these thoughts danced in our heads.

The parallel ORWAC course for commissioned officers also began at Ft. Wolters, but because the treatment of WOCs (Warrant Officer Candidates) and commissioned officers was vastly different, our training programs would be separated.

Our bus passed through the main gate at Ft. Wolters, the largest heliport in the world with 1,200 helicopters. Helicopters were mounted on stands on either side of the gate. Our aviation adventure was about to begin. We couldn't wait to get our hands on the controls.

The bus rolled to a stop where several Tactical Officers (TACs), were waiting for us. The TACs were Warrant Officers, with shined helmets, gleaming boots and immaculate uniforms, the role model, by appearance at least, of what we wanted to become. Their harassment had a new flavor, less profane, more refined, quick and cerebral machine-gun-staccato challenges and questions that had no right answer to let us know right away we could not win, and that the purpose was not to correct but to make us sweat, to harass.

We were informed our new title was "Candidate."

When they gave the order, "Fall in!" we did it quickly, having learned how to form a platoon of four squad lines, each line an arm-length behind the other, each man one arm length from the ones on either side, standing squarely behind the man in front. We even knew how to stand at attention. But our TACs had new and higher standards for us to meet, and we would soon learn to do what we thought was impossible, and what seemed irrelevant.

The chief TAC asked if any of us had maxed the PT test in Basic. I and others who did said loudly and proudly, "Yes, Sir!" He instructed us to retrieve from our duffle bag the little trophies given for maxing the PT Test, and to hold them high. He said, "For the next week you Candidate shitbirds will have these trophies with you at all times. Everywhere you go you will take them with you and you will not set them down!" I should have listened to the Drill Sergeant in Basic.

We were the brown-hat 5th WOC, meaning the 5th class of Warrant Officer Candidates in this 21-week cycle. Each of the 10 WOCs had different colored hats. There were 330 of us at the beginning of flight school

Preflight Phase

OUR FIRST FOUR WEEKS were called pre-flight, classroom education to teach eager young men about basic helicopter aerodynamics, flight controls, instrumentation, map-reading, navigation, weather and a host of other subjects. The curriculum was broad, the pace quick, the classroom questioning intense, and we learned when we answered a question that a whole bunch of *attaboys* could be wiped out by a single *aw-shit!* The standards were tough and deliberately washed out the weak.

The TACs used this time early in our cycle to harass the hell out of us to identify and wash out weak personalities who could not take the pressure. If they could break us and wash us out then we would be headed for Advanced Infantry Training back at Ft. Polk, something we feared! Over the course of flight school over 60% washed out for one reason or another.

Our TAC Officer, Mr. Webb, was low key but very demanding to keep us striving for perfection. At the beginning he would wake us up at 4:45 AM and walk through a preliminary inspection while we scrambled to shower, shave, dress and get our area ready for inspection in 15 minutes. At first it was impossible, but we learned to adapt by preparing many things the night before to make it happen. Eventually the impossible became routine, and that was the plan. I think one of Mr. Webb's favorites after a while was his soft voice announcing through the squawk-box from his chair in the orderly room next door at 4:45 AM, "Good Morning, Candidates" as our wake-up call, and the barracks immediately came to life, Candidates suddenly standing at attention in their spot in their skivvies to reply in unison "Good Morning Sir!" even though they knew he was not there. And then Mr. Webb was likely to softly say, "Now drop down and give me 10 to properly start the day," and the barracks would ring with the shouts from front-leaning-rest pushup position as one voice, "One-Sir! Two-Sir! Three-Sir! . . . "

Soon that early morning wakeup would precede our company formation outside at 5:00 AM, just 15 minutes later, our individual cubes ready for inspection.

We marched to the mess hall, where we formed the normal single file line outside, but formalities were not relaxed for meals. We were required to march up to the door, stop to open it, march in and gently close it. If the door slammed then we were made to do an about face and apologize to the door, saying "Mr. Door, I am sorry." Inside the mess hall we stood silent at attention except to pick up our tray of food. We marched to our table and stood behind a chair. We stood there holding our tray until there was someone standing behind every chair, at which point the last one to arrive would say "Ready, seats!" We would in the most military way possible take a chair and sit ramrod straight. We leaned over the table only to take a bite of food, and those of us with PT trophies held them in our free hand the whole time, careful to not let the trophy touch down anywhere.

After transferring food to our mouth, we put the fork back on the table and leaned back to chew and swallow, then repeat the process. We all sat at the table at attention until the last Candidate finished his meal and said, "Ready, up!" and we stood in unison, bent over to gather our trays in unison, executed a right or left face in unison to get pointed in the right direction. We marched single file, in step, and placed our trays on the conveyor, marching our way out, mindful of the rules applied to the door. We stood in formation outside the mess hall and were given our fair share of harassment before marching back to the barracks.

And so it went with all of our meals. We eventually got rid of the butterflies that can't be good for digestion.

We went to meals, classes and everywhere else in formation, at double-time. That means running, not walking, but the good news is the Airborne shuffle is a quick-time short step that a man in condition can do nearly all day long. Amidst ever-tighter discipline, the Candidates were looking better, we began to move as one, and even something as simple as double-timing to breakfast could make a little spot of pride glow inside as the company moved like a well-oiled machine.

The TACs assigned our bunks with two men to a cubicle. I was in the upper bunk while Terry Garlock was in the lower. He was as skinny as me back then, no kidding. We learned how to make those beds every morning so tight, with no creases, that a quarter might have bounced. The TACs inspected daily and judged with a demerit system, demerits marked on the sheet we kept available on our desk, punishment for microscopic infractions: bunks not perfectly made, clothes on hangers in closet not evenly spaced, chair not straight, dust on top window trim, the smallest imperfection earning the gig of demerits. If our brass belt buckle had not been taken apart, stripped of the protective coating and freshly shined *inside*, we received demerits, and so on. We stood inspection each morning and evening at attention, in front of our bunks. Once our TAC got down on hands and knees to eyeball the toe-line of shoes lined up under our bunk to check that the line was straight. We were not allowed to leave the Company Area for the first month without a TAC. There was a possibility of a weekend pass at the end of the Preflight Phase if we did not have too many demerits, and since Terry and I had wives impatiently waiting off-base, we were, well, eager to excel. We would learn the meaning of the phrase, "If the Army wanted you to have a wife it would have issued you one!"

The physical side of our training was less demanding than Basic. However if one man in the company formation sneezed then everyone was dropped for 10 to 25 pushups. In my first few days there I did more pushups than I could count. This was their way of team building; if one man screwed up then everyone paid the price, just like in combat except for the deadly result.

In the Preflight Phase we learned how to play the game and as our numbers began to dwindle from wash-outs, we knew we must stay sharp to succeed. We were down to around 250 at the end of Preflight, and we were ready and anxious to fly. We even thought we had learned what was coming from our classroom sessions, but we were in for just one of the surprises of our flying career.

The TACs divided us by height into different "Flights" since different height ranges were best fits for the three different types of helicopters used for training. Apparently the Hughes TH-55 helicopter, the one we called the Mattel Messerschmitt because it looked like a toy, must have been least costly, because the Army was buying a lot of them, but there was a modest height limit for the pilots. I was tall and would be flying the Bell OH-13, with bubble canopy and erector set tail boom, recognizable from the movie and TV series *MASH*. We also had one Flight assigned to fly the Hiller OH-23D, as large as the OH-13. Our third Flight was within the height limits and flew the TH-55.

In Preflight, the classroom subject we most eagerly devoured was about helicopter flight controls, our fingers itching to give it a try. Since so many stories from the Vietnam War involve helicopters, I will test your patience by describing how the controls work.

Helicopter Flight Controls

CYCLIC - THE MAIN ROTOR system of a helicopter is controlled with the cyclic stick, the long stick coming up through the floor between the pilot's legs like a commode plunger stuck to the floor with the handle moveable in a circle. The cyclic actually controls the tilt of the main rotor plane, which you can think of as a dish that tilts one way or another. To move the aircraft forward that dish must tilt down slightly in the front, accomplished by moving the cyclic forward very slightly. If we wanted to turn we moved the cyclic in the direction of the turn and the dish tipped in that direction.

The key word for the cyclic is *sensitivity*. Our first attempts with the cyclic would be crude movements that jerked the aircraft around, and we would soon learn to keep the cyclic, and the rotor plane, quite still and make tiny cyclic movements to make the aircraft move in the desired direction. If we wanted to turn left, we learned to just *think* left and the aircraft would turn. Learning a sensitive control touch would be essential in Vietnam because the heat and density altitude ate away at the aircraft's lift, and when lift was strained, unnecessary movement of the rotor plane let precious lift power leak away, wasted. Sometimes a new pilot couldn't get a heavily loaded Huey off the ground because he was moving the cyclic too much, whereas a more experience pilot could take over and nurse it off the ground with a tender cyclic touch that kept the cyclic plane steady, preserving every ounce of lift.

COLLECTIVE - THE COLLECTIVE IS like a lever, rising at an angle from the floor in the left rear of the pilot's seat, up and forward to the pilot's left hand. The collective's purpose is to increase the pitch, or angle, of the rotor blades when pulled up to *ascend*, and reduce pitch when lowered to *descend*. The collective had a motorcycle type twist throttle attached at the handle, controlled with the left hand. When we were pulling up on the collective, called *pulling pitch*, we had to simultaneously twist to advance the throttle to increase power in order to maintain rotor RPM – more pitch to increase lift requires more power to keep the rotor blades turning at the same standard speed. To *descend* we lowered the collective, thereby reducing the pitch or angle of the blades, as we simultaneously twisted the throttle the opposite direction to decrease power and maintain RPM.

PEDALS - IF YOU LOOKED down at a helicopter from above, the blades turned in a counterclockwise direction, the rotor blades turning in front of the nose from right to left, and the laws of physics made the aircraft fuselage try to rotate in the opposite direction. This is where the pedals come in. The foot pedals actually control the pitch or angle of the tail rotor blades. As power was *increased* in the main rotor system, the nose tries to turn right. To keep the aircraft aligned we would apply left pedal, which increased the pitch and the thrust of the tail rotor. When we *decreased* the power or pitch of the main rotor system then we pushed on the right pedal to reduce tail rotor thrust and keep the fuselage aligned with the direction of flight, keeping the aircraft *in trim*.

AUTOROTATION - HAS TO BE part of any discussion on helicopter flight. When a helicopter engine fails, autorotation is a maneuver we performed to fly the helicopter downward in a rapid descent without power and with pitch in the blades dramatically reduced. Think of it as trading altitude for airspeed and the power to control the aircraft just enough to execute a soft landing. Even though there is no power turning the blades, the air forced up through the blades in descent keeps the blades turning, which lets the pilot control the aircraft airspeed and direction, and builds the kinetic energy in the turning blades so the pilot can *pull pitch* at the bottom to cushion the landing after he flared the aircraft by pulling back on the cyclic to slow it down. In a real autorotation you get just one chance to get it right!

Here's how autorotation works. When the helicopter engine quits, the clutch in the transmission would remain engaged unless we decreased the collective pitch immediately, the rotor blade RPM would bleed off quickly and the aircraft would drop like an anvil. If we did lower the collective immediately when the engine quit, then the clutch disengaged the main rotor from the engine, allowing it to freely turn with flat rotor blades. As the aircraft descended rapidly, the air rushing up through the rotor system with this reduced pitch kept the rotor turning. A long and steep descent might even make the rotors turn too fast and we used the collective to pull a little pitch to slow the descent and keep RPM at the desired rate. We would descend rapidly while maintaining airspeed, with one chance for survival by making quick and proper control movements.

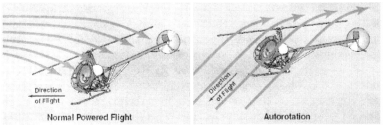

Public domain image, no rights claimed

Whether in autorotation or a normal landing with power, we always wanted to land facing the wind because we needed to keep airspeed up to control the aircraft, but also needed to slow down groundspeed to reduce risk; wind in our face meant groundspeed slower than airspeed. When the engine quit we must do two things immediately; lower the collective and turn into the wind. At all times, a helicopter pilot has to know the wind direction and potential emergency landing spots in addition to everything else he was doing.

When the engine failed and we had lowered the collective and turned into the wind, and we had selected a landing spot and lined up, we had to prepare carefully for landing because we would use up all the blade pitch to cushion the landing, there would be no lift remaining. Autorotation meant one chance to get it right. When we approached about fifty feet of altitude we would pull back on the cyclic to flare and decelerate. As the aircraft continued to descend at a slower rate, when the aircraft started to settle to the ground we would pull the collective in what we called *initial*. This would be pulling about one-half of the full travel of the collective. This rapid movement up on the collective would further slow our descent with a cushion of air created by the increased blade pitch, rotor RPM decreased and left pedal needed to be applied to keep the aircraft facing straight forward. As we settled to a few feet off the ground we would slowly bring the collective up all the way until we could basically feel it in our armpit. Ideally that would happen with little to no groundspeed and a gentle touchdown and we hoped for very little ground run since sliding the skids on open ground begged for rolling it over. Actually a skilled pilot can land a helicopter just as softly without power as with power, but he has only one chance to get it right when the engine fails.

So far we had learned these concepts in the classroom. Putting them into practice would be . . . difficult.

FLIGHT TRAINING: PHASE 1 – Preflight was over and we started eight weeks of Phase 1 flight training, during which we would actually begin to fly. Our instructors were contract civilians, the Flight Leader an Army Captain. On the day we reported to the flight line for the first time and entered our classroom we were assigned an instructor pilot and a table. While most instructors had two students, ours had three, but in the second week one washed out.

I think all of us remember our first flight with the IP, hands in our lap but impatient to get on the controls as the IP lifted the aircraft to a three foot hover ever so smoothly, making us think this wouldn't be that hard at all. The first takeoff was exhilarating, sitting exposed in that Plexiglas bubble at a hover over the takeoff pad, then with tower radio clearance the IP nudged the cyclic microscopically forward and with our gathering forward motion we dropped a foot or so and then leveled off and began to climb, and I knew from the classroom we had just gone through *translational lift*, in which a helicopter loses the air cushion of hovering while beginning to fly. Pleasure pumped through my veins as the world slowly slipped away below and we left everything behind in our little bit of freedom, nothing to concern us but what we could see in the sky, the mild roar of our engine and rotor blades and the quiet in between radio talk. Flying at last!

Our IP started us off by trying one control at a time once airborne to get accustomed to how the aircraft responded to control movement. First the cyclic, then the collective, then the pedals, and slowly he let us try all three controls for a brief moment, trying to maintain straight and level flight. We were all over the sky! Who knew the controls were *that* sensitive? Over time we began to get the feel of the controls and could keep it straight and level, graduating to turns, climbing and descending, and even trying to hover.

Our IP, Mr. Starkey, had been a bomber pilot in WWII and had flown all his adult life. He had been an IP at Ft. Wolters for several years and had two midair collisions with other helicopters, one resulting in burns and hospitalization for six months. Mr. Starkey was a screamer. He called me names I had never before heard. It is a wonder that I made it being taught by this bitter old man who took his frustration out on his students. Somehow both students made it through.

Once we learned a crude control touch, our IP would let us try all three controls at once, turning over the controls while flying straight and level. Once we took the controls, straight and level disappeared. It would start descending so we'd pull pitch and the RPM would bleed off while the nose swung right, so give it some left pedal and RPM would bleed further so twist the throttle to increase power and the nose swung right again and now we're ascending and maybe turning left while slowing down and . . . it's exhausting when you have to think about each movement and each movement affects all the others. Think of learning to drive a stick shift truck, then multiply by ten! IPs took their life in their hands every day with new students.

Learning to fly a helicopter involved intense concentration as we continually decided how to move the controls that interact with each other, an exhausting process. *Flying* happens when you don't have to think about moving the controls, you just do it instinctively, like driving a car. One day the lucky ones wake up and just do it! The others washed out.

Once we were able to keep the aircraft sort of steady in the air, the IP would let us try the far more difficult task of hovering, which requires intense focus on any changing movements in pitch, roll and yaw. Things are slightly different in the hover mode. Power and pitch has much the same effect on the main rotor system. The difference in hovering is that we had to make very tiny adjustments on the cyclic to remain over a specific spot. Visualize placing a quarter on top of the cyclic and not allowing the center of the stick to move outside that small ring. The collective was handled the same way as in flight except when the skids were three feet off the ground it took tiny adjustments to keep the altitude constant. Pedals reacted the same with power changes as in flight but if we wanted to turn the helicopter around at a hover then we applied more pedal than normal in the direction of the turn. As we turned to the desired direction we pushed slightly on the opposite pedal to slow and finally stop the turn.

First attempts at hovering resembled a giant yo-yo, twisting and turning as we bobbed up and down with the IP grabbing the controls to prevent a crash now and then. First it took a football field size area to be semi-safe, then a parking lot, until finally, after sweating in the cockpit spread over a number of sessions, we learned to keep the aircraft in a small area with tiny movements back and forth, left and right, up and down, but safe. Sort of. Keeping the aircraft still in a hover would come over time as we developed our *control touch*.

Takeoff was from a hover and landing was to a hover, so only after learning to hover were we ready to begin learning takeoff and landing. Once we learned these basics along with adjusting all the controls and flight patterns for wind direction and velocity, and proved enough proficiency that we *might not* kill ourselves, the magical day came when the IP would tell us to go fly by ourselves, our very first solo, a major rite of passage leading to a ritualistic toss in the pool at the local Holiday Inn.

By the time we soloed, we had begun to practice autorotation in our training flights. While we had the controls our IP would surprise us by turning down the throttle at least once during every flight to simulate an engine failure to test us while we were flying. He always seemed to pick a time when we were looking the other way or preoccupied with something like a radio call. If we had a *constant* awareness of wind direction and possible landing spots then we had a good chance of lowering the collective, turning into the wind, lining up our landing to a good spot, then the instructor would take over, turn up the throttle and recover the aircraft while still at altitude, returning to normal fight as he told us whether we had earned an *attaboy* or an *aw shit!*

Eventually we would practice autorotations all the way to the ground at the airfield. A perfect autorotation had very little ground run, because in a real engine failure we would probably put the aircraft down in some kind of grassy field where a long ground run on the skids might roll it up into a fireball. So our autorotation competition with the instructor and with one another was on shortest length of ground run on the asphalt runway. Sometimes we would flare, pull initial, pull final and set it down on the runway still going too fast and slide on the skids a hundred feet or more. Sometimes it turned out right with a very short run, and the practice would serve us well.

Throughout flight training we practiced hundreds of autorotations, many of them all the way to the ground. It was a fun maneuver after you got over the initial shock of "Holy crap, we're going to die!" and it was good training that saved many lives.

WE HAD COME A long way in a short period of time, both as individual Warrant Officer Candidates and as a cohesive unit. That was the plan, of course. Team building is the foundation of the military, and we were continually challenged to act more as a team – even in unorthodox ways. Our TAC Officer, Mr. Webb, was a short man who proudly carried a swagger stick, a metal and leather stick about 18 inches long ostensibly crafted to encourage your horse to run faster. His swagger stick was a fitting completion of his uniform, always immaculately neat, black TAC helmet gleaming, boots freshly polished with the toes like black glass, tropical worsted khakis elegantly tailored and the brass on his uniform shined to perfection. He expected no less from us, but our commitment to such perfection as we sweat our way through flight school needed some . . . encouragement. Mr. Webb used his swagger stick to emphasize certain points in a way that was obnoxious but he didn't touch us with it. He would rant and rave in his low key manner and generally look for imperfections in our dress or demeanor to give him an excuse to drop us all for pushups or award demerits for microscopic infractions. That normally happened if he looked long enough for a reason, so we could count on some exercise to start the day.

One day at the mess hall, one of the candidates swiped his swagger stick from the coat rack and hid it. It was all polished, brass and chrome. That evening the word spread quietly from cubicle to cubicle to go to the latrine. There we found the swagger stick had been placed in one of the urinals. You can imagine what happened.

The next morning the candidate carried the moldy looking and smelly stick out to the formation area well ahead of our formation time, leaving it propped against the curb. We were all called to formation and sure enough Mr. Webb was on a rampage. He wanted to know who had taken his swagger stick and where it was. After about five minutes of this, one of the candidates in the front row of the formation said "Sir, is that your stick on the ground there?" Of course it was. Mr. Webb knew immediately what had happened as he reluctantly picked it up and ranted more about punishing whoever was responsible. No one said anything. We were marched to the mess hall for breakfast and we knew from that point on that we were a cohesive unit. One for all and all for one. The TACs knew it too and were secretly grinning at their accomplishment of bringing us to the point of covering for one another.

On another occasion a few practical jokers among us moved Mr. Webb's desk and chairs out of his office and into the latrine. When Mr. Webb entered the barracks and someone yelled "Attention!" two rows of candidates appeared at the foot of their bunks standing at attention in their skivvies while he slowly walked down the row looking over each one. When he arrived at his office there was a stifled giggle or two from candidates. But true to his low key demeanor he emerged from his now-empty office as if nothing had happened. He slowly walked back down the row and turned right into the latrine. When the door closed behind him we couldn't hold it any more and laughed our asses off, and had to stifle again when he emerged from the latrine and we tried to stand straight at attention without movement. Slowly walking down the row toward the exit he couldn't hide a little smile as he softly said "Put it back" on his way out. Some of the bullshit actually had a purpose, like minimizing our individuality and driving us to work together.

Some of the pressure was laced with humor. Demerits given for a dead fly on a windowsill may have described, "Improper treatment of deceased aviator," followed by Candidates carrying out a semi-formal burial ceremony, but those same demerits might have prevented one of us from spending a weekend with our wife, with Vietnam on the horizon. Yessiree, flight school was a far more mature setting than basic training.

By the end of Phase I we had lost fifty more candidates. Our number was now two hundred.

FLIGHT TRAINING: PHASE 2 – Now it was time to apply our new flying skills to navigation, cross country flying and tactical area landings, which included pinnacle landings and confined area operations where trees limited the landing space. This phase required a new airfield and all the IPs were military and Vietnam veterans. The training seemed to go easier now, as we were treated more like adults and pilots.

Flight school was fun for a little while as we learned these more useful applications of flying. We were challenged in the cross country portion, often with two candidates flying together. One would navigate and the other would fly in a "buddy cross country." It was amazing that anyone found their way and many of us got lost now and then, but we were loathe to make radio calls for help, admitting we got lost, and we always seemed to solve the problem before we ran out of fuel. Night cross country flights presented unique problems. Meeting the challenge was invigorating.

Local ranchers in the north Texas area rented landing rights to the US Army flight school so that our practice landing area was scattered far and wide. These landing areas were marked simply with a tire painted either white or yellow or red depending on the degree of difficulty. White tires meant a low degree of difficulty, yellow was more difficult and we had to have IP clearance to fly yellow tire areas solo, while red was more difficult or even dangerous and required an IP with us.

We started by landing in confined areas. Our helicopters back then could never take off straight up because they were underpowered. High elevations or hot temperatures made it worse since these factors affected the *density altitude*, meaning the force our rotor blades could take out of the air. As a result, although we landed in an area, we might not get out if the angle of ascent was too steep for the power we had available. Our procedure was to find the longest axis of the landing area that allowed approach and takeoff into the wind. On the ground we would tighten down the controls at an idle, exit the aircraft and walk off the area counting paces, because we had to back up to give us maximum takeoff room while protecting the tail rotor from trees. This was good practice making judgments about risk.

There is a critical stage of helicopter flight in the transition from hovering, which depends on the ground effect for the air cushion, to flying, where the air is drawn through the rotors and pushed to the rear. That transition is called *translational lift*. From a standard three foot hover, the pilot pushes the cyclic slightly forward and as the aircraft picks up forward speed it will momentarily dip toward the ground and shudder with vibration as it passes through translational lift, then it gains altitude and begins to fly. Hovering requires more power than flying. If the aircraft did not have enough power to hover, we could apply enough power to get it light on the skids, start forward movement to slide the skids on the runway and if we reached the speed of translational lift it might take off. The same applied in reverse when landing, and learning these lessons would serve us well in Vietnam. For example, C-model Huey gunships in Vietnam were overloaded when taking off with fresh fuel and ammo, and they often had to do a running, sliding takeoff on the runway. Since they were lighter when returning from their mission, having consumed fuel and expended ordinance, hovering at landing might not be any problem.

We didn't practice sliding takeoffs from unprepared areas because there was too much risk of snagging a skid and rolling the aircraft up into an unfriendly ball of spare parts. But we did have to judge whether there would be enough length to hover forward, gathering speed to translational lift in order to take off. It was often nip versus tuck, very close, and we sweat a lot until the trees were cleared. If we misjudged an area and were unable to attempt the takeoff after we landed, we would be stuck and have to call for help; we knew if we survived the aftermath we might forever be stuck with some nickname raising the image of calling mommy for help. So, we were careful.

We practiced landings and takeoffs in confined areas, and pinnacles as well, in the next two months. Pinnacles were small areas on the tops of hills that were leveled off for our training. Some red tire pinnacles were no more than a skids length long, though most were larger than that. If we overshot the pinnacle landing spot we risked trying to land on the sloped side of a hill instead of on top. As landings became more difficult and we struggled with the controls we had to avoid getting *behind the power curve* where we exhausted our power and settled uncontrollably to the ground, not a good thing on a pinnacle approach.

Nursing the aircraft, gently rationing the power with the controls was good practice for Vietnam, because we would learn there that when we were operating at the edge of the aircraft's power supply, clumsy control movements would make the aircraft settle toward the ground whereas very gentle and steady control movements conserved the air cushion and made it possible to find some lift.

By the end of flight training at Ft. Wolters, we were moderately competent as a pilot in our assigned aircraft, but far from proficient. We gathered skill and competence daily in tiny and unnoticeable bits. But now that flying had become nearly routine, the revised object of our desire was the Huey we would meet at Ft. Rucker in Advanced Flight Training.

FT. RUCKER IN ALABAMA was surrounded by small towns, principally Enterprise and Daleville, towns that catered to what soldiers needed to buy. A major change at Ft. Rucker was that married candidates could live off-base; Terry and I each lived with our wife in the same trailer park, the amenities we could afford.

While we were anxious to fly the Huey, we first had to endure the torture of instrument training for eight weeks.

So far we had learned to fly by VFR – Visual Flight Rules, meaning we could see clearly. Because weather is a major factor in all flying, because weather was a particular problem in Vietnam, and because combat required conquering the weather when possible, we would learn how to fly our aircraft on instruments or IFR – Instrument Flight Rules.

We learned to fly the helicopter from takeoff to landing short-final without visual reference outside. This involved a lot of classroom time learning how to use instruments to navigate. We used the altimeter to track how high we were flying, the attitude indicator or artificial horizon, and the trim ball-and-tube, to tell us what the aircraft was doing since we were flying under a hood that prevented us from seeing outside. The instructor had a clear view of the sky for safety.

We learned that vertigo, losing your sense of movement, happens naturally to everyone and while the aircraft is in a climbing turn to the right the pilot might be convinced by his feel that he is in a left diving turn. We learned that when a VFR pilot flew into clouds that obscured the horizon, if he used instruments to fly he had a chance of surviving, but if he relied on his feel of what the aircraft was doing and then making corrections on the controls, he had maybe 30 seconds before he would lose control of the aircraft and end up making a smoking hole in the ground. We learned to trust the instruments and mistrust our own instincts.

We learned to use a map with navigation aids and the aircraft's compass with built-in ADF (Automatic Direction Finder) that would point in the direction of a radio signal when we dialed in the radio station frequency. Two different ADF frequencies would help us triangulate our position, complicated by the fact we were always moving and had to recognize whether the ADF needle movement of each station was in the right or wrong direction as we homed in on the target turning point.

There were other instruments connecting us to airports to help us detect the proper glidepath for a descent to landing.

For some of us this was the most difficult stage of flight school. We earned a *Tactical Instrument Ticket* from this training. We would discover that meant we didn't have enough training to be IFR rated, but we had enough training so an over-eager commander could order us to fly when the weather was bad enough to make us wish we were back on the ground.

Completing the instrument course was doubly refreshing because we had to think too hard, and because we were eager to get our hands on the controls of a Huey.

THE BELL UH-1 Huey was the workhorses of the Vietnam War. We learned to fly the Huey in a month-long phase we called *Contact*. The Huey was easier to fly. The throttle was automatic. After rolling the throttle full open a governor would maintain proper engine RPM regardless of collective position. It was bigger, turbine powered and seemed like a dream to fly and hover. There were many new systems to be learned so proper corrective action could be taken in the event of their failure.

We flew takeoffs, landings, traffic patterns, autorotations, cross-country, nights, etc. until we were not too dangerous as Huey pilots.

TACTICS WAS THE TERM applied to the final four weeks of flight school, applying more real situations to the real aircraft most of us would fly in Vietnam. In this phase we would learn how to nurse the Huey into and out of confined areas and pinnacles.

We would also learn formation flying. Now that was scary. The idea was to keep approximately one rotor diameter away from the helicopter in front of you. We practiced being lead as well as wingmen. If the lead did not have a slow, gentle, steady touch on the controls, especially in turns or climbing or descending, the movements were exaggerated with each position so a long formation might expand and contract like a slinky, sometimes in dangerous ways. The worst for me and most others was night formation.

While learning formation flying we also had field operations, living in tents for a few days and flying in formation to PZs (pickup zones where troops are loaded) and LZ (landing zones where troops are offloaded) to simulate combat assaults.

Watching a flying formation from the ground, it all looks smooth and easy to do. In a helicopter it is much like an accordion in the air, with constant power, pitch, roll and altitude changes to stay in the ideal slot. We learned simple tricks like keeping the rotors of the aircraft just in front on the horizon, meaning we were a few feet higher just in case our rotors overlapped. We learned to pick out a particular visual reference on the aircraft we were following and keep it at the same angle and distance. Like flying in general, formation flying became a matter of experience and feel. Many of us would soon learn to do this well, in ever-tighter formations, even while being shot at close to the ground.

I was not fond of formation flying and as luck would have it, I never had to do it in Vietnam.

SOMEWHERE IN THESE FINAL days we all got our orders for the next duty assignment after flight school. We were constantly graded on classroom and flight performance, and I turned out to be fifteenth in the class, in the top ten percent. We had all filled out what we called "Dream Sheets" months before indicating our top three choices of assignment as well as what kind of aircraft we would prefer to fly when assigned as a pilot. The top candidates were assigned further training in what we called a *Turn-around School*. The options were to transition into CH-47 helicopters (Chinooks, tandem rotors), AH-1G gunships (Cobras), attend Maintenance Officer School at Ft. Lee, Va. for another 16 weeks, or Medical Evacuation Training (Dustoff/Medevac).

I had requested Cobras first and Medevac second. I had been talked into believing I would really like to blow some of these bad guys away in a gunship! So naturally I was assigned to Medevac! And with this school assignment came the orders for Vietnam.

To wrap up our final days in flight school we attended an escape and evasion exercise. There was a class to teach us the unnecessary concept that if we were shot down we would be wise to make every human effort to avoid capture by a very brutal enemy. We learned many types of bugs and snakes and plants we could eat to survive in the jungle. We were taken to a mock POW Camp with enemy interrogators prepared to put us in an underground pit or a small box or a cage, and they demonstrated the water torture where the victim feels like they might be drowning. We were informed that during the evasion exercise when, not if, they caught us we would be brought here for interrogation, and it would be our job to resist giving up anything besides name, rank and serial number.

We were dumped in the woods at dusk in teams of four with a map, a compass and instruction on our checkpoint objective. There were *aggressors* out there looking for us with full knowledge of our objective, troops who knew the woods and were rewarded for catching us with beer and 3-day passes. If discovered by the aggressors we would be taken to the mock enemy village for interrogation.

I was not captured and we never found our way to the checkpoint. There is nothing worse than a pilot trying to navigate on the ground. We had a hard charger leading our group who had no better idea where we were than I did. At 2:00 A.M. the exercise was over and all that had not made it to the checkpoint were to go to the nearest road and would be picked up and transported to the checkpoint. When we were picked up we were informed we had to take a remedial class the next week in ground navigation, but I was satisfied that we evaded capture, which to me was the main point.

When I returned home from the exercise at about 4 AM, I discovered in the shower that I had ticks all over my genitals. I used a cigarette to touch them at which point they would immediately release. I must have had twenty of these little critters attached to me and I remember Beth laughing at my circumstance. I did not burn myself badly, but it would be enough to get any man's attention! I wonder if there were ticks in unmentionable places I could not see?

WASHOUTS CONTINUED IN INSTRUMENTS, Contact and Tactical. I lost the exact count but we were told the washout rate was typically about 65% and just 35% of the original complement of young men would be there for graduation, which was rapidly approaching.

We were given more free time, freedom to come and go as adults. We could drive to classes instead of being bussed. On graduation day we flew a *flyby* over the parade field, four flights of eight helicopters each in formation. It is thrilling to see this many war machines in close proximity.

In a ceremony later that day a friend, family member, a girfriend or wife would pin on our new Warrant Officer bars, and pin the US Army wings of a pilot to our chest. We were proud to have accomplished a great deal, relieved at having survived the gauntlet and so glad it was over. We were Warrant Officers and hot-dog pilots!

Little did we know we new pilots were just raw material, finally qualified to begin learning the things that might help us come home from Vietnam alive.

Postcard to America

WE WERE TOLD AS *we departed Vietnam to return home that our welcome in the US could be a little rough and we might want to wear civilian clothes after we were processed thru for our flight home. We were also warned not to tangle with protestors or it would be the military guy who suffered consequences.*

Being the good Airborne Ranger that I was, I decided to dress in my class A uniform, bloused boots, and all my decorations on full display for the whole trip.

When I arrived in the San Francisco Airport for my flight to Philadelphia I was confronted by three hippies, one man and two girls, who spit at me but not on me. As I was getting myself in deep trouble by decking one hippie who poked me in the chest and insulted me, a strong arm draped my shoulder and a soothing voice said "Captain, come with me."

He was an older man with a kindly smile who had spotted my 101st Airborne patch. He directed me into the nearest bar and told me he had landed with the 101st in Normandy in WWII, and he wanted to buy me dinner. We had a great time together and got a little drunk before my flight home. I have never forgotten that kindness and have bought many drinks recently for men and women sporting the Screaming Eagle in airports.

Over the years I avoided conflict by not advertising my Vietnam service much outside my small circle of friends and work colleagues.

I did receive accolades for my Vietnam service in an unusual place. I was in Russia shortly after the fall of communism, where my Russian hosts warmly referred to me as their friend Jim who fought communism in Viet Nam. Later, in 1991 while in Moscow, after some conversation the conductor of the Moscow Orchestra gave me a bottle of North Vietnamese vodka he had been given while in Hanoi during the war. I still have that bottle, unopened.

I received gifts and congratulations for my role in the war from Russians that I would never receive at home in the US. Go figure.

James William Oram, Jr.
Berwyn, PA
1969-70 101st Airborne, Camp Eagle, remote firebase, company XO, CO then Bn S3

So Others May Live
Ron Current
Mableton, Georgia

SCARFING MISSIONS WAS A wicked custom in our Dustoff unit. If we were flying and heard the call come in for a mission, we would change our course and max out airspeed trying to beat the other crew to the location to make the pickup. It kept us focused on fast response to pick up our wounded brothers.

At lunch time one day while on *first-up* standby, I asked Jay Brannon to cover for my co-pilot and I while we grabbed chow at the mess hall, and to call me so we could come running if we were scrambled before returning from lunch. I told Jay we had a pending mission from the morning, in the same area where I took rounds the day before.

While we dashed to lunch the mission finally came in but instead of calling us Brannon grabbed a co-pilot, took my crew and the Huey that I had preflighted and had ready to go in five minutes. He flew the mission. I returned to discover Jay scarfed me and had been shot down during the pickup. I immediately got another crew together and scouted around for a flyable Huey to go after them. The Operations Sergeant said Major Rose was already on his way to make the rescue in the only remaining flyable aircraft. Jay was a good friend; if he was dead or hurt I would feel partly responsible, and if not I wanted to kick him in the ass for scarfing me.

I waited by the radio very anxious to hear whether my friends were alive. It turned out that Jay was shot down on final approach, lost control and went in upside down. The main rotor separated, cartwheeling off to parts unknown, and the rotor mast buried three feet into the ground. It did not sound good for a while, then the radio chatter said all the crew got out of the wreckage and away from the aircraft, still on the ground. One had a broken leg and the rest were just cut and bruised. There was no fire, amazing luck. Gunships were pounding the area and the crew were keeping their heads down. Major Rose circled over their position thirty minutes when he finally decided not to wait any longer for a break in enemy fire; he went in to make the pickup under heavy fire and with lots of gunship activity. They did find an LZ to land in and made not only the pickup of the crew, but the patients that the original mission was called in for. They were back in the unit area about an hour later after taking the wounded, now including the crew chief, to the Cam Ranh Bay hospital.

I met them on the flight line and they looked pretty haggard but Jay was grinning from ear to ear, ready to dig me forever with pride about scarfing me and surviving that mission. I gave him hell but I would have done the same thing. There were enough stories from that incident to last several nights at our club as we pushed the bad stuff to the back of our minds with the help of banter and booze. I got to listen to Jay for days about his big scarf.

If scarfing missions sounds like the sordid competition you might see between ambulance companies racing one another to the scene of a car wreck, well, it isn't because . . . you see . . . well, OK, it's the same damned thing! But that was the fun part.

I AM PROUD TO have been a Dustoff pilot. Our motto was "So others may live." To the men in danger in the remote bush, the knowledge that if wounded they would likely be in a hospital within an hour was a huge factor in morale. The quick evacuation of the wounded to a hospital meant a marked increase in survival rates, and contributed to the effectiveness of operating units in the field that could quickly hand off their wounded and focus on their mission. Of the wounded who survived the first 24 hours, 99% lived to tell their tale.

After graduating from flight school I attended the four week long course *Essential Medical Training for Army Aviators* at Ft. Sam Houston, Texas. The course concentrated on anatomical training and films of actual treatments and surgeries in field hospitals. We learned the Priority System of categorizing patients as urgent - would lose life or limb within two hours, priority - could wait at least eight hours, or routine - good for 24 hours. We also prioritized patients as US military, US civilian, Foreign Military such as the ARVNS (Army of the Republic of Vietnam) or ROKs (Army of the Republic of Korea) and finally foreign civilian.

Part of the training at Ft. Sam Houston was a visit to the *Burn Ward*. Most of the burn patients were air crewmen. That gave me a sinking sensation because burning was and still is the only real fear I have. I never worried about dying so much as burning. There were many real heroes there in the Burn Ward, guys who had saved other's lives and were shot down or crashed and burned.

A captain who talked to us had injuries from his crash-and-burn in a scout OH-6. He lost both ears, his nose and several fingers, and told us of his struggle to regain normal life after spending nine months on the Burn Ward with all the reconstructive surgery and pain he had to deal with. He had been given false ears that were sort of glued on. He was accustomed to stares, but one day walking downtown San Antonio, near the Alamo he noticed more stares than normal. He finally looked at his reflection in a store window and saw that one of his ears had slid down and was now in the middle of his neck! He plucked it off and went on his way. He was a living testament that no matter how bad things are, life goes on and someone else has it much worse! He helped us put things in perspective.

THE AIR AMBULANCE OPERATION in Vietnam was part of the 44th Medical Brigade, which controlled Dustoff and hospital units all over the country. Our callsign was Dustoff and our aircraft had red crosses on the nose and side doors but no M-60 machine guns like other Hueys. Each crew member had a rifle and a pistol for the protection of our patients and ourselves in the event we were shot down.

There were two Medevac units outside the control of the 44th Medical Brigade: the 1st Calvary Division had their own air ambulance helicopters known by the call sign Medevac; they also had the red crosses but they did have M-60 machine guns and used them to shoot their way into and out of LZs. The 101st Airborne also had their own air ambulance helicopters with M-60s; their callsign was *Eagle Dustoff* since their unit insignia was an eagle.

When I arrived in Vietnam I was assigned to the 498th Medical Company (Air Ambulance), a unit of the 44th Medical Brigade. The next day a helicopter with the red cross on the nose and doors arrived from our unit, piloted by Lt. Anderson, to take us to our assignment. The trip took several hours including a refueling stop in Nha Trang, and I was struck on that flight by the country's beauty. In II Corps near my unit there were some of the most beautiful mountains I had ever seen and some great beaches. The countryside was pock-marked with bomb and artillery craters that looked like scattered round mirrors from the air, filled with reflective water from recent rains. As we approached what would be my new home, the crew phonied up an enemy attack in process on our airfield, all for my benefit so when we touched down to drop me off I would low-crawl and scramble with wide eyes of fear, which I did. The helicopter crew and the ground crew laughed their ass off. I may have been a newbie, but at least I was good for a few laughs.

My first afternoon at the 498th I met CW2 John Siverd, a guy with a number of scratches on his face and walking with a limp. He said that he had turned a helicopter upside down in the river. Everyone got out but they all had cuts and bruises. I later discovered he had been flying aggressive low level, caught a skid on a rice paddy dike and flipped over. The dikes were normally only three feet high so he definitely was low!

The first few days in my new home were spent getting over jet lag and being indoctrinated to new surroundings and AO as well as the unit's mission. We staffed *standby* Dustoff missions at remote locations including LZ English and LZ Uplift in support of the 173rd Airborne Brigade, An Khe in support of the 4th Infantry Division, and Tuy Hoa in support of mostly ARVNs. These standbys required at least one helicopter and crew on call around the clock for up to ten days.

On the third day I had my in-country checkride, a *Contact* check ride to see if I could really fly. We practiced something new to me, hovering autorotations from fifty feet in the air. In flight school we never did them from higher than three feet. The higher we were at a hover made these maneuvers more exciting and far more dangerous. This maneuver was vital to us since we would be doing *hoist* missions, hovering above the jungle, winching the wounded up on a cable.

The remainder of the checkride consisted of flying around our AO to get familiar with landmarks, unit locations, enemy hotspots and the helipads we would use to drop off certain categories of wounded. There was the 24th Evacuation Hospital in Qhin Nhon and dispensaries located at LZ English and An Khe. The closest large hospital besides the 24th was in Pleiku, used by our standbys at An Khe when the dispensary could not handle the severity of the wounds encountered.

My new home at the 498th was just outside a village named An Son, close to the large town of Qhin Nhon on the coast with a beautiful beach to its south. The flight line was called Lane Field, where the *first up* crew handled all the incoming missions first on their 24 hour duty day, prepared to be airborne within five minutes of a mission call. *Second up* was the backup crew used in the event that *first up* was out on a mission when the call came in for an evacuation. We had a *third up*, but used them mostly for back hauls, transporting stabilized wounded from dispensaries to larger medical facilities for treatment.

The *first* and *second up* crews stayed in a hooch on the flight line, eating and sleeping a few feet from the door to their helicopters. The field phone in the hooch was directly linked with a strung wire to Operations. The missions were radioed in to Operations and they called to scramble us on the wired phone. When the field phone rang in the hooch, even at 2 AM and ruining a deep sleep, I knew someone was hurt bad and we were going to fly, bringing help to our brothers on the ground.

LIKE OTHER NEW CO-PILOTS, I flew the safest and routine part of trips; the Aircraft Commander took the controls for the hard part while I watched and learned. It normally took three or more months in country to qualify for AC, a goal we lusted after more and more as time went by.

Daytime LZs were marked with smoke to help us pinpoint the spot amidst an ocean of jungle. We always tried to take off and land into the wind because that slowed down our ground speed and gave us more control of the aircraft. The smoke also gave us a visual reference on wind direction and speed, an important part of deciding our approach and takeoff path.

Since the enemy monitored radios, as we approached we radioed the ground troops to "pop smoke" or the ground crew would tell us "smoke out" on the radio without identifying which one of our five smoke colors. When we spotted purple smoke from the smoke grenade, we would say, "I have purple smoke," and the ground crew would either confirm or deny. Sometimes the bad guys popped smoke of different colors, hoping to get one right and lure us into a false LZ for a helicopter kill.

ONE OF THE ACS that impressed me, CW2 Terry Zinger, was with me on my first standby at An Khe. I listened to him and tried to learn as much as I could. He had been in-country a year and was on a six-month extension, an option in those days that would get him thirty days of extra leave and then released from the Army when he hit the states on his way home. Most of us would have two years left to serve after a year in Vietnam.

Zinger was a very good pilot and as I came to learn why he was on his extension, I admired him as a person, too. In the 10th month of his first year he crashed a helicopter on a hoist mission. All of the crew was hurt and his crew chief died from his injuries. Zinger felt obliged to come back on an extension to somehow try to make up for the crash.

Zinger showed me the ropes of our mission, very tolerant of my new-guy mistakes and learning curve. The standard approach we learned in flight school, as in flying at 500 feet and then decelerating slowly and smoothly as the helicopter descended into the landing zone, would get us killed in Vietnam because it gave the enemy too much of a target for too long. We customarily flew above 1,500 feet because lower than that was in the *kill zone*, within the effective range of small arms fire. A couple of miles out from the LZ we would drop down to just a few feet above the trees for our final approach, close to the trees at a hundred knots and following the dips and rises of the terrain, always a thrill, clipping tree branches now and then with rotors or skids, and the enemy had a hard time hitting us with fire since we flew over their heads so fast. Very near the LZ, the pilot would drop the collective and kick a pedal either left or right, depending on which seat you were flying from, and lay the aircraft over on its side in a flare to decelerate rapidly. This *side flare* maneuver brought us to a stop in about 50 yards from a speed of 100 knots. As the helicopter approached a hover we kicked the opposite pedal to straighten it up and normally do a *button hook*, turning the aircraft 180 degrees as we came to a landing. If it all worked properly, just as we completed the turn our skids touched down in a level landing.

The button hook was intended to put our tail toward the bad guys since there was a lot of metal to shoot through in the tail boom and fuselage, before it would get to us. Whether the LZ was hot or cold, we used this approach as practice for the real thing when the LZ was hot. This maneuver required guts and skill since we were flying *on the edge* with main rotors very close to the trees.

Sometimes we used a *hammerhead* entry to the LZ. The hammerhead started with a *cyclic climb*, pulling the cyclic back toward our belly so the aircraft traded airspeed for altitude by jumping up in a steep climb until we zeroed out the airspeed with the nose to the sky. Then we dropped the collective and used the pedals to flip the tail to the sky and nose at the ground in a 180 degree pivot, and the aircraft would drop rapidly until we added power to accelerate to the landing spot or to a hover for the hoist. One purpose of the hammerhead was to stop quickly. The other purpose was to show-off and play; who said flying a helicopter in a war zone can't be fun?

One day I flew standby with John Siverd out of LZ Uplift. John flew low level at every opportunity and this time the crew chief was sitting on the floor of the helicopter with his back to my seat on the right side, his foot hanging out of the aircraft. As we flew close to the ground, dodging and weaving between the scrub brush in the flat lands, his foot was hit with a bush at 100 knots and nearly broke it.

New guys inevitably make mistakes; we just hoped they were not fatal mistakes. I was flying with Zinger on a mission about dusk that involved a routine pickup along a road. I had the controls and I approached the LZ that had been marked with smoke, moving too fast, low on the road between the trees on both sides. We flew past the LZ at about 50 knots as they probably wondered where the hell we were going. To recover, I completed the side flare maneuver with the button hook and proceeded to hover fifty yards back to the LZ site.

Maybe the guys on the ground thought it was cool and didn't realize a green pilot had screwed up, but Zinger and I knew. He didn't say a word and didn't try to take the controls; he was smart enough to let me learn my own lesson. From that day forward I never flew by an LZ again. It takes a lot of peripheral vision and a confident control touch in the aircraft to do these things properly.

WO Ron Current at the Phan Thiet RTO shack
Photo courtesy of Ron Current, all rights reserved

WHEN THERE WAS NO clearing for us to land to pick up the wounded, we used a hoist. We hovered just above or even down amongst the trees with a winched boom out of the side of the aircraft, 250 feet of cable and a weighted hook on the end attached to a jungle penetrator, a three legged affair shaped like a bullet when the legs were retracted so we could drop it to penetrate the jungle. The three legs could be pulled down to allow one soldier to sit on the device with a strap around his back to be lifted through the jungle. We preferred an LZ to land in, but we did what we had to do with so much jungle cover. Hoist missions were, uhm, well, exciting! While we hovered, unable to move with our own wounded coming up slowly on the hoist, we presented a large and irresistible target to the enemy, who used the protected red crosses as targets.

We all kept track of the number of hoist missions we performed as some macho measure of our courage, or perhaps a small celebration that our luck was holding to count how many dangerous missions we had survived. I flew my first hoist mission with Terry and he became quite a mentor to me. I knew that I had *arrived* after I rode through 15 hoist missions before I flew one on my own. The hoist missions were the most critical of all things that we did since our tail rotor and main rotor were very close to the trees, requiring all the crewmembers to be on *hot mike* to talk without pressing buttons to tell the pilot ". . . left two feet. . . forward three feet . . . " to protect the main and tail rotors. The lower we could get into the trees the less of a target we presented to the bad guys, so we would try to nestle the aircraft down in the trees with just the main rotor above the tree line. We had to constantly watch the tail rotor because its small size and high RPM made it fragile to contact and if we lost our tail rotor we went down immediately. The job of the medic and the crew chief was to help the pilot flying to keep his tail rotor out of the trees. The crew chief had the added responsibility of operating the hoist, lowering the jungle penetrator down to the people waiting below and keeping the pilot aware of where the aircraft was in relation to the trees and the people waiting to hook up. The pilot not flying would help by keeping his side of the aircraft clear and giving similar instructions to the pilot when he thought a hazard might exist from hitting trees with the main rotor. The landing light and search light were controlled by the pilot in the right cockpit seat, so that was an added responsibility for him if he happened to be flying the aircraft during a night hoist.

There was non-stop hot-mike chatter among the crew on a hoist mission hover. Add a little wind into the equation or rolling fog and you have a high-risk and tense maneuver. Ideally, once the penetrator hit the ground then it would take thirty seconds to a minute to get the wounded loaded. Time was of the essence, particularly if the enemy was in the area, which was likely since most wounds that required us to be there were the result of contact with the enemy. The less time the helicopter was in this *sitting duck* position, the better chance for a safe mission.

During the hoist missions while the penetrator was being lowered, the crew chief was often partly outside the aircraft with one or both feet on the skid, secured by a *monkey strap* in case he slipped, a 10 foot-long strap attached to the transmission bulkhead. Some called it a *sissy strap* and would not wear them. Fortunately nobody fell out on any of my hoist missions.

POLICY REQUIRED GUNSHIP SUPPORT for medical evacuations resulting from enemy contact if there had been enemy fire within the last 15 minutes. Sometimes that worked and sometimes in outlying areas it was not possible since the gunships might be 30 minutes or more away from the LZ. The AC's decided whether to wait for gunship support or not. I tended to involve the entire crew in those decisions since their neck was as precious to them as mine was to me. In most cases we decided to go in regardless of gunship support unless it was obvious that the ground unit was still in contact with the enemy.

There were many times gunship support was too far away and we flew the mission anyway.

I SPENT A WEEK in LZ English with a *first up* and *second up* crew there since there was plenty of activity for us. There was a 155 mm unit 50 yards from our hooch. The 155s would fire every night starting at about two AM for two hours straight, shaking our hooch so the beds would move across the floor. For the first and second nights I did not sleep after they began firing, but by the third night I had grown accustomed to the blasts and fell asleep. After a while we could sleep through anything, which was good because we could get an occasional *combat nap* for twenty minutes or so between missions, which were no longer than forty minutes in the air, since we were close to the action by design.

I woke up one night to a gunshot blast in the hooch, grabbed my pistol from under my pillow and rolled off the bed onto the floor. While my body struggled awake to catch up to my mind I searched for the enemy, ready to fire. In a few seconds the lights came on; one of our pilots had fired at a rat that had been running across the rafters but slipped and fell onto his bed. The rat scurried to the footboard of the bed and sat long enough for this pilot to pull out his 45 caliber pistol and blast him. There was a bed abutted to his footboard occupied by another very lucky pilot since the round went a foot over his head, and he was pissed!

Our idle time on standby would often be spent playing cards in the small area that had a table and chairs and a cooler for our cokes. We played hearts or spades, waiting for the field phone to ring. Many days we would fly six to eight hours in 30 to 40 minute increments. The most I ever flew in one day was 14 hours, but the intensity of concentration required in combat flying, especially in contact or in weather or at night made an hour of flying equivalent to about two hours of strenuous manual labor. At the end of a long day of flying we were beat!

There was a lot of hustle involved with these trips and a real sense of urgency to get our wounded out and back to safety. Our AO was mountainous and covered with triple-canopy jungle, three different levels of trees, making it impossible to see the ground from the air and requiring the hoist. Flying in the mountains brought other challenges like *geographical turbulence* created by the winds forced upward by the terrain.

This triple-canopy deep cover was enemy territory, and LRRP territory as well. LRRPs were vulnerable because they were small teams and lightly armed, and if one was wounded he not only couldn't do his specialized job but also posed more risk to the team that had to care for him. So we picked up LRRP teams frequently.

LRRPs always whispered on the radio, either because the enemy was that close or the discipline they had ingrained against breaking their cover. When we were nearly overhead the noise of the rotors drowned them out on the radio. We normally hoisted LRRPs out. Sometimes when we picked them up in the mountains we kept the aircraft at a hover with one skid touching the steep slope and the other in the air. It was tricky with the main rotor dangerously close to the ground on the uphill side, requiring intense concentration to keep the helicopter stable while these guys were jumping on with their gear.

We got a call one night to pick up LRRPs in the flat area north of Phan Thiet, some twenty miles away, them whispering on the radio as usual. As we approached their location on this dark night, they told us that they would mark the LZ with a strobe light. They put the strobe in a *steel pot*, a combat helmet, to prevent the light from being seen on the ground but visible from the air. We located the light and set up an approach to the area. We came in rather steep and *blackout*, without any lights turned on the helicopter, until the very last minute. We lost the light several times on our final approach because of the depth of the steel pot, but with a little increase in collective we picked it up again. We turned our landing and searchlights on at the very last minute and safely landed. When we were down we turned all the lights off again, so we wouldn't give the bad guys too much of a target. We sat there with the helicopter running for what seemed like hours, really just minutes, until they showed up, running to the helicopter. I don't remember the nature of their wounds but after the whole team was on we immediately pulled pitch and got the heck out of there.

AN KHE HAD SEVERAL Red Cross volunteers in a building where the men could come in and play pool or ping-pong and have coffee or sodas, but no liquor. These centers were run by American ladies we called *Donut Dollies*, a big hit with the troops since these were the only *round eye* women they might see for a year. Their living area was a couple of double-wide trailers surrounded by high fencing to protect them from Vietnamese sappers that might break into the compound, and from American troops who might become overly amorous after a couple drinks at the local club!

One of our pilots spent his free time with his special *Donut Dolly* while on standby. One night he was still in the Donut Dolly compound when his crew was scrambled. The Peter Pilot cranked the helicopter, hovered over the compound and hoisted the AC into the aircraft. Asses were chewed later.

The routine after coming back to Lane Field from standby duty was to have that night and the next day off. Most went to the club and tied one on with their friends that night. John Siverd came back from one standby and proceeded to get really plowed. He didn't like the unit XO, a major, who had an air-conditioned room quite a bit larger than the standard for the rest of the pilots, and a refrigerator, which was unheard of. After John got enough liquor in him he popped a purple smoke grenade and tossed it in the XO's room about midnight. The sleeping XO and everything in his room were rendered . . . purple! When the CO and XO ranted and raved at us the next morning, searching for the guilty party, John looked like hell, appearing for all the world to have been stabbed in both eyes with an ice pick. Despite his hangover, he took his medicine by confessing and cleaning up the mess.

ONE DAY FLYING WITH Zinger I made a pickup in the An Lo Valley, just east of An Khe, a routine mission with no contact. I had the controls and made a low and fast departure, accelerating to 100 knots, maybe 50 feet off the trees. I was approaching a ridgeline and decided to do a cyclic climb. About 100 feet from the ridge and 200 feet below the top of it I pulled back on the stick and instinctively pulled up on the collective for some added power. I was already close to the maximum power of 50 pounds of torque when I did this and ended up at 56 pounds before the maneuver was over. We did climb over the ridge, a thrilling maneuver.

Zinger noticed that I had over-torqued the aircraft. We got back to An Khe, dropped the patient at the dispensary and hovered to the POL point for a *hot refuel*, meaning we would keep the aircraft running while the crew chief got the fuel hose that was attached to the bladder. He refueled us in five minutes, we repositioned to our helipad and when we shut the aircraft down, Zinger mentioned that we would have to inspect the head of the rotor system since I had over-torqued it on our flight. We completed the inspection and of course I was embarrassed about doing something so unprofessional when there was no reason for the extra power except to *cowboy* the aircraft. The word got out to the rest of the unit and I was known to some from that day forward as *High Power*!

JUST WHEN I WAS settled and getting confident controlling the aircraft, the Army needed me elsewhere to help start a new unit, the 247th Medical Detachment. They had six helicopters, based at Phan Rang Air Force Base in southern II Corps. On the flight down we flew the coastline down all the way past beautiful beaches and mountains with picturesque waterfalls. We passed Nha Trang, a sought-after assignment since their Dustoff detachment lived downtown in villas. We passed Cam Ranh Bay and the big Air Force Base and, finally, into the valley where Phan Rang was nestled. It had high mountains to the west and a nice beach about two miles to the east.

The AO was not as hot, less enemy activity, and the terrain 30 miles to our west and 5,000 feet high at the airport was breathtakingly beautiful. We flew standbys at Phan Thiet, an Army airfield on the coast about 50 miles south of Phan Rang, and one in Bao Loc, inland some 30 miles and 30 miles south of Dalat. We had two aircraft on continual standby at Phan Thiet and one aircraft and crew at Bao Loc. At Phan Thiet we would alternate between first up and second up. We had a hooch all to ourselves there with bunk beds for all of the crew, a table and blood cooler used to keep our sodas cool. We ate C rations most of the time, although there was a mess hall down the road for an occasional hot meal.

We had a real bug problem at Phan Thiet. We could hear big cockroaches moving across the wooden floor at night as they headed to our table where someone might have left some food open. We ambushed them one night after all the lights were out using an opened can of Spam for bait. When we surprised the cockroaches by flipping on the lights, one crew chief standing over the table sprayed them with lighter fluid lit by squirting through the flame of a lighter. He got them all with this homemade flamethrower, over two hundred kills that night. We had a burned spot in the middle of the table, and the warm glow of revenge for creeping us out at night. We repeated our ambush for three nights in a row and put a small dent in the roach population.

DUSTOFF MISSIONS WERE INHERENTLY short by design; we were stationed close to expected action for quick extraction and return to a medical facility. By the time I became AC I had over five hundred missions and had logged some 300 hours in the aircraft. I was confident of my ability to do anything to complete any mission. I had flown over 50 hoist missions and had not lost a crewmember or helicopter yet. The helicopter was beginning to feel like an extension of my body. I didn't fly it, I wore it. I could maintain the aircraft within inches in altitude of where I wanted it to be even at one hundred knots. As I look back at it now, I was bordering on dangerous because of constantly flying *on the edge*.

I grew a mustache while I was there, a common practice by Warrant Officer pilots despite strict limitations on the length of a mustache as well as hair. I was normally within the haircut policy but way out of the regs on my mustache. Mine was a handlebar mustache that you could see from behind my head. My wife sent me mustache wax from the states. I got away with it.

I went to the O Club with everyone else and if I was not flying the next day, I just might have too much to drink. I never did drink before I went to Vietnam, but . . . close contact with so many dead and wounded did have an effect on all of us. Drinking was our escape, deadening the senses so that, for a little while, we could forget what we had seen recently.

ONE DAY WE WERE supporting the *Whitehorse Division* of the ROK on a hoist mission in mountain jungle. We were told on the way that the troops were still in contact with the enemy, but that there were two men near death. We made several low passes over the troops on the ground, did not draw fire, and decided to go in for the pickup even though we had no gunship support. We made our approach to the one hundred foot hover and came to a stop with the nose of the helicopter some twenty feet from the side of the mountain.

The crew chief began lowering the penetrator and we began to think it would be a routine mission. Suddenly Plexiglas was breaking everywhere. My chin bubble was shot out and our radios were gone. My Peter Pilot on this mission was WO1 Jack Parks, whose voice raised an octave or two as he said, "Taking fire, taking fire!" I lost radio communication with the crew chief but decided to leave since there was no one on the penetrator yet. I turned the helicopter away from the hover spot and lowered the nose to accelerate. We were dragging the cable and penetrator through the trees and as a result the boom of the hoist knocked the crew chief down on the cargo floor of the helicopter. I told Jack to cut the hoist. He did nothing, still paralyzed by the rounds tearing through the aircraft. I finally reached up to the switch on the console and cut the cable myself. We landed the helicopter ten minutes later in the company area, so riddled with bullet holes, we would learn later, that it had to be sent back to the states to be rebuilt.

We switched helicopters and flew back out there to find the gunships had arrived and were softening up the area with rockets which sounded like the side of the mountain was falling through the rotor system while we hovered. We finally got our patients and returned them to Cam Ranh Bay at the large hospital there. That area was hot for a long time.

THE LAST TIME I took rounds on a mission, I thought I was shot. I was operating out of Phan Thiet and I was short, just one month to DEROS. We had an urgent hoist mission in the mountains to our west without gunship support and no time to wait. We nestled into our hover about one hundred fifty feet above the ground and all of a sudden we were taking rounds through the aircraft. I felt something hit my left arm and thought I was hit in the elbow. I thought to myself it didn't hurt very much for a gunshot wound. Since I was concentrating on the controls I had no time to look at my elbow.

The patient was not on the penetrator yet and the fire was intense so we cut the hoist cable and made a quick departure back to Phan Thiet. The crew chief on this particular mission was Salvador Vargas, the nephew of the guy who drew the *Vargas Girls* in Playboy magazine. Salvador was wearing his survival kit on his leg that day and in that survival kit was a small flashlight in a side pocket. What had hit my arm was the lens of his flashlight after it had taken the bullet and broke up. Salvador was not hurt, very lucky. We all were. I had a bullet hole through the door post of my door just inches behind my head. To make matters worse, we were scarfed! While we were en route back to Phan Thiet, a Huey arrived on the scene and brought the patient back to the dispensary and he survived. We were installing the spare hoist on the Huey when we got the word the patient was en-route. I didn't like being scarfed but was glad that they got him out.

AS IN ALL OTHER locations, we had a field phone connected to the Operations area. I dreaded hearing the phone ring in the middle of the night. We routinely would fly four to five hours a day in support of US and ARVN units in the area, most of our missions lasting just 30 to 40 minutes and would end with us bringing some poor souls back to the dispensary for medical miracles.

We provided a lot of Dustoff support for the local civilians, too. It seemed that we were always taking pregnant women with complications to our dispensary. I wondered to myself what these people did before helicopters arrived on the scene in Vietnam. For the most part the women were very hardy there. Most would work in the rice fields until they delivered their babies and go back to work within a day or two after the baby was born. They would carry their newborn in a backpack affair while they were working. I was impressed with such industrious and hard-working people.

NOT ALL OF OUR flights were strictly business. While on standby at Bao Loc, surrounded by rice paddies in a very flat area, we were called to pick up a patient that was pretty bad off and took him for treatment to Phan Thiet, some four thousand feet high nestled in the mountains. We refueled and started back, flying low level on a river flowing out of the mountains down into the Phan Thiet basin, not the safest maneuver since bad guys were often along the river hunting wildlife for food. As we followed the winding course of the river, having fun with our rotors just above the trees, we passed a couple of deer on the river bank but they somehow weren't spooked by the helicopter noise. We immediately turned around to see if we could get a shot. One deer ran but the other stayed put long enough for the crew chief to hit him with his M-16. He fell into the river and floated down several hundred yards until he finally hung up on some rocks below the surface of the water. We took the chance to hover just above the deer, keeping the skids out of the water by a few inches while our crew chief got out on the skid and lowered the cable from the hoist, wrapped the cable around the deer's neck and hoisted his body out of the water, reached under him and with the help of the medic pushed the whole animal inside the helicopter and we took off. It was a large deer, probably two hundred pounds. Back at Bao Loc some troops butchered the deer and we all had a great meal that night.

(L-R) Scotty Davidson, Malcolm Hartman, Robert Deaderick, Ron Current
Photo courtesy of Ron Current, all rights reserved

WE FLEW A NUMBER of Dustoff missions for Special Forces based just outside of Dalat, and they invited our unit to a party and BBQ at their place about 5,000 feet above sea level. I was the AC in charge of flying our unit personnel that could be spared. We put a dozen of our troops on board plus the crew of four and we were overloaded. The aircraft did not want to hover. I pulled 48 pounds of torque getting it off of the ground when we would normally only take 30 pounds or so. 50 pounds was the maximum power that could be pulled out of the engines before the rotor rpm would start slowing down. I decided to take it that way anyway; after all I was bulletproof.

I hovered just inches above the ground and slowly gained speed and passed through translational lift. When that happens the helicopter will tend to descend a little and this day was no different. I did not contact the ground though. On this day my skill far exceeded my judgment.

We made the flight from Phan Rang at sea level to the party area and suddenly I realized that I wouldn't be able to hover at this altitude with this load." Fortunately for us they had a level, grassy area where we could do a running landing. It took less power for a running landing and the maneuver turned out just fine. Our guys had the chance to get out of the unit area for a few hours, which was a treat for most since they were stuck back at the base doing maintenance on the helicopters most of the time. I realized after we got there that I would have to make two trips back to take everyone home on a safe flight. I did just that.

FINALLY, FOR ME IT was over. I was to fly out of Cam Ranh Bay Air Force Base on a contract airline after a day of out-processing. Our flight would take off at midnight with some three hundred soldiers on that flight. One of the enlisted men in the out-processing section told me that I would not be allowed on the aircraft with my mustache since it was completely outside of regulations. Is there anything quite so stupid as a zero tolerance policy that relieves people of the responsibility to think and apply judgment on what is important? I cut the darned thing off.

The flight home from Vietnam to the US was a long one. I was sitting near the front of the aircraft when we first spotted the lights of the good old USA. Someone yelled, "Hey, there it is guys!" bringing a loud and long cheer from the troops on the plane. We were finally home!

We landed at McChord Air Force Base outside of Seattle. It was dark and very early in the morning when we landed, all dressed in jungle fatigues as required on these contract flights. We went through the customs area to retrieve our bags, then everyone immediately went to the closest bathroom to change into TWs (tropical worsted), our tan short-sleeved uniforms. The floor of the bathroom looked like an Army Surplus store, fatigues discarded everywhere.

Since we landed at McChord we had to grab a taxi to Seattle-Tacoma Airport for our commercial flight home. My flight to Des Moines was not scheduled to take off until 8 AM. At 4 AM we were waiting for shops to open.

At 6 AM the bar opened and there were some twenty guys waiting to get in. The first three guys through the door dropped a twenty on the bar and told the bartender that anyone with a uniform should get drinks until the money ran out. Twenty bucks bought quite a few drinks back then. I had never had alcohol early in the morning until that day . . . but back in Vietnam it was the cocktail hour! We had a little time to chat about our experiences in-country. It was a happy time for all of us, safe and home and ready to continue our lives outside of the gunfire and daily hazards.

Finally I boarded my flight. I sat behind a pretty girl from Seattle to Denver where I would catch a connecting flight to Des Moines. She reminded me of someone I knew and during the flight she turned around and asked, "Did you just get out of basic?" I shouldn't have been puzzled by the question. I felt older but was still just 21 and I do have a slim face, and my uniform was clean of decorations other than my wings and Warrant Officer bars. I could easily be mistaken for a trainee. I didn't even wear the Bronze Star ribbon I received in Vietnam, which was pretty routine for helicopter pilots. I thought medals were overrated since some units were tight with awards while others gave them out like popcorn.

I answered, "No, I'm on my way home from Vietnam." That seemed to end her desire to talk.

I could have taken the time to tell her that I flew Dustoff in Vietnam, and that I was quite proud of having mastered flying the aircraft in difficult conditions, that I could now make that helicopter do things, even while being shot at, that new pilots could only dream about. I could have told her I was damned proud of being part of an operation that saved a whole bunch of lives. I could have tried to find the right words. If I did that, do you think she would have understood what an important part of my life the last year had been?

Neither do I.

Photo courtesy of Ron Current, all rights reserved

After Vietnam Ron Current taught Basic and Advanced Instrument courses to helicopter pilot students at Ft. Rucker. He transitioned to fixed wing and had Army aviation assignments in Thailand, the US and Europe. He retired from the Army in 1988 and continues his love for flying as Chief Pilot on an Astra Jet for a private company in the Atlanta area.

Postcard to America

FROM 1968 THROUGH LATE 1971, as a young Lieutenant Naval Aviator assigned to VRF-31 at NAS Norfolk, VA, my job was ferrying all types of aircraft to many points in the world including Vietnam. In this capacity, about every two months, I and other pilots were tasked to TRANSPAC - fly across the Pacific - from one to four new or reworked aircraft to Da Nang AFB in Vietnam. This involved several days of flying, with multiple mid-Pacific air refueling by our peers in the Air Force and stopovers in Hawaii, Wake Island, Guam, or the Philippines. The return to the US from Da Nang was usually accomplished by hitching a ride on the next scheduled departure, generally a USAF C-141 or one of the frequent charter flights. The latter caused us to jump for joy due to the added comforts and amenities of the civilian jets.

On at least two dozen occasions over that time frame I had an opportunity to see first-hand how the hostile welcome by the anti-war crowd developed and changed at the California airports, particularly San Francisco. From my very first return trip to the US in the summer of 1968, I dreaded having to run the gauntlet of hateful hippies and college students who were waiting for us when we debarked the Air Force bus which brought us from Travis AFB to the SFO airport for a commercial flight out to our next aircraft pickup point.

I believe these organized groups had informants on the base who alerted them when the flights were inbound from Vietnam. They were always waiting for us at the airport with hateful signs and chanting slogans as the bus left, calling foul insults to the military personnel on board.

The worst part was having to thread through them on the sidewalk, making our way to check in for our outbound flights at SFO, having been warned that altercations were to be avoided. The average returning GI only had to endure this nasty treatment once as they returned from their tour in Vietnam, while I got to do this around two dozen times over several years as the anti-war hostility became progressively worse, adding to the insults spitting, tossing urine from jars, chicken blood in flimsy packets that would break and even feces in bags. Sometimes there were physical attacks.

After leaving Active Duty in 1971, I visited my small hometown a few miles west of Boston. My father and uncle belonged to the local VFW Post and I had looked forward to the time I could join the VFW, too. I went down to the Post and asked the Commander for an application. He looked me straight in the eye and said; "We don't want any baby killers in our Post!"

The anti-war radicals did a fine job of lying and propagandizing the citizenry into vilifying a whole generation of good people like me, who only wanted to do our duty, for our country and for them.

Drew Johnson, Rancho Santa Fe, California
Captain - FedEx-Retired
US Navy Reserve-Retired

LRRPs and Small Men
Larry Hogan
Rome, Georgia

WE HID BY MAKING ourselves invisible, blending our camouflaged clothes and faces into the thick, steamy jungle by holding every muscle still while the enemy walked by close enough to touch, searching for us with flashlights, probing the trees and brush, trying to flush us out like deer while we barely breathed. I feared they could hear my loudly thumping heart, and I kept my eyes turned away from their flashlights to preserve my twilight vision and to keep the whites of my eyes from reflecting the light which would reveal our LRRP team's position. Somehow our covert recon mission was compromised; something had alerted the enemy that we were in the area.

They knew we were hiding; they called us "the men with painted faces."

When our LRRP team was compromised the pucker factor was high because there were only six of us deep in enemy territory, lightly armed, sometimes out of the reach of artillery support, and our helicopter lifeline was usually at least a half hour away. We were on our own for a while and far away from our PZ, in the middle of a large enemy force. They continued looking for us while we stayed very still and hoped that, when we could risk calling on the radio for help, the voice on the other end of our radio lifeline was a man who knew LRRPs, a man we could trust.

WHEN I THINK BACK to these times we flirted with the enemy and death, it seems combat drove us closer to our comrades with a powerful force, and that it magnifies who you are.

I never really thought about combat magnifying who you are until I encountered a *small man*, my term for those rare individuals who are focused on making themselves look good no matter the consequences to others, swaggering fools full of their own rank and selfish purpose, quick to use other men without regard for their welfare. These characteristics, too, I think, are magnified by the pressure of combat situations. "Small" has nothing to do with size, and everything to do with something ugly within them where virtues should reside.

I'll tell you about LRRPs, and I'll tell you about one of these small men, a major whom we would never have accepted among us on a LRRP team.

AMERICAN UNITS BEGAN DIRECTLY engaging the enemy in Vietnam in 1965, elevating the need for intelligence on enemy movement and activities in remote areas where deep jungle made recon from the air or satellite impossible. Special operations teams were required for stealthy observation deep in remote country for dangerous four-or-five day missions. These men were volunteers, graduates of the toughest military training, Ranger and Recondo schools. They were called Long Range Recon Patrol, (LRRPs, pronounced "lurps"), later shortened to Long Range Patrol (LRPs, pronounced the same way).

LRRPs operated in six man teams, lightly armed so we could move fast. Each team member had a specialty, like radio, explosives, medic, etc., and each cross-trained to handle any position in case a team member was wounded or killed. LRRPs used extreme camouflage and we made it a practice to communicate with each other by hand signals and whispers in the field. Sometimes we didn't speak a word for days. In the Mekong Delta with low vegetation and little natural concealment, LRRPs were forced to sleep during the day and move at night. In the jungle cover of the rest of the country, we moved mostly during the day, but sometimes at night. Stealth was crucial to our mission, and to our survival.

Larry Hogan with the company pet, Nugen Van Dog
Photo courtesy of Larry Hogan, all rights reserved

LRRP missions were inherently dangerous as we moved quietly amidst far larger enemy units. Our *box* was an assigned five to six square mile area in which we would move and observe, sometimes so far away from established US bases that radios did not reach and a radio relay station was required. When our team was compromised and found by the enemy, we ran while trying to drop artillery in our footprints by radio fire direction, calling for an emergency extraction while we ran. One of our team always fell back to cover the team's rear, just a reactionary and voluntary judgment by one guy that no one would ever argue about. A half hour to helicopter extraction was a lifetime when we were running from an enemy hot on our heels.

LRRPs were very effective at assessing bomb damage, observing enemy presence and movement, snatching enemy prisoners for interrogation, wiretapping enemy communication lines, confirming enemy unit strength, identifying enemy units, pinpointing targets, observing and adjusting artillery fire missions, finding helicopter LZs for planned troop movements and even maintaining LZ observation during a helicopter combat assault. Sometimes we were an ambush patrol, or even a decoy with a large rifle company on standby to handle a large enemy force that attacked us when we were intentionally compromised. LRRP kill ratios were very high.

The enemy knew we were effective, too, and placed a bounty of about $1,000 on the head of any LRRP. That was a lot of money in Vietnam in the 1960s and 1970s, strong motivation to find us. They looked hard to find and kill us when we were compromised, and we knew if the enemy caught us when we were wounded there would be a bullet in the head – or worse - instead of a doctor or hospital.

A LRRP TEAM USUALLY consisted of six men, five Rangers and one ARVN, all volunteers, living together in a hooch and doing everything together. We showered together, ate together, read each other's mail, and shared personal experiences we would not even tell our own family. Our bonding was very tight and it crossed over to other teams because we often filled in with their team and they filled in on ours. When a member from our team or another team was killed, there was not a dry eye in the company.

The LRRP team would be inserted by helicopter during a red-herring game of leap-frog where two or more helicopters would touch down at many LZs to confuse an observant enemy trying to determine where the LRRP team had been inserted. Most LRRP missions were not intended to result in enemy contact; instead we tried to move undetected, spying on the enemy in their own back yard. Ideally, they never knew we had been there.

When we were inserted, the team and supporting helicopter crews were tense until we were confident that we had not been compromised; then the helicopter crews went on a standby mode during our mission. If we were compromised and calling on the radio while running to our PZ, the helicopter would scramble to pick us up while gunships provided cover. Sometimes, when the enemy discovered our presence we would go to ground while they searched for us.

For a light six man team on a four-to-five day mission, we would normally be equipped with two radios, two spare batteries, food and water. Our weapons of choice were the Car-15, a shortened version of the M-16, an M-79 grenade launcher, eleven magazines of ammo, four grenades, one smoke, one white phosphorous, a claymore mine, and one tear gas canister distributed among the team. Each individual carried about eighty pounds.

Some point men preferred to carry a pump 12-gauge shotgun with each shot sending about 12 pellets down range, very effective to about 30 meters. Charles "Jolly" Haussler often carried this weapon. Jolly, at 6'4", weighed about 250 pounds and towered over every other man in our unit. Walking point, the lead man is in the most vulnerable position when we were on the move. I never knew of anyone who had a problem walking behind him.

LRRP team, Front L-R, David "Little Mac" McLaughlin, Gary Brown
Back L-R, Jerry Howard, Larry Hogan, Allen Caprio
Photo courtesy of Larry Hogan, all rights reserved

People back home, the media and even most in our own military had never heard of LRRPs. The enemy actually knew more about us than our own troops did; the NVA had reaction forces that were constantly looking for LRRP's. They would often wait until after nightfall to try to take out a LRRP team, when extraction and directing gunship fire is more difficult. We spent many nights with the enemy thrashing the brush, beating on pots and pans, yelling at us, trying to make us react and reveal our position. These were long nights and if we did sleep the dreams were never very peaceful.

Chopper pilots knew us best. We depended on their slicks to play the leap-frog game to insert our team and to pick us up as planned, but we fell in love with them when they pulled us out in an emergency extraction with the bad guys closing in on us, doubly so when they had to use ropes or ladders because there was no room to land the helicopter and they took enemy fire while holding the chopper in a still hover for us. We depended on medevac helicopters to pick up our dead and wounded, and we depended on the gunship pilots to put the rockets and minigun fire *danger-close* when the enemy was right on top of us. Sometimes we spooked our helicopter pilots as we directed their fire, because we whispered into the radio, sometimes out of habit, and sometimes because the enemy was so close and the gunship pilots worried they would hit us.

LRRP patrols were always carefully planned. Before insertion, the team leader would make an over-flight and select an LZ, route of travel and a PZ for extraction. Every member of the team had maps and was fully briefed on the mission.

Insertions were always an adrenalin rush, never knowing what was waiting just inside the tree line. The tension lasted until we were confident we were in control of our situation and well away from the LZ. With gunships flying close support around our position, most dinks wouldn't take a pot-shot at us for fear of giving up their position to those bad boys but you never know. Extraction was always exciting, too; soon you would be carried to safety, but we always knew the enemy might make a last ditch effort to inflict casualties.

Other than the normal combat gear, food, water, compass and maps, each member carried a medical kit that included bandages, tourniquets, a blood expansion unit to be used if a member had severe blood loss, and a pill kit. The pill kit included salt tablets to help retain water, and water purification tablets that made the water taste so bad that you would rather have the malaria. A pill was included that would prevent you from taking a crap for days, though I never knew of anyone using the no-crap pill. I could not imagine six constipated guys trying to stay alert. Better to crawl back into the brush, dig a hole and cover it up. A Librium pill was included to calm down a team member who was in a panic mode. A morphine shot was included for the seriously wounded and dextro-amphetamines to be used if the team was totally exhausted. We only took this dextro-amphetamine drug as a last resort because it would increase your heartbeat, make your mouth dry and your hands sweat and make you wide-eyed and alert. It would also make some people see imaginary things at night. A seasoned team leader would often break one tablet in half or quarters and distribute a piece to only the most physically distressed team members. At this point the team had better be pulled out the next day, because when the drug wore off the user would have a hangover far worse than the fatigue it relieved.

When we were not able to watch trails that would lead the enemy toward our position, we might set booby traps as an early warning device. Using a fragmentation grenade, we would remove the pin and replace it with a safety pin, so it would slip out easily, and cover and camouflage the grenade on the side of the trail, attaching a low hanging limb or trip wire to the pin. That would give us at least a five-second warning.

Sometimes life teeters on the most inane of balances. One of our members recovered his booby-trap grenade when we were moving out but he failed to replace the safety pin with the original pin. The safety pin worked itself out as he walked and set the grenade off in the bag hanging on his web belt. While he was still alive he was screaming, "Charlie! Charlie!" as if the Viet Cong had hit us. His intestines were blown out of his body, hanging in the branches of trees around him. When the team realized they had not been hit by the enemy, they pulled his intestines out of the branches and placed them back into his dead body. They did that because he was their brother, and he would do the same for them.

YOU MIGHT SURMISE THAT, because of the inherent risks of LRRP missions and the value of the intelligence they gathered on the enemy, that LRRP units were given high priority on equipment and other resources. If you believe that I would have to conclude you were never in the Army.

When the 196th LRRP began in late 1966, there was a shortage of camouflage fatigues, known as Tiger fatigues. Of course they were important to our stealthy mission, so we wore regular fatigues in the company area, reserving wear of our scarce Tiger fatigues for missions, when it mattered. When we saw a guy we didn't know wearing faded Tiger fatigues in our LRRP unit, we knew he had paid his dues.

When we were at the Chu Lai airstrip, we saw soldiers on bunker guard duty wearing brand new Tiger fatigues. The Army would send us on missions 30 miles into the boonies where our lives depended on blending in with the jungle, but the geniuses in the distribution chain gave Tiger fatigues to guys on guard duty a few hundred yards from their hooches. You had to wonder.

We had an old three-quarter ton jeep with broken springs, and we figured our CO, Cpt. Gary Bjork, deserved better. So, knowing the regular channels would disappoint, we decided to hunt for a couple of jeeps for our company area, and some 45 caliber pistols, like the one the motor pool Sgt. carried as if he might come under attack from the mess hall. We needed them for our missions. Some of our guys would go to the NCO Club near division headquarters, pick out some targets, buy them a few drinks to get them loaded, lift their 45s and drive away with their jeeps. We changed the numbers on the jeeps and got away with it. Necessity is the mother . . .

IN DECEMBER, 1966, THE Recondo School in Vietnam was established for the purpose of training LRRPs and other highly specialized teams. Getting into Recondo School was like getting into Harvard, very difficult. It required combat experience, a minimum six months left on your tour, top physical condition and determination of steel. It was an honor to be selected for Recondo, a name derived from *reconnaissance* and *commando.* Honor notwithstanding, many would drop out of each class.

Recondo School provided a three-week course, taught by Delta Force soldiers in Na Trang, one of the toughest schools the military had ever devised. I have never met a Ranger or any Special Operations soldiers from the Viet Nam era who did not know of it and every one of them said that this was the best school the Army had ever put together.

When the first week of training began, we quickly learned that these instructors were different from any we had experienced. Their expectations were quite specific and if you failed any part of the training you would be immediately returned to your unit.

The instructors had two or three tours in Vietnam in Special Forces. What set them apart from the ordinary sergeant instructors in the Army was that they did no yelling. They would spend time with us, one on one, to make sure you understood the instructions, then we were on our own.

Our day started about 4:00 a.m. with long, intense training that could break us physically and mentally. The first morning started with enough physical drill exercises to leave us trembling in formation on wobbly legs despite our strength and good condition. Next we would put on all our web gear, with our weapons and rucks filled with 30 lb sand bags for a one-mile run. Every day they would increase the number of exercises and add another mile to the run until the seventh day with a seven-mile run in the suffocating heat and humidity of Vietnam.

On day one, a couple of the students in front of me pulled their bayonets out to stab the bottom of their rucks to let the sand dribble out, to lighten the load. It was a clever adaptation but at the finish line, in full view of the students and instructors, each sand bag was weighed and the offending students were out. The first day sent a lot of the students back to their units. By the time we had finished our seventh day, and the seven-mile run, our class size had dropped to the point the instructors became more personal. They would spend more time with us, one on one, according to the need of each student.

In the classroom one day, one of the instructors asked me where I was from. Noting that he was wearing a Ranger patch, I told him that from my home in the states I could see a mountain that he had repelled from, Mt. Yonah near my home in Cleveland, GA, just a few miles from Dalonega Ranger Camp. He smiled and said, "There ain't nothing there but car thieves and insurance adjusters." He was referring to the era of muscle cars that were stolen and stripped to provide parts for souped-up 1955-57 Fords and Chevys. We understood each other perfectly.

Week one was also a lot of classroom training that included map reading and technical training such as calling in artillery, air strikes, directing fire from gunships, handling medical emergency situations such as sucking chest wounds. We took blood from each other to learn how to hook up a blood expansion unit if a team member had severe blood loss, and learned how to pump our own blood directly into a wounded team member.

At the end of the first week the five members sent to Recondo from our unit had been reduced to two: my friend, Jerry Howard, from Seattle, and me. Week two continued with the calisthenics and running increasing every day with faster times required. Repelling training began from a high wooden tower. It was as if all the open blisters on our feet, ankles, and shoulders were not enough, we needed more on our hands. No matter how much instruction on how to handle the guiding hand - the breaking hand - nature would take over and we would blister. By the time we advanced to helicopter repelling we had the proper procedures. Jerry Howard still has scars from the rope burns on his right hip from these exercises.

If we had not gained confidence in the repelling rope by this time, we sure would after the emergency extraction exercises. With three 120-foot nylon repelling ropes hanging from a helicopter, we sat in a 2" wide sling strap with a noose to slip our right wrist through in case our butt slid out of the strap. Our left arm was interlocked with the student next to us while the helicopter climbed to 1000 feet. When the helicopter made a banking turn it seemed that the ropes would never stop stretching. Damn, I loved those ropes.

Those of us who survived the regimen so far were required to make a one hundred meter swim into the ocean around a raft and back. If you touched the raft you were out. This seemed like a simple task after all we had been through and certainly soothing for all our blisters in the salt water, but the relief was illusory. We made the swim in our boxer shorts. By the time I was half way out I was thinking I might drown in the South China Sea in a stupid training exercise. At this point I could only think, "Mother and Dad, I love you, and please God don't let me touch the raft. If I can make it back to shore, I might have a chance to beat the odds!" I made it.

Nearing the third week we were taken on a survival mission to a wooded area with a stream next to the wood line and a rice paddy to our front. The objective here was survival. We had to forage from our surroundings to eat. Jerry Howard and I decided there might be crawfish in the stream and, while neither of us had ever eaten crawfish, it seemed that the tails might be like shrimp. Within an hour, we were boiling crawfish tails in a canteen cup, eating better than anyone else in the camp. Later we saw an ARVN walk up the creek out of our range, then a few minutes later we heard muffled blasts that we knew were grenades exploding under water. He came back with a bunch of what looked like small black catfish strung on a stick. Another ARVN had been on the other side of the creek digging up bamboo sprouts. The fish and sprouts were dumped into an ammo can and boiled into a delicious meal for us and the others who had been eating grasshoppers and worms.

The last week of Recondo school, the deadliest school the Army had ever put together, was known as "You Bet Your Life." It was an actual patrol into enemy territory in which we would try to locate the enemy and observe their troop size and movement. Contact was made on these missions at times and the teams took casualties. This was for real. The enemy would kill us if we screwed up and gave them the chance. With a Special Forces instructor and five other team members, we were briefed on the mission then dropped into our assigned AO. Each member would rotate as team leader.

Those who completed the course and the patrol mission received a certificate and an arrowhead-shaped Recondo patch. That might not seem like much, but the pride that went with it is hard to describe. Was this school worth the agonizing effort? Damn right, it was! What we learned there helped keep us alive on LRRP missions. After four years the school closed on December 31, 1970.

WE RETURNED TO OUR units and we resumed LRRP missions.

While on one observation patrol we had moved down the side of a ridge so we could observe a trail alongside an abandoned rice paddy. We set up in the brush in a position overlooking the trail from about 200 meters above it. The trail led to an abandoned hooch another three hundred or more meters up the trail.

Thinking this was going to be a boring mission, we sat watching when, suddenly, one of my team members pointed to one lone VC coming up the trail with a rifle slung over his shoulder. Knowing there might be more to follow, we waited. He made his way to the hooch and went inside. An hour later we were still waiting for a larger group to join him so we could hit them all with artillery. After using the map to calculate the exact coordinates for a direct hit we waited.

A FAC "forward air control" plane flew near our position. I got on his radio frequency, gave him our location coordinates and told him about just one VC in the hooch. He made a couple of passes, firing rockets into the hooch, then said he was going to redirect a pair of jets from another bombing run to finish this job. I could not believe what I was hearing . . . all this for one VC? But his rockets had left the hooch standing without the dink trying to escape.

Suddenly two jets screamed into our area. The FAC fired a violet colored smoke rocket on the hooch to mark their target which was followed by two bombing runs that reduced the hooch to a smoldering pile of rubble.

The FAC pilot came over the radio and told me to report that the enemy was killed and that he was out of there.

I made the call to report one enemy KIA and sat back with the other team members, grinning and whispering back and forth about the big display that the FAC and jets had provided. A few minutes later, when everything had quieted down, one of my team members nudged me and pointed toward the hooch. The smoldering straw was moving. Suddenly a man who seemed a little dazed emerged from it and ran down the trail. He no longer carried a weapon, only the clothes on his unscathed body, and he set a blistering pace, gone from sight in seconds.

We didn't take any shots at him, we were too busy pondering this problem as only young men can. We should have been moving out quick since we were far away from any support, in the boonies where every man within hearing of the blasts would love to have a chance to kill us. But we were busy debating how the VC lived through the bombs. We could understand the rocket shrapnel missing him, but the bombs? The hooch was too low near the rice paddy for any tunnels which would fill with water, but many of their living quarters would have a grave-like hole underneath their bed. When the shit hit the fan, all they had to do was roll out of bed and into the hole. Maybe that's how the luckiest VC I ever saw survived, but his bell must have been ringing for days!

After this fascinating puzzle, we spent two boring days waiting for the enemy to show up. They never did. I wonder why?

IN LATE FEBRUARY, 1968, I was assigned to take a patrol into the Tam Ke area. This was my only LRRP mission without a fly-over for mission planning. We would walk out from a firebase controlled by the First Cav. We would have to depend on them for support and we didn't know if they knew how to support a LRRP team.

My team for this mission was mostly new members with the exception of my ATL (Assistant Team Leader) Sgt. Laszio Kiraly, who had served with the 101st Airborne on his first tour. In my pre-mission briefing with Intelligence at Tam Ke, I was told that our infantry had a Regiment of NVA surrounded. My mission was to determine if the NVA were escaping over the mountain ridge on a route of about four miles along the ridge. They informed me that they had flown a *sniffer* flight over the area but had not picked up any human activity. If I could confirm that the enemy was crossing this ridge then they would pull my team out and send in reinforcements.

We would be traveling light for a three-day hike, although a two-day run might be a more appropriate description for this one.

Upon arriving at the fire base by helicopter, I was met by an Army major that I remember as being overly cocky, and, while I don't remember his name, I thought of him as *Major Ego*. He had a cactus drawn on the side of his camouflage helmet cover and I wondered if he was from Texas or Arizona, but it was obvious that we were not going to get that personal. He instructed me that if we should need any support that he could give it to us with his mortars. I ignored my better judgment and did not tell this know-it-all officer that if we were in trouble we would need fire placed with far more precision than mortars could possibly provide. When we started our patrol out from the fire base, which was located on a point of the long ridge line, I felt that we were pretty much on our own.

We stealthily worked our way along the ridge line which was heavily wooded with thick cover. About one mile out we heard a single rifle shot about 200 to 300 meters ahead of us. This happened about three times within 15 minute intervals. These might have been warning shots that we were approaching; we might have already been compromised.

We came to a large opening covered with elephant grass. From our viewpoint we could see a clearing to our left in the valley below and five NVA crossing the clearing. I called back to the fire base to direct mortar fire on them. The first round hit so far away that the NVA did not even realize they were being fired on. I tried to adjust fire by radio directions to walk the rounds toward the enemy but it became obvious that the mortar crew at the fire base would probably hit us before hitting the enemy. At this point I knew we were in deep trouble, either because we had lousy support or that I did not know where I was on the map, which was not likely.

We continued on the open trail through the eight feet tall elephant grass until we reached the wood line. Suddenly, everything seemed so quiet that our instincts told us something bad was about to happen. No birds were chirping, not even mosquitoes were buzzing our ears. Something was just not right. It was getting late so we elected to backtrack into the elephant grass to set up for the night. We dropped off to the right side of the ridge deep into the grass opposite from the side we had seen the NVA. We rotated the watch as we did every night in the bush.

When we awoke the next morning we heard movement in the tree line to the right of the ridge below where we had set up for the night. I took one of my team members and crawled as close as possible to the enemy, intending to toss as many grenades as we could before they knew where we were. It turned out to be a group of baboons. We re-grouped with the rest of the team and proceeded to move out to the wood line, re-tracing our path from the night before. When we reached the wood line we dropped off the main trail into the wood line to the right. We avoided trails by practice for two reasons: the trail is where booby traps awaited, and trails are the foot-highway of our enemy. Our mission was to observe them, not to engage them. We found dug-in fighting positions and fresh footprints leading away, making it clear to us the enemy had set up to ambush us. We had stopped the night before within 50 meters of being ambushed and we had slept that night about 200 meters from the ambush site. Our no-noise skills certainly paid off.

This observation put my team on high alert. We had been so close to being ambushed the night before. It is quite likely that the baboons had saved our butts by distracting us, making us move out a bit later that morning.

We continued down the ridge line into a saddle back on the well-traveled trail. The canopy was so open at the saddle that we could be compromised from either side of the ridge. We continued up the rise from the saddle toward the next high point on the ridge line where the trail ran between some large boulders at the peak. There it intersected another trail at the peak. To the left it dropped off steeper than at the saddle. To the right was heavy vegetation with the large trail through it. These were the largest trails I had ever encountered in the mountains, wide and well-maintained.

I considered our options. We knew we were within 200 meters of a clearing of elephant grass that would be our closest LZ for extraction, but that was on the edge of a free-fire zone which could at any moment be hit by artillery or air strikes. Otherwise, we would have to backtrack to the baboon clearing, through or around the ambush site, and to avoid the ambush site we would have to drop well off the steep rise into dense cover, chewing up hours to maneuver safely back.

Along the trail we found a small paper with a lot of Vietnamese scribbling on it. Our new ARVN team member had not yet been proven, and my experience told me to suspect any ARVN until they had proven their loyalty to us instead of the enemy. He translated the writing with the aid of his translation booklet since he did not speak English, and we concluded that it was a supply list to be taken to a certain NVA unit. I kept my suspicion of our ARVN quiet, thinking sooner or later I would know whether I could trust him.

We turned to the right on the trail toward the free-fire zone, toward the edge of a clearing on the right side of the ridge peak. We were in heavy fog with our sight distance limited to less than 100 meters. We stopped at the edge of the clearing, relieved to find another LZ.

Suddenly NVA soldiers started standing up in the grass and calling to us. Their own visibility was limited and they obviously thought we were some of their own. More and more of them started standing up in groups of five or six in each group. They started calling back and forth between themselves and toward us trying to identify us. At this point there were probably five groups of five or six men each with the closest group to us about 40 meters away.

We hastily moved back several meters along the trail around a sharp bend in the heavy canopy where I told my team to get the hell out of here to the right of the trail. My point man covered the trail toward the enemy and I covered the rear until everyone was off the trail. As I turned back to see if everyone was off the trail, I saw my point man, armed with a beehive round in his M-79 grenade launcher, looking in my direction. Suddenly, two NVA stepped around the bend of the trail about 20 meters away. They had their AK's or SKS's pointed toward the ground as if they were looking at our footprints. Immediately I turned my CAR-15 toward them but could not fire because my point man was between me and my target. He fired the M-79, the two NVA fell and he jumped into the thick foliage to our right where the rest of the team had gone. Since the M-79 was a single shot weapon and he needed to reload, I immediately took the job of rear security. I stepped off the trail inside the thick cover, scared as hell, trying to keep my feet from running away with me. I needed to give my point man a head start.

Moments later while I was running down the side of the ridge, my point man stepped out from behind a tree, naked above his belt. In his rush to escape, his ruck had become entangled in the brush and he had jerked loose from everything leaving his ruck, ammo belt, the M-79 and his shirt. It was too late to worry about a lost ruck and weapon; I needed to get my team the hell out of there.

Our escape route would lead us into the area that the NVA regiment was supposed to be, but it was the only choice that provided heavy cover. We moved through heavy canopy crawling most of the time until darkness started to set in and set up for the night in the thick underbrush. At daylight we would plan our next move. We were all exhausted from the adrenaline rush and our escape. I called in our position by radio as we did every night in case the enemy found us and we needed support. Someone shared their poncho with our point man, who now wore only his pants and boots.

Late that night I fell asleep while on radio watch. After missing our scheduled communication check, artillery rounds stated falling close to our position. I awoke from my dream that we were being mortared. I immediately made our communication check and received a verbal reprimand from some SOB at the fire base reminding me to stay awake while on radio watch. Being admonished in our situation, by a man who was safe and comfortable, rubbed me the wrong way.

When daylight seeped into the next morning, we could see an abandoned hooch near where we had set up for the night. On further investigating we found that it had been recently used but no one was around. We wondered if we might have slept that close to the enemy for a second night.

At this point our mission had been completed as far as I was concerned. The NVA were crossing the ridges. I called to report on our observations and to request extraction. As part of my report I mentioned we had lost the M-79 in our escape from the enemy. Unfortunately, the man on the other end of the radio was Major Ego, who confirmed that he was a small man when he ordered me to take my team back to the site of our contact to recover the M-79.

Holy shit! Was he out of his mind? We had to follow orders of superiors, but this guy was willing to risk six men to retrieve an unimportant weapon.

Major Ego said if the NVA has the M-79, it could be used against his troops, so go back and find it in the brush or find the NVA and take it back. The stupidity of that thinking is boundless. There was no mention of support from them except that if we confronted the enemy again we were to hit them hard then back up and call in artillery. I realized that Major Ego had no knowledge of how little ammo we carried and I guess he had forgotten there were only six of us.

At this point, I had the very lonely feeling we were in the core operating area of the enemy, close to the river, and there would be no extraction area except back at the baboon clearing or the free-fire zone clearing, both involving exposure to the enemy.

We worked our way back up to the top of the ridge where the large boulders were on both sides of the trail. As soon as we came to the top of the ridge my ARVN turned and started running back toward us, wide-eyed, in fear. A couple of others on my team and I moved up to see what had scared our ARVN and saw just before us about 15 VC sitting 30 meters away, apparently eating lunch, dressed in black pajamas instead of NVA uniforms. At this point we were so unorganized I knew we had to re-group to plan our next move. We still had to find the M-79 or Major Ego would leave us there. At this point the most important thing wasn't killing the enemy, it was recovering the M-79 and getting my team out safely. We quietly moved back down the trail toward the saddle of the ridge for about 150 meters and dropped off the side of the trail where we set up our defense and possible ambush on the enemy if they moved toward our location.

I radioed to the fire base about our situation and requested gunships. Major Ego responded that they had a company pinned down by a sniper, implying the six of us did not matter much at this time. A few minutes later we were informed that the gunships were on the way. I gave the radio to my ATL for him to direct them in at a certain azimuth over our position and to fire rockets first then follow up with mini guns twenty meters past the yellow smoke.

My ATL made the call and we popped the yellow smoke to mark our position. As we waited the choppers were approaching very close and I knew that our call was risky with the enemy at 150 meters but the closer the rounds were to our position the less likely the enemy would know we were that close. It was risky, but we did this all the time.

As the gunships closed in, I suddenly realized that the wind had drifted the smoke out of the trees behind us. Before we could contact the gunships, rockets hit the trees almost on top of us and the mini-gun fire started, making the leaves dance just a few feet in front of us. By the time I could call a cease-fire, it was over. I quickly looked over my lucky team and everyone was OK except our ARVN, who had been grazed in the head and was lying there with his eyes rolled back. My deep, dark instinct was that I should shoot him because I did not trust him and he might be a hindrance to our survival. Before anyone could help him, he rolled over, searched through his ruck sack, jerked out a bandage, placed it over his bleeding head, then opened his pill kit, and popped a couple of Librium pills to calm himself down; I knew at this point that he had Recondo training, that he was one of the best, that I should trust him, that we needed to get him to medical attention and get our asses out. I called for a medevac for my wounded man knowing that under other similar situations, when we were being controlled by a man with common sense, the team would be pulled out also. We should have been pulled out the day before. But we still had to find the M-79. We carefully moved back up the hill and found that the VC had moved out before the gunships arrived; our gunships shot up some empty jungle while alerting all the bad guys in the area that we were nearby. Could this mission get any worse?

We found the spot where my point man had been entangled in the brush and, to our complete surprise, we found the M-79, ruck, shirt and everything only about three meters off the trail. We checked it out carefully to make sure nothing was booby-trapped. My point-man put his shirt and gear back on and we backed up into the brush about 10 meters off the trail, where we sat waiting for the medevac chopper.

Minutes later a group of five NVA walked past us. I signaled my team not to fire because we were going to be getting out of there. Heavily camouflaged in the thick brush, we would be hard to see. A few minutes passed and then a single NVA walked around the trail, turned and looked straight in our direction. All of our guns were on him but every one held his cool, staying very still in cover, knowing that he might be a decoy. This might have been the luckiest NVA ever. I also thought that there were so many NVA crossing this ridge that if he had seen us he might have thought we were part of his own regiment. He just moved on.

Suddenly, we heard choppers moving in our direction. We moved out to the edge of the clearing. I was expecting an extraction but it turned out to be one medevac chopper supported by two gunships. I informed them that we would pop smoke to mark our position and for the gunships to be prepared to spray the tree line around our position with mini-gun fire, as if I needed to tell them what to do. The gunships circled the clearing right on top of the tree line but never fired a shot. As soon as the medevac chopper touched the ground my whole team climbed aboard and someone came over the radio to tell me I could not bring weapons on a medevac chopper. I screamed back over the radio for them to get the hell out of here and I was not throwing any weapons off the chopper, especially since we had to deal with Major Ego.

When we landed safely back in Tam Ke, none other than Major Ego, with the cactus on the side of his helmet, was there to greet us. As soon as he saw that my wounded man was an ARVN, he stepped up in my face and told me that if he had known that it was a gook that was wounded that my ass would still be up there in the mountains. I told him I knew that was probably what he would do, implying I didn't tell him on the radio because I didn't trust him. This made him furious but he turned and walked away. He didn't seem interested in attending my debrief with Intelligence.

When we arrived back at our chopper pad and walked down the hill, there were twenty-four of our LRRP members ready to move out. SSgt. Earl Toomey and Sgt. Franklin Williams were there to lead nine-man heavy LRRP teams back into the same area we had patrolled.

Photo courtesy of Larry Hogan, all rights reserved

In the debriefing from Intelligence, I was told that we had been close to a POW camp. A Marine had been released or escaped and had helped them to identify where the prison was and that there were American pilots and others being held there. I sat down and talked with "Jon-Jon," Sgt. Ronald Jonsson. I had been on two other missions with him and we had learned to trust each other in the field. We had housed together in base camp for months where he and Allen Caprio gave me hell about being a Georgia hick. They taught me to say "you guys" and sometimes I might catch them saying "y'all."

Jon-Jon told me their mission was to get the POWs out. He would be a part of the primary nine-man team and they had a twelve-man team back-up if they got into trouble. He showed me that he even had an ice-pick in his ruck sack, ready for a close encounter. James Davidson, another of the nine men, had come into the unit with Jonsson and me. I told the team members about our experiences and that if they go to the two boulders they will have to make some kills because the enemy will be there every few minutes.

That was the last time I saw Jon-Jon and some of the others.

Within the next twenty-four hours, one team killed two NVA. SSgt. Toomey sat on a punji stake, feces-dipped bamboo sharpened to a razor edge, carefully placed by our enemy to surprise, wound and demoralize American troops. Infection set in and his scrotum swelled to the size of a grapefruit. He didn't want to leave his team, but he had trouble walking and his team convinced him to be medevaced. One other member had become very sick and was also medevaced with him.

The mission to get the POWs out was scrubbed for some reason, and the two teams joined up in the bush. During the early morning of March 3, 1968, in the heavy fog in the mountains, a large force of NVA made their attack.

Apparently the enemy had already pinpointed their position. They moved in right on top of them while most of the men were still asleep, tossing grenades into their position and opening up with automatic weapons fire. Terry E. Allen, James R. Davidson, Ramon S. Hernandes, Ronald B. Jonsson, Edward M. Lentz and Jose E. Torres were killed. James E. Kesselhon died from his wounds on March 21, 1968. Franklin Williams, Bob Wheeler, John Guntrum, Rob Campau and others were wounded. Bob, who had a shrapnel wound to his forehead, told me of the team's panic to keep throwing grenades back out of their position.

I always wondered whether the small man I knew as Major Ego left these heavy LRRP teams at risk the way he did my team. I suspect some LRRP deaths might have been sacrifices to his chest-thumping.

SO FAR AS I know, the POWs were never rescued; Frank Anton, a gunship pilot, who wrote the book *Why Didn't You Get Me Out?* was most likely one of the pilots in that POW camp.

So far as I know, throughout our missions the officers who were on the other end of our radio lifeline almost always understood our precarious circumstance, the risks we took, and that our lives rested in the balance of their decisions. But now and then, when the enemy was closing in and we called for help or extraction, I had the bone-chilling realization that I was talking to a small man, and wondered if we would live through this session of his stroking his own ego.

Larry and Carolyn Hogan
Photos courtesy of Larry Hogan, all rights reserved

Larry Hogan lives in Rome, GA, with his high school sweetheart and wife, Carolyn, who keeps him straight in their lumber business. Larry's family has a proud history of serving their country. His father, Glen, served in WWII and earned the Silver Star, Bronze Star and Purple Heart. Larry and his brothers Ralph and Wayne served during the Vietnam War. When Larry's brother Rex became of interest to the draft, Mrs. Hogan visited the draft board to inform them they had quite enough of her sons. They didn't have a chance.

Postcard to America

WHEN I RETURNED FROM Vietnam in June, 1969, there were protestors outside an eight-foot-tall chain link fence when we deplaned at Ft. Dix, New Jersey, at 11:30 PM. A row of MPs kept us from responding to the protestor insults as we trudged toward the terminal building. Thank God I was from the Midwest and avoided the confrontations faced by returning vets in major airports on the East and West coasts.

However, I couldn't help but notice the total lack of compassion my old high school and college classmates had for my sacrifices in Vietnam. They didn't want to discuss the war or my participation in it. It became an "unmentionable" and I soon found it wise to put my experiences behind me. Although I never lost the pride I felt for my service to my country, I accepted the fact that it had to be private if it was to survive at all.

Socially, I was considered "damaged goods." I put it all in a private place and never looked back for the next 17 years, until my comrades in arms reformed at our first reunion since the war. That reunion was a good thing . . . a redemptive thing . . . enabling all of us to openly express the pride of service we had suppressed since the war.

Our nation betrayed us by blaming us for a long, unpopular war. We had been unprepared for the rejection and many of us let it adversely affect our behavior. Now, in our declining years, some of us have finally begun to heal but for so many the healing has come too late.

I hope our nation has learned its lesson well.

Gary Linderer, Branson, Missouri
Company F (LRP) , 58th Inf., Company L (Ranger), 75th Inf.
101st Airborne Division, Vietnam 1968-69

Magnet Ass
Andy Burleigh
Greer, South Carolina

THE DELTA REGION OF Vietnam, IV Corps at the southern end of the country where the mighty Mekong River splits into countless smaller rivers and canals before its waters join the South China Sea, building more so-called land constantly as the silt settles and shifts, is an intensely green, flat, hot, wet, stifling place where rice flourishes in flat, square green paddies knee-deep in water, separated by dikes just wide enough to walk on. Viet Cong moved at will in this region by rivers and canals at night until they were decimated in the battles of their own Tet Offensive in 1968. After that, the VC were weak and the NVA from North Vietnam sent their soldiers to reinforce the enemy effort to conquer South Vietnam.

The delta was a miserable place for westerners not accustomed to the constant stench of rotting vegetation, huge bugs that might gang up to carry you away for dinner, oppressive heat that shimmered and rose in visible waves, humidity and dampness that literally rot clothes off our soldiers' backs in a few days and an elusive enemy who usually decided when and where we would fight. There were also booby traps trying to kill or maim ground troops like punji pits, bouncing Bettys, grenades triggered by items that would attract a GI's attention and trip wires to detonate trailside devices or even artillery shells suspended overhead. In the flat delta there were no enemy tunnel complexes and there were no foxholes to hide in at night because when you dig in the delta you hit water fast.

Waterways were the highways in the delta. The people fished and took their goods to market by boat. Families visited friends by boat. The VC moved their tools of war by boat, at night, and the US Navy patrolled and hunted them, trying to stem the tide of terror waged on the South Vietnamese people by the north. At night the people stayed home because they knew violating the curfew on boats was a deadly game.

In this lovely locale I became one of maybe a thousand other guys in Vietnam to earn the nickname *Magnet Ass*. You can guess why.

I was a helicopter pilot, a Loach-driver, a Scout. I didn't walk the ground with a rifle, but I covered many who did and, boy, was I glad to be up in the air with the wind blowing instead of down there with the heat and humidity and snakes and bugs and booby traps, never mind the enemy.

The Navy hunted the enemy on the rivers and canals. Our job was to find them and kill them where the ground was at least semi-solid.

BIEN HOA WAS BRIGHT and sunny when we landed after 20 bone-weary hours in the air flying from California to our war. As the cabin door was opened we were hit by our first and overwhelming sensation of Vietnam, a complex aroma of diesel fuel, jet fuel, decaying foliage, unfamiliar foods cooking, sulfur, and death carried by a blanket of humidity you could almost wear.

As we stepped out to squint in the sunlight, we were greeted by many guys going the other way, on their way home, their tour done. They had leering smiles on their faces, faces that appeared to have weathered many miles in a short time. We heard joking comments about FNGs and many other words of derisive encouragement.

We were all transported to Long Binh where we would wait several days for our assignment. My fellow classmate, Tim, and I were taken to Di An, not far away across the river, squadron headquarters for 3/17th Air Cavalry where we waited two days for our assignment to one of the troops. A Troop (Silver Spur) and C Troop (Charlie Horse) were both here at Di An. B Troop (Stogie) was at Dong Tam on the Mekong River about 40 minutes south of Saigon by Huey. As we were sitting in the orderly room of HQ awaiting assignment, I picked up the latest issue of Stars & Stripes newspaper. "Tim, I know where we are going." On the front page was a Navy Seawolf Huey gunship lying on its side, a victim of the explosion of the ammunition storage facility at Dong Tam just the day before. Dong Tam and B Troop - here we come!

We were picked up by Captain Mike Valardo in a B Troop Huey and given a lesson in low-level flying around the west side of Saigon and south to Dong Tam. I knew we were low when I looked down at the M-60 machine gun in front of me and saw a blade of grass wrapped around the barrel. OK, so the elephant grass was very tall.

Carved out of the flood plain just north of the Mekong River, Dong Tam was rather bleak. If something was metal it was painted olive drab. If it was wood, it was brownish-gray, or was that grayish brown? Everything else was varying shades of brown and tan including the so-called potable water.

We settled in quickly, housed in two-story billets of unpainted wood with half-screen exterior walls to allow ample ventilation. When I looked across the street to the compound nearby, I noticed a large POW painted on the roof. As I looked back, I realized our own compound looked almost the same, with less concertina wire. We were checked out in the Huey and were flying missions as peter pilot soon after.

B Troop 3rd Squadron 17th Air Cavalry was organized into four platoons. First platoon was Scouts, with a full complement of ten Loaches, our slang for LOH, or Low Observation Helicopter, the Hughes Cayuse OH-6A, a highly maneuverable little bird with a four-bladed main rotor, a 285 horsepower turbo shaft engine, and a cruising speed of about 140 mph, 125 knots. The Loach gave a very responsive low vibration ride. Ours were usually equipped with an electrically powered 7.62 mm six barrel mini-gun mounted on the left side of the fuselage. The ammunition container in the back of the cabin held belts of about 4,000 rounds, enough to last just one minute if fired continuously on full speed. That's over 66 rounds per second with every fifth round a tracer, an arced bright-orange river of lead. We learned to fire two-or-three-second bursts at half speed to conserve ammo and to avoid overheating the barrels.

Second platoon was slicks, the lift platoon, eight H model Hueys used for various missions including personnel transport, medevac, resupply and command and control.

Third platoon was guns with nine Cobra gunships.

Fourth platoon we called the Blues, our infantry. They were tasked to perform ground work usually in conjunction with our air operations.

Andy Burleigh in Vietnam, 1969
Photo courtesy of Andy Burleigh, all rights reserved

WITHIN THE FIRST MONTH I had the opportunity to join the Scouts and check out in the Loach. After a short transition school in Di An, I returned to Dong Tam to become a real Scout. This called for flying as an observer in the right seat with an experienced Scout pilot, and then as pilot with an experienced Scout pilot in the observer position. Once checked out, we would fly with experienced observers for a while.

B Troop flew in a hunter-killer team consisting of a Cobra up at about 1,500 feet orbiting over the Loach, while the Loach was down low and slow, looking for signs of the bad guys. When we found a target, we identified it either visually or with a smoke grenade for the Cobra to roll in and blow it away. Targets we looked for varied from people to bunkers to food or weapons caches or an enemy shooting at us. Many times signs of something would be so subtle we might not be sure why it came to our attention, just that something seemed out of place. Camouflaged sampans in a small canal, sunk into shallow water, were a sure sign there was something amiss. Bunkers with the appearance of recent use or footprints where we didn't expect them were evidence of enemy activity.

A Scout's job was dangerous, flying low and slow to snoop out signs of the enemy, even to goad the enemy to fire at us and thereby reveal their position. When the enemy fired at us we skedaddled out of the kill zone as fast as we could while the Cobra rolled in to cover our butt and kill the enemy.

MY FIRST CONTACT WITH the enemy was while sitting in the observer seat. We were supporting what turned into a big operation in the hamlet of Tan Tru, near the BoBo Canal. As we came roaring in low, my experienced Scout pilot instructor sensed something wrong and abruptly banked away just as I noticed a number of sparkles in a brush line. I wondered what could be sparkling in the brush when I suddenly realized someone was shooting at us – those were muzzle flashes! And the fire was on my side of the aircraft. This sudden contact with the enemy grew into a significant fight with a regiment-sized group including suspected NVA as well as VC. This large enemy force used the BoBo canal to infiltrate from Cambodia, staging for attacks in the local area or ultimately Saigon, the capital city. We seriously interrupted their plans that day, and their resistance shot up seven helicopters. When we later found the holes in the bottom of our Loach, it became obvious that my pilot instructor's abrupt turn saved me from being hit, possibly many times. I didn't kiss him, but I should have.

A few days later I was flying with Mark "Rock" Furman as his observer in support of ground troops. During our low-level observation, an explosion went off underneath our aircraft. After the initial confusion from the concussion, we checked for damage but found none except blood seeping through a hole in my glove. I had been initiated into the world of Scouts by a small piece of shrapnel, a reminder of how close we flew to the action on the ground.

TEN DAYS LATER, LARRY "Duck" Simpson and I were flying low and slow, looking for campfires, fresh trails or any other sign of the enemy when we spotted a fresh footprint. As I hovered overhead, looking at the footprint, I noticed the grey color of the inside of the depression in the wet clay-like soil of the Mekong Delta. In this area, the top of the ground was generally brown, but when the surface was broken we could see gunmetal grey clay. After a few minutes, the depression of a footprint would fill with water and the brown silt from the water would change the color of the footprint. When we spotted the fresh grey of the clay we knew the foot that made this print must be very close because the print was made just a moment ago. Was I a helicopter pilot or a detective?

We found this footprint in the middle of a clump of Nippa palm about twenty to thirty feet in diameter. With no evidence outside this clump, we suspected the bad guy was still there. Nippa palm looks like giant blades of grass ten to twelve feet tall and grows in dense clumps, an ideal hiding place.

Had we not spotted the footprint, we would never have suspected someone was there. I hovered lower and slower so the downwash would spread out the growth of the Nippa palm while Duck leaned out to look straight down. Suddenly we heard the rapid pop-pop-pop, of an AK-47 from underneath us and I felt the rounds hitting the bottom of the Loach. I pulled pitch, nosed over, and hollered on the radio to the Cobra overhead. "Taking fire, taking fire," and realized a second later I had been hit in my butt. I transmitted, "I'm hit!" The Cobra called back, "Are you hit or is the aircraft hit?"

Painting "Eye of the Tiger" by Joe Kline [www.joekline.com], customized for Andy Burleigh
Photo of painting courtesy of Andy Burleigh and Joe Kline, all rights reserved

I yelled into the mic, "I'm hit, God-dammit!" It's funny how getting hit will piss you off and make you think irrational thoughts like, "Jeez, doesn't that guy realize he could kill somebody?"

Duck, my non-pilot observer, looked over at me as though I might drop dead any second, and he grabbed the controls while staring at me. The nose pitched over and the ground got bigger, but Duck didn't realize he was about to kill us. To keep it in the air and preserve Duck's piloting confidence, I took the controls again, lifted my hip up and told Duck to see how bad it was. He figured I was OK, just a cheek wound, and I kept the controls. In the meantime, the Cobra and ground troops dispatched the enemy who shot at us, presumably the same man who made the footprint. We never saw him.

We flew back toward Tan An and the dispensary to check my ass and the helicopter for damage, but before we arrived I smelled fuel and had Duck look out to the rear to check for leaks. Sure enough, he reported fuel streaming down the belly toward the exhaust. Great! Now that we had escaped one enemy soldier shooting from point-blank range, we were going to turn into a torch!

I landed in a large, open, dry rice paddy. As I set the aircraft down, two Hueys landed beside me, one on each side, with grunts on board to secure our position. We shut down and made our way over to one of the Hueys, but not before I retrieved my thermos of iced tea from the chin bubble. While troops guarded the Loach until it could be picked up by sling for transport and repair, we hitched a ride to the dispensary, where they patched the bullet hole in my ass.

That day I acquired permanent membership in the *Magnet Ass Club*.

ONE DAY WE FOUND evidence of recent enemy activity, tracked by grunts we were supporting. They found and dispatched the owner of the footprints, but not before one of their guys sustained a gunshot wound to the chest. Having no communication with Medevac, we decided to take the wounded soldier to the Tan An aid station ourselves since he was critical and we could get there in less than ten minutes. Duck jumped out and loaded him into the back of the Loach and we hurried off to Tan An as fast as we could fly. I was hollering on the radio as soon as I pulled pitch to let them know we were inbound with a sucking chest wound and they met us on the helipad. The GI was alive when we got there. I have always wondered whether he lived or died. If he lived, was it because we did the right thing? If he died, was it because we did the wrong thing? It would be great to know we were successful. We never knew.

WHEN STEVE BROWNELL'S LOACH took enemy fire one day, he had a complete engine failure in hostile territory. He landed without damage but was rather anxious to get the hell out of there. His Cobra came in to pick him up but had no place to put him. Steve opened the ammo bay door, just a ledge barely wide enough to sit on. He climbed on, with only the support wires to hang onto. The Cobra pilot could not see his condition but assumed Steve was badly injured and felt he should hurry. Well, Steve would have preferred a slow and gentle flight while he desperately balanced on his precarious perch, but at 140 knots he was scared to death and damn near blown off the ammo bay door! There was plenty of discussion about Cobra medevac that night back at Dong Tam over rusty cans of Hamms or Pabst and glasses of Blue Nun or Mateus.

WE USED WILLY PETE (white phosphorous) grenades to mark targets for the Cobras because the brilliant white explosion was not only highly visible, it lingered. Of course if Willy Pete touched you, it would burn the hell out of you.

Leroy C. Laine, Jr. was teaching Bill Holt, a slick driver, how to be a Scout, like when you mark a target considered dangerous, you "make one pass and haul ass!" Leroy, seated in the left observer seat, pulled the pins on two Willie Pete grenades while holding the spoons to keep them from going off. Leroy was dressed in full nomex flame resistant flight suit with gloves and helmet with visor down. Bill was in tailored jungle fatigues, open at the collar, with sleeves rolled up, and gloves rolled down. He wore his helmet with visor up and sunglasses. He was dressed for comfort, not for safety.

As they approached the target, Leroy, with a grenade in each hand, tossed the one in his right hand first. It hit the d-ring handle in the door frame, fell to the cockpit floor, and rolled between the pedals into the chin bubble. With no chance in the world to reach the grenade, Leroy put his feet together and over the grenade to shield from the blast, threw the grenade in his left hand out, and put his hands over his face.

When the grenade exploded in the cockpit, it created a very dense cloud of white smoke with a Loach in the middle of it! This immediately disoriented Bill, who was flying, and before he nosed over and crashed he must have instinctively squeezed the minigun trigger. The Cobra was above, watching, with a crew chief in the front seat, who managed to get a spectacular but distant photo of the Loach crossing a canal before impact in a very wet rice paddy, minigun rounds stitching a line out in front of them, a bright orange glow in the nose of the aircraft, and a dense plume of bright white smoke trailing behind as the aircraft pitched over, hit the ground, and rolled up into a flailing ball of mud, smoke, water and spare parts.

Loach crashing with exploded white phosphorous grenade, pilot squeezing minigun trigger
Photo courtesy of Andy Burleigh, all rights reserved

We assumed if they lived they would be seriously injured. Once the mud, water, and smoke settled down and all the flailing parts stopped moving, out of the muck came Bill and Leroy slogging out of the rice paddy, each bubbling and smoking from the white phosphorus which was still on them. After we knew they were OK, we laughed our asses off. Leroy was burned only on his exposed neck. Bill had many little spot burns all over. None were serious, all were painful.

AFTER ABOUT A YEAR, I returned to the US to teach Vietnamese pilots to take our place. I often wonder what happened to all of those young men we sent back to defend their country.

Arriving home was notable for its insignificance. While I was proud of my service in Vietnam, nobody else seemed to care. Wearing the uniform I was proud to wear seemed the best way to ensure I was ignored by the public, as if I were invisible or inviting clerks and waiters to deliver poor and grudging service since the young and immature had the most difficulty concealing their contempt for our own military, contempt they had learned by watching TV and listening to their genius buddies.

Looking back on my Vietnam experience, I'm pleased to have served in that war for three reasons. First, I'm proud to have served my country when I was needed, when wearing the uniform was not fashionable. Second, by flying helicopters in Vietnam, I became a member of an informal club of guys whom I admire for their courage, service, loyalty and fierce patriotism; we are friends for life. Third, flying helicopters in that war was my introduction to an aviation career I wanted since I was a lad.

I grew up on a farm, about two miles west of the end of the Dobbins Air Force Base runway in Marietta, Georgia. When the airplanes flew over, I always looked up with wonder. Sitting on my bicycle in my back yard in the 1950s, watching the F-84s, F-86s, B-47s, and many other aircraft, I imagined myself in the cockpit, at the controls.

As I reflect on how I got here, I am sitting in the captain's seat of a passenger jet at 37,000 feet with all passenger seats full, flying from Guadalajara, Mexico, to Atlanta, Georgia. I am here after many miles, much work, and a great deal of luck . . . not all of it good.

If I could change one thing about my past, it would be the way my brothers and I were treated when we came home from Vietnam. On the 4th of July, 2006, I was marching in the Washington, D.C. parade with the Vietnam Helicopter Pilots Association. In front of the US Archives Building, someone in the crowd stepped out into the street, grabbed my hand to shake it and said "Welcome Home!"

It sounds like such a small thing, but it meant a lot to me, the first welcome home I heard from someone outside my family in the thirty years and three months after my return.

*Andy Burleigh with wife, Margie and grandchild
Photo courtesy of Andy Burleigh, all rights reserved*

After Vietnam, Andy Burleigh spent a year teaching Vietnamese pilots how to fly. He flew F100 and F105 fighter jets for the Air National Guard for eight years, and was an air traffic controller, one of many fired by President Reagan over the strike issue. Andy flew for Atlantic Southeast Airlines out of Atlanta for 23 years covering the continental US, Canada and the Caribbean until retired by FAA rules in 2006 when he reached the age of 60. Andy and his wife, Margie, live in Greer, South Carolina, near their youngest daughter and her first daughter, settling into the grandparent mode.

Postcard to America

I SERVED IN THE Marine Corps as a machine gunner in Vietnam among other duties. I was wounded very near the end of my year and spent two months in traction, then two months in a body cast. That summer I was able to come home for periods of time from Great Lakes Naval Hospital, which was close to my home in Minnesota. On weekends I would hitchhike home as money was tight, wearing my uniform since I thought that would help my chances of getting a ride. Boy, was I wrong! People would drive by me on interstate 90 and throw cans of pop or other various objects at me.

When I was discharged I went to our local VFW to join. At that time the people at the VFW told us returning vets we were not welcome to join.

At the local college we had a guy in charge of filing our GI Bill papers who hated vets. He intentionally screwed up the documents for years, causing endless problems with veterans waiting for GI Bill benefit checks for tuition, books, etc.

I dated a class mate of mine from high school, but it didn't work out, maybe because her mother told her since I was a Vietnam veteran I wouldn't amount to anything.

David Crawley, Rochester, Minnesota
Kilo Company, 3rd Battalion, First Marine Regiment, Vietnam 1968-69

Shipmates Forever
Alex Nides
Tyrone, Georgia

A FEW YEARS AGO, nearly four decades since my military service, while at dinner in a restaurant with my wife, Fran, my mind drifted to July 29, 1967, when a man near me lost his life simply by losing his balance. During a major fire emergency on the aircraft carrier USS Forrestal, I was part of a small group below decks on the forward starboard elevator and had just pushed a plane out of Hanger Bay TWO, through Bay ONE and onto the elevator. I was vaguely aware of a sailor standing somewhere to my left and a bit behind when suddenly everyone around me was yelling "Man Overboard!" I knew in an instant the sailor was gone, and I knew instinctively the entire crew must remain focused on the fire threatening the ship and the sailor had little chance of surviving the fast turning screws of both the Forrestal and the destroyer Mackenzie which was running close and playing firehoses on our starboard side. With that vivid recollection in mind, as we sat at our restaurant table I dropped a few tears on the tablecloth while my wife wondered what the hell was happening.

Memories don't always wait for a convenient time and place.

I HAD BEEN IN the Marine and then Navy Reserves since 1962. When I was activated for two years of active duty, I reported to Charleston Naval Base, South Carolina, the day after I submitted the final draft of my master's thesis. Two weeks later I received orders to COMCARDIV TWO along with plane tickets to Atlanta and on to San Francisco, then by bus to Travis Air Force Base. All I was told was that I had "staff" duty. On I went, calling my parents from the Atlanta airport, not knowing where I would wind up. After processing at Travis AFB, I was called for my flight to Clark AFB, Philippines, for transport to the giant naval base at Subic Bay. There, at the Subic receiving station, I learned that COMCARDIV TWO meant Commander Carrier Division TWO and that the aircraft carrier, USS Forrestal (CVA-59) was his flagship, then in transit to Subic from its homeport at Norfolk NOB, Virginia. After I enjoyed a few glorious weeks at Subic Bay, Forrestal tied up at the Cubi Point Naval Air Station across the bay and I reported aboard with the staff of COMCARDIV TWO. The Vietnam War, which had been a public and an academic issue for me, was now getting closer.

I was an E-3 Seaman with hopes of becoming a yeoman which, for you flatland tourists and pollywogs, is a sort of clerk-typist. The staff seemed to have one yeoman too many. I was sent to the ship's Public Affairs office along with an ensign to take care of any work that needed doing for the admiral, COMCARDIV TWO, who would assume command of *Yankee Station*, the US Navy's task force of aircraft carriers and supporting destroyers waging air war against North Vietnam.

On July 25th, 1967, Forrestal joined the other two carriers in air operations against targets in North Vietnam. Those days were exciting and hectic. The ship was a constant buzz of activity twenty-four hours a day with the launching and recovery of aircraft from our Air Wing. We were busy in the Public Affairs Office interviewing pilots for press releases and putting out a daily ship's newspaper.

Public domain photo of USS Forrestal, no rights claimed

On the evening of 28 July, we were rearmed from the ammunition ship USS Diamond Head (AE-19) with extra ordinance including 1,000 lb bombs that I later learned were pulled out of an ammunition depot in the Philippines. Some of those bombs dated back to 1935. The bombs were rusty, filthy and leaking. The ship's weapons officer alerted Captain Belling who tried to refuse delivery of the old bombs. The Diamond Head's captain refused the pleas from Forrestal; this ordinance was to be used for a maximum effort Alpha Strike against North Vietnam the next day.

Rumors circulated that the pallets with the 1,000 pounders still had jungle vines enmeshed between the boards. Despite the reservations of the officers of Forrestal, the orders for the next day were carried out. After recovering the first strike, the second strike was readied on the flight deck. Early warning radar aircraft (E-2Cs) and tanker aircraft (KA-3Bs) for in-flight refueling were first to be launched. The A-4E Skyhawk and F-4 Phantom II strike aircraft, armed with a variety of weaponry including sixteen of the suspect 1,000 pound bombs, were arrayed in a horseshoe around the after-flight deck, their engines coming to life, started by auxiliary power generating yellow tractors. As the pilot of one of those aircraft, an F-4 Phantom II, started up and switched to internal power, a series of events began that came within a hair's breath of our total destruction.

There was a constant scramble aboard an aircraft carrier in a war zone to launch and recover aircraft because there were two critical bottlenecks: one was the catapults to launch aircraft, and the other was the flight deck to recover landing aircraft. There were many things to do to prepare aircraft for a catapult launch and economy of time was important to avoid the necessity for early refueling once aloft.

Under these pressures, squadron officers were alert to opportunities to save time if it could be done safely. One way they found to save time was to arm the weapons, like rockets and bombs, before the aircraft engines were started, relying on redundant safety measures to protect against accidental triggers.

When that F-4 pilot switched to internal power, an electronic surge leaped past safety measures and fired an air-to-ground MK-32 Zuni five-inch rocket held in a four-rocket canister under its wing. The Zuni shot across the flight deck, hit a sailor in the shoulder, and struck the fuel laden drop tank of an A-4E Skyhawk in the pack of aircraft waiting its turn for launch, piloted by Lt. Cmdr. John McCain. The A-4E fuel tank ruptured and ignited, spreading flaming fuel across the deck and throughout the horseshoe of aircraft on the flight deck aft. The 1,000 pounders were also knocked off their racks. McCain escaped the flames by climbing out onto the nose of the aircraft, jumping into the fire below and scrambling clear.

The fire sped from one plane to another, armed and waiting their turn to be shot off the flight deck. Chief Aviation Boatswains Mate Gerald Farrier walked toward the first of the 1,000 pounders engulfed in the flames, spraying it with a fire extinguisher attempting to cool it off to give the pilots a chance to escape their planes. Ninety-four seconds after the fire started, the first of the ancient bombs exploded. Chief Farrier disappeared in the blast and the first trained team of firefighters was wiped out. In all, nine of the 1,000 pounders detonated, rending huge holes in the armored flight deck down which thousands of gallons of flaming jet fuel poured into the heart of the ship.

The Public Affairs Office was located on Second Deck, immediately below the cavernous hanger deck. The first we knew of what has happening above us on the flight deck was when the Boatswains Mate of the watch on the bridge got on the 1MC - the general ship wide communications net - and rang a bell followed by words that are ingrained in my memory.

"Fire, Fire, Fire . . . Fire on the flight deck aft!"

He began again:

"Fire, Fire [short pause] *General Quarters, General Quarters, All Hands Man Your Battle Stations!*"

We rapidly exited the PAO and ran aft on the port side to get to our battle stations. Mine was in the Admiral's Flag Office immediately below the flight deck. That was when the first of the 1,000 pounders detonated. I felt the ship pushed down into the Gulf by the force of the explosion. At the same time something very unusual happened: the passageway was showered with particles and dust shaken loose from the overhead. This was memorable to me simply because Navy ships are immaculately clean. That either was dust hiding around various conduits lacing the overhead or flecks of paint shaken loose by the force of the explosions. That's when I knew that it was going to be a bad day.

I cut across the ship at the first passageway and ran forward to a ladder which would take me up four decks to my battle station, beginning a surreal hour or so which I still relive. I didn't notice as I ran to my battle station that I was the only sailor on the ladder or in the passageway to the Flag Office. When I swung into my battle station from the passageway the Flag Office, normally a hive of paper handlers dealing with all levels of classified materials, was wide open and deserted. There I was, new to the staff and the ship, never having been to General Quarters in my life, in an empty office with all hell taking place over my head on the flight deck. Scared? Oh, yes! I expected the flight deck might implode into my battle station. However, I felt moderately secure knowing that the Navy had developed contingency plans for any conceivable scenario. I knew there must have been casualties on the flight deck, but expected things would soon be under control. I was wrong more than once that day.

Battling the fire aboard USS Forrestal
Public domain photo, no rights claimed

Our ship was in deep trouble. The bombs blew holes through seven or eight decks.

Flaming jet fuel poured through holes blown by bombs, starting new fires below. Crews in the hangar deck were desperately unloading missiles off jets to remove them to safety, forced to do their work by touch because they couldn't see through the dense smoke.

Below decks, off-shift sailors were sleeping in their bunks, unaware of the developing emergency above them. When the bombs began to cook off, one compartment of off-duty sailors from the air wing was cut off from an escape route and 50 were killed, some still in their bunks. Nearby, also immediately below the flight deck, another 41 were killed in their berthing compartment.

Over many years of combat casualty experience, the Navy has developed damage control into a near science. In the first few minutes of the fire essential actions were taken that, combined with the firefighting actions of the crew, saved all of our lives and the ship. Among these actions taken by individuals was to stop pumping jet fuel to the flight deck and drain all the fuel lines. The oxygen generator was turned off and the storage tank drained into the sea. These events, along with those taken by Commander Merv Rowland and his crew in Damage Control Central, saved Forrestal.

The ship stood in danger of capsizing as thousands of tons of salt water from fire hoses accumulated and the ship took a dangerous list to port and aft. Cmdr. Merv Rowland began pumping fuel and oil to the starboard tanks to counterbalance the list and reduce the risk of capsizing.

Off Hanger Bay THREE, Robert Clark was the sole sailor in the compartment with the ship's oxygen-generating plant and a tank of 800 pounds of highly explosive liquid oxygen. Clark was sealed off from the rest of the ship while the fire heated his space, peeling the paint from the hot bulkheads. Clark was slowly draining the tank as instructed, using a one-inch hose stretched to an outside sponson, an open partial deck outside the hull, as he dodged flaming debris falling from the decks above.

*Destroyer USS Rupertus steers dangerously close to
USS Forrestal to spray seawater on the flames
Public domain photo, no rights claimed*

The "bomb farm" on the flight deck stored hundreds of bombs on the starboard side of the towering island where the command and control functions were located. The crew pushed and rolled bombs and other weapons over the side, and they used tractors to move aircraft forward away from the fire.

The USS Mackenzie (DD-614) on the starboard side and USS Rupertus (DD-851) on the port side, both destroyers, maneuvered dangerously close to the Forrestal, dodging bombs and aircraft that were pushed over the side, to direct their fire hoses to hard-to-reach burning areas. USS Tucker (DD-875) and other ships picked up crew and bodies in the water and chased off small Vietnamese boats that were closing in on sailors in the water. The carriers USS Bon Homme Richard (CVA-31), aka "The Bonnie Dick," and USS Oriskany (CVA-34) ferried the seriously injured to their medical departments while the hospital ship USS Repose (AH-16) raced from Da Nang to meet up with the ships of Yankee Station.

Smoke from the fires raging on the flight deck of the listing ship could be seen for hundreds of miles. Recently, I met a former sailor who was on the flight deck of the carrier Oriskany who averred, "I never knew smoke could go up that high!"

In the Flag Office I lost all sense of time. I received one phone call from the Flag Bridge telling me not to come up to the Flag Bridge. That's all. After some time, several sailors and a red-headed ensign collected in a passage way. The immediate threat to the ship was over and those of us who did not hear instruction to abandon the deck below the flight deck were collected and brought to Hangar Bay THREE where five or six gangs of sailors were manning fire hoses prepared to rush into Hanger Bay TWO when called upon. We were organized into a party to go into Hanger Bay TWO and manually push aircraft out into ONE and onto the forward elevator where a like party would push the aircraft onto the forward part of the ship.

It was during the movement of aircraft to the elevator that a sailor fell overboard. I never knew what happened to him. I've often wondered and I know he must have been killed or drowned as there were two destroyers running along side us playing their fire hoses onto the flight deck. The rushing water between two ships traveling along side each other at speed and the high turns of their propellers all make for an extremely hazardous place to be in the water.

There were real heroes on the ship that day. I am told 130-pound Lt. Otis Kight lifted and threw overboard a 250 pound bomb; the explosive ordnance demolition officer, LTjg. Robert Cates, defused and jettisoned one 500-pound and one 750-pound bomb that were on the flight deck smoking from the heat; a team of men with fire hoses charged the intense fire and were vaporized as several 500-pound bombs cooked off, leaving just cut-off hoses spewing water and huge holes in the deck. There were many selfless acts as men desperately tried to help one another and save the ship. Three sailors, James Blaskis, Ronald Ogring, and Kenneth Fasth, manned their battle station in the port after steering compartment, to steer the ship in case control was lost from the ship's bridge. All three were seriously wounded and yet they stayed at their station as fires closed in on them making escape impossible. Still in touch with Damage Control Central, they managed to execute the last order from Cmdr. Rowland, who listened as they died, one by one.

A very tired crew conquered the deck fires in a few hours. Fires in various places below continued to burn and took several days to control. Several sailors died as salt water and battery acid combined to create chlorine gas in the confined areas where they fought the flames.

That evening, we were met by the hospital ship USS Repose and transferred many of our wounded and dead along with one of the ship's dentists carrying dental records for identification.

Two days later, under our own power, we pulled into Subic Bay with fire alarms sounding. I thought to myself that that ought to impress the folks waiting on the pier. Late that evening, on a tarmac close by, I watched a C-130 transport, its engines humming, ready to depart the base with the bodies of my shipmates aboard.

We lost 134 killed, 161 badly injured and 18 missing or unidentifiable. Twenty-one aircraft were destroyed and forty damaged. It was the worst US Navy disaster since World War II when crew of the carrier USS Bunker Hill (CV-13) saved their ship after being struck numerous times by Japanese suicide pilots.

Forrestal returned to Norfolk in September, 1967, for repairs and never returned to the Vietnam War. Her service in that war had lasted just four and one-half days.

We were scheduled to return to Yankee Station aboard our new flagship, the aircraft carrier USS America (CVA-66) in April 1968. I was sent to work in the Air Intelligence Office of the admiral and attended a short intelligence course taught at the Norfolk NAS. In a brief few months, I went from graduate student with my nose in books on British foreign policy and Argentine history to a fully-accepted sailor of the staff. It became my new home and, although I didn't quite appreciate it at the time, on these two ships I served with the finest young men I have ever known. As planned, we departed Norfolk on my 25th birthday for another deployment to Yankee Station. I won't say that this deployment was uneventful but it is enough for this accounting to say that we did our part in the war and *America* brought us home.

BEFORE ACTIVE DUTY, AS a graduate student I read all I could about the war, consuming daily newspapers as a sort of mental exercise. I thought I was well-informed but I learned far more from the intimate view I had and the real-time knowledge I picked up in the CARDIV TWO Air Intelligence Office. I was a minor cog, like a fly on the wall, a participant and witness to the activities of the naval forces operating off the North Vietnam coast.

My observations and later studies taught me the sophistication of our ships, aircraft, equipment and weapons, the skill and courage of our young men, stood in stark contrast against the utter stupidity imposed on us from Washington.

One thing revealed to me in discrete bits was the Rules of Engagement (ROEs) and the powerful effect they had on the conduct and outcome of the war. These self-imposed limitations were meant to show the Chinese and Soviets that we were not being "provocative," that the US was fighting a "limited war" in Vietnam to avoid escalation that could lead to a nuclear confrontation. Our ROEs were also meant to reassure the world we were withholding our considerable power to limit civilian casualties.

That may seem perfectly reasonable on the surface, but a war has two players, not one, and like a chess game, each side uses its different pieces to take maximum advantage and use them in unique ways to surprise and overcome the enemy. In Vietnam our enemy quickly recognized the weakness we imposed on ourselves and took full advantage because they were playing to win at any cost.

We established rules against crossing the border of Cambodia and Laos and so that is where our enemy built their supply trails, training camps and rest areas. We gave them their own sanctuary.

Our bombing targets were limited by political considerations, the fear of provocation and the dimwitted hope that our restraint would be returned in kind.

Our pilots sometimes risked their lives to repeatedly bomb the same worthless targets, like road intersections, while their requests to hit prime targets, like enemy airfields, were denied.

We prohibited our pilots from bombing in populated areas of North Vietnam, including the main port of Haiphong harbor for fear of provoking the Soviets by hitting their ships as they delivered a constant stream of weapons to our enemy; the Soviets mounted anti-aircraft guns on merchant ships and shot at our planes when we were in range.

About the power of the lever as a tool, Archimedes of ancient Greece said, "Give me a place to stand, and I will move the Earth." Our own ROEs gave our enemy a place to stand, a fulcrum point against which their lever moved the world's greatest military power to a point of capitulation. Archimedes would have been proud of our enemy, disappointed in us.

The reality of all this came together for me when USS America was in Hong Kong harbor for a few days of R&R. I had missed the liberty boat and happened to run into Lt. Edwin Schippel, the man I worked for in the AI Office. We took a local small water taxi called a *water buffalo* to fleet landing. As we plied the waters of the harbor, Mr. Schippel pointed out merchant ships at anchor, loaded with war materiel, destined for Haiphong, the main port of North Vietnam. Some of these ships were flagged from the *communist bloc,* others were neutrals, and some were our so-called allies. I knew we kept an eye out for merchant shipping to the North and that we periodically received MERSHIP reports but this was the real thing. What they carried would kill our pilots and troops in the south. This scene would not make one of those great paintings to proudly hang over your mantle.

Thinking back to what I saw aboard ship, I was struck by a targeted road intersection that our planners decided could be easily interdicted because of its relatively low lie and potential for making it impassable during the rainy season. All we had to do, it was reasoned, was churn up the soil by bombing the hell out of it. This place was identified in our "Bombing Encyclopedia" –

renamed "Basic Encylopedia" for public relations purposes - as *Nui Moi Vulnerable Highway Segment*

When I noticed the target, it was bare of any growth that could be seen in recce photos. There were bomb craters filled with rain water and a winding road. In the next series of recce photos, the road was indeed hit but a new road wound around the craters. The bombing went on for some time, giving birth to a new road each time. Although Mr. Schippel never pronounced on this thunderstroke of military genius, he did keep a close eye on it and shared particularly illustrative photos.

One day, Mr. Schippel's objectivity was broken with the receipt of a report of a POW interrogation. In telling of his travels to the south, a captured NVA soldier told of a *political officer* illustrating the stupidity of their enemy by pointing out the bomb craters in a low, muddy area. He told them that American planes continued to bomb the place and that "heroic people" maintaining the road simply redirected traffic around the craters. "Here it is," Mr. Schippel declared for all to hear, "Nui Moi."

Sometimes you don't appreciate the significance of what you know until years later, long after the event has passed from memory, like the folly of conducting an air war against a wily foe and at the same time trying to show the world our reasonableness and our humane conduct of the war. I have had many occasions to remember the words, "Nui Moi."

Shortly before our second deployment aboard USS America, President Johnson ordered a halt of all bombing activity above the 19th parallel, leaving untouched all enemy activity in its major cities and ports. Supplies and munitions were free to enter the North, off load unimpeded, cross bridges and enter the trail system in Laos and Cambodia, destined to kill our troops in South Vietnam. With our restricted target geography, the air war was even more unfocused than the critics allege.

Our pilots well knew that Laos and Cambodia were off limits to the US, unless of course they were assigned a covert mission to fly there and bomb there, like hitting an enemy unit on the Ho Chi Minh Trail when our intelligence provided such information. Sometimes, walking this diplomatic tightrope got a little silly.

On the first day of USS America operations off the north, a two plane flight from Strike Fighter Squadron VA-82, flying A-7 Corsair 2s, was passed from one FAC to another FAC who directed them to a target over Laos. The lead pilot, Lt. Jerry Fields, rolled in and attacked a river crossing where a large barge was tied up off-loading enemy supplies. On the first run, they hit the target on the nose. Contrary to the Air Wing's ROEs, Jerry thought it useful to make a second run to hit the unloaded supplies and vehicles. By this time, a hive of AAA was brought to bear and he was shot down. Rescue operations commenced. Before the saga was over, the Air Force flew 189 sorties, lost seven planes, four pilots ejected, one pilot killed, and another captured and remained a POW until the end of the war. Lt. Fields was on the ground for forty hours evading the NVA and Pathet Lao guerrillas searching for him. When Jerry was finally returned to the ship, the Air Wing Commander (CAG) began proceedings to have him stripped of his wings for violation of the Air Wing's ROEs. CAG was finally dissuaded because it would make for terrible publicity, but the incident does make one think our pilots faced more than one enemy.

As USS America departed for home months later, all bombing of the North was halted in the hopes of achieving a negotiated end to the war. It was not until 1972 when President Nixon unleashed all our airpower on the North and mined their harbors that they agreed to seriously negotiate.

Alex Nides aboard USS America
Photo courtesy of Alex Nides, all rights reserved

IN DECEMBER, 1968, I left the Navy after returning from Yankee Station. While I was aboard ship, with the exception of a few really bad days, it was as though it was just the pilot's war. When I got home, the war became mine.

Having learned to revere the men who served in WW II, I was not prepared to face the wall of silence that greeted me.

I returned to the academic world, met my wife, Fran, and began a career teaching history. The revisionist, politically-correct stereotypes and clichés in our high school textbooks about the Vietnam War made me sick. It felt like the country was taken over by some group that had no connection with me or my experience. When I wrote curriculum guides for the school system, I made sure our story was fairly told but teachers will teach what they believe to be the truth and I fear I did little good convincing others.

I never experienced any overt hostility about my service, but the silence I sometimes felt from others spoke volumes, conveying disapproval and discomfort. I always believed a nation should honor the young people it sends to do dangerous things, and I was deeply disappointed in how America treated Vietnam veterans. I tried to put it out of my mind. Nineteen years later I suddenly realized that not one person in my own family asked me what I did in the war.

Perhaps the most egregious part of coming home was having to deal with the academic world's version of the war. At conferences and meetings, in textbooks and curricular materials, curriculum guides, and in personal conversations with colleagues, I had to put up with a skewed version of the war. At its most subtle form, it was bias by selection with dramatic photographs leading the text. In other cases, it was a celebration of the worst kind of self-congratulation by and about the leading members of a generation who reveled in their rites of passage at the expense of the interests of our nation. The practice of denigrating our warriors and their sacrifices had almost become a sport to many. In recent years, I saw much of the left's response to the Iraq War as having been conditioned by the views formed in opposition to their nation in the Vietnam War.

On the campus of the University of Dubuque, a wonderful small college in Iowa, is a black stone memorial to four of our own that were killed in the Vietnam War. Carved in the stone are their names, rank, branch of service, the date of their deaths, and "Republic of Vietnam." Below their names are the following lines taken and translated from another war memorial in Greece.

> "GO TELL THE SPARTANS, THOU WHO PASSETH BY,
> THAT HERE, OBEDIENT TO THEIR LAWS, WE LIE."
> THERMOPYLAE, 480 BC

I had a hand in the erection of that stone. It was the best thing I've ever done.

I am frequently reminded of just how close we all came to being annihilated on the Forrestal. So many things could have happened, but didn't. I think of those events almost daily. I am grateful for all these years I was given by the courage and sacrifice of my shipmates and by God's helping hand, of whose gift I am always mindful. Every July 29 I pause to pray for the souls of those sailors who didn't return home from the sea.

In July, 2007, the staff of Commander Carrier Division TWO gathered for our only reunion at Arlington National Cemetery in Arlington, Virginia, on the 40th anniversary of the Forrestal fire. We visited the Forrestal Memorial where, behind the Tomb of the Unknowns, the remains of our unidentified sailors rest together. There we honored our shipmates who lost their lives.

Our hair is thinning and turning grey, our waistlines have expanded and we have begun and ended careers since seeing one another. We have had children and grandchildren since the last time we spoke. And yet, for some of us it was like picking up a conversation after a brief pause as the years fell away and the day of the fire with its sounds, smells, adrenaline and fear bound us together as close friends for the rest of our lives. Our frantic teamwork on the most intense day we ever lived came back to us as if it were yesterday.

Maybe the others were thinking the same quiet thought I had: "Damn, it was good to serve with such fine men." At dinner one of the guys stood and summed it up nicely for all of us by saying, "Friends are for life but shipmates are forever!"

Shipmates forever, indeed. I wonder if all of them get a little drifty now and then as I do and feel themselves transported back in time in a snap as they hear with startling clarity the alarmed voice from the Forrestal 1MC: *"Fire, Fire, Fire . . . Fire on the Flight Deck Aft!"*

Shipmates (L-R): Richard Carr, Yeoman; Skip Keller, Yeoman; Bill Crawford, Quartermaster; Alex Nides, Air Intelligence Yeoman; Matthew Hoy, Quartermaster; Terry Wagner, Photo Interpreter. Photo courtesy of Alex Nides, all rights reserved

Dr. Alexander G. Nides grew up in Astoria, Queens, New York. After service in the US Navy he continued his graduate studies while pursuing a career teaching high school history and ultimately earned his PhD degree. Now retired, Alex lives with his wife, Fran, in Tyrone, Georgia, just south of Atlanta, where he enjoys creating original water colors and digitizing them to make prints.

Postcard to America

WHEN THE DUST SETTLES from a recent war, it is the victors who write the war's history. Even though we did not win the war, there were victors in America. The victors were those who opposed the war, the good people who objected as well as the radicals who waved the enemy's flag in protest, those who burned their draft cards and fled to Canada, those in the left and the media who worked against US efforts in Vietnam, the professors who pushed their anti-war agenda on campus and perpetual students who used academic draft deferments to avoid serving. Ask yourself through what prism these Americans viewed the Vietnam War as they took history's pen in hand to write their version in our schoolbooks.

Think about it. The people who avoided service feel the need to rationalize what they did. Whether consciously or subconsciously, when they demonize the war as immoral, when they insist the US had no strategic interest in Vietnam and improperly intervened in a civil war, when they treat Vietnam veterans as either villains or victims, when they distort the war's history by emphasizing the negative and suppressing the positive they are legitimizing the actions they themselves took long ago. Whether they have seeds of regret about not serving their country or remain committed to their anti-war course of action many years ago, it is perfectly understandable that they twist the history of the war and in so doing they reassure themselves.

Frank G. Pratt, Jr., M.D., Psychiatrist, Rome, Georgia
US Army, West Point, LRRP detachment CO, Vietnam, 1966

The Snake and the Ox
Terry L. Garlock
Peachtree City, Georgia

I didn't like the order I was given: "Kill him!" Oh, man! I didn't want to shoot him. The command came through my helmet from the back seat of the Cobra, from the Aircraft Commander, my boss for this mission. From our low altitude of about 300 feet we just assumed my target was a *him*, but maybe not.

334th Playboy Platoon Cobra
Photo courtesy of Graham Stevens, all rights reserved

"Use your thumper," said the AC.

Thumper was our lingo for the grenade launcher in the chin turret directly under my feet, and I aimed the turret weapons left-right and up-down by my moveable sight with a lighted dot bull's eye we called a *pipper* projected on angled glass in the sight, with adjustable range marks. The grenades were 40mm, about the diameter of a fat banana and about five inches long with a rounded nose containing the fuse and detonator. The fuse armed the grenade by centrifugal force of spin in flight after it was fired. We carried belts of 150 or more of these grenades, but I didn't like to fire them close to our troops for fear of hitting them instead of the enemy since firing the grenades was prone to inaccuracy. These grenades were easy to aim in an arc when fired by experienced ground troops from an M79, like a short shotgun with a big kick. In flight, though, it was a different story. The projectile flew a relatively slow arc, making them hard to aim while moving and even more difficult to adjust fire because by the time I watched the point of impact we were in a different position, and our turning and climbing or descending made it worse.

The AC knew all this, and he wanted me to use the thumper since no friendlies were nearby, a good time to practice with the weapon.

I aimed the turret and pressed the thumper button for about a second, enough to fire two rounds, hoping I would hit the ox in the head to get it over quickly. I missed him as we circled, but detonation was close enough to hit the ox with shrapnel and he was clearly in distress. I adjusted and fired a few more, hitting him in the side with one grenade but it just blasted a hole in his flesh that looked twice as big as my fist. Now he was down and kicking.

I couldn't take any more of this, so I switched to my minigun, an electrically driven 7.62mm six-barrel gun that fired 4,000 rounds a minute, 2,000 a minute on the half-speed I always used in 1-2 second bursts to conserve ammo and reduce the risk of jams. Whether full or half speed, the minigun fired so fast that it made a loud burping sound and rained a thick stream of hot shell casings on anyone below. Every fifth round was a tracer, so aiming the minigun was like aiming a garden hose but with a brilliant stream of tracers instead of water. I told the AC I was switching to the minigun and gave him a moment to back off from the target a little to prevent ricochet rounds from hitting us. I fired a bunch of rounds into the ox to put him out of his misery, the tracers careening wildly but away from us as some of the rounds bounced. I also sprayed the cart the ox was tied to, attempting to damage the weapons or ammo hidden under the tarp.

We flew away leaving the dead ox in a growing pool of blood. I felt like a blue ribbon asshole for killing him, but we both knew it had to be done.

It was necessary because the ox cart was part of an enemy supply train in a free-fire zone, an area of repeated confirmed enemy activity, out in the middle of nowhere in jungle-covered hills far from any civilians, who had long been on notice to stay out of this area. The enemy used trucks, people, bicycles, ox carts and any other means on their perpetual supply train moving the weapons and ammo they would use to fight us and terrorize civilians. Their supply line stretched from North Vietnam where arms and ammo arrived in the Haiphong harbor by Russian and other ships, down the Ho Chi Minh Trail through Laos and Cambodia, winding through jungles and over mountains through hundreds of remote trails into every corner of South Vietnam.

When we spotted the ox cart from the air we dropped down low in the kill zone to circle and check it out, with our wingman providing cover from a safer altitude. A large enemy unit might take on a pair of snakes, and since we had received no fire this was likely a small and well-hidden enemy team, or a large unit that did not want to reveal their presence and position.

Maybe you believed the nonsense that *free-fire zones* in Vietnam were places where we shot people, kids, animals and "anything that moved" with bloodthirsty abandon. The truth is a *free-fire zone* merely meant relaxed rules of engagement applied, that we could use our own judgment when to fire on the enemy without calling higher-ups for clearance. The rules of war that made indiscriminate killing of non-combatants a war crime still applied in a free-fire zone. A single man, or a small unit, might cross the line somewhere, sometime, to commit such a crime, and they would be prosecuted if their crime was discovered, never mind the fiction spread about free fire zones by the anti-war left. For Pete's sake, the whole of Europe was a free fire zone in WWII.

My ox was unlucky that day when we caught him in one of the few open spots on the enemy's supply line, left there exposed by NVA or VC who surely heard our aircraft noise and scrambled to cover in the jungle. We took the ox away from them, just as we would destroy a truck, making their infiltration just a little harder.

It has been over 40 years since I flew in the snake and killed the ox in 1969. If I had magical powers, I would wish the ox alive and well, still living a long life because I love animals, too. But I would kill him again if I had to, if it was necessary, because we didn't have the luxury of indulging our sensitive side. We had to deal with harsh reality.

WHEN I ARRIVED IN Vietnam in 1969 I was 21 years old and had been in the Army just long enough for basic training and flight school plus a month of Cobra school. I joined because my draft number was coming up and joining gave me choices; I wanted to fly instead of pounding the ground with a rifle.

The war was controversial when I joined and the anti-war movement was forcefully sweeping the country, but I never once considered refusing to serve. My Dad served on an escort carrier in the Pacific in WWII, and even if he had not, how could I look him in the eye again, how could I look at myself in the mirror again, if I didn't answer my country's call? So I did my duty and learned to fly helicopters even though much about the military, like marching and saluting and the strict command structure, has always made my head hurt. I graduated from flight school near the top of my class and thereby had a choice of a special school before Vietnam. I chose the Cobra, otherwise known as the *snake*. I wanted to fly the first helicopter designed to be a gunship.

At the time, Cobras were relatively new in Vietnam and in high demand because they flew faster with more firepower than the Huey B and C models modified to be gunships. There weren't enough snakes and while the Huey gunship pilots continued to serve with distinction, a few specialty Cobra companies served the needs of other units that didn't have them. I was assigned to one of those Cobra companies, the 334th Attack Helicopter Company, 145th Combat Aviation Battalion stationed at Bien Hoa in III Corps near Saigon. Our platoons were the Dragons, Playboys and Raiders. I was a Dragon. Gunslingers, we thought of ourselves.

We flew a variety of missions, farmed out daily to other units who needed our fire teams. I'll tell you about some of those missions.

We usually had one fire team of two snakes on emergency standby at the 334th, with the preflight inspection done and the aircraft ready for startup and takeoff in five minutes. When a ground unit needed gun support fast, they radioed for help and when 334th Operations received an emergency mission, they scrambled us. When our scramble radio buzzed, the AC sprinted to the aircraft to start up while the co-pilot took mission details over the radio, then the co-pilot sprinted to the aircraft, climbed in and was shutting his cockpit door and putting on his helmet while the AC hovered to the pad and took off. The co-pilot told or motioned to the AC the general direction, or wrote the location in grease pencil on the canopy where the AC could see. The co-pilot also wrote the ground unit FM radio frequency, and their callsign so the AC could dial in and listen to what was happening while he also dealt with departure control on the VHF radio and talked to other aircraft on the UHF radio. The co-pilot quickly pinpointed our destination by coordinates on his map, gave details to the AC and the wing ship, identified every artillery base whose turf we were flying through, looked each one up on our list and dialed their FM radio frequency to call for *arty clearance*, making sure we didn't fly through the gun target line on an active fire mission and get shot out of the sky by our own artillery. The co-pilot would tell the AC how to divert if necessary to skirt arty problems. Then things settled down, the co-pilot stayed focused on the map and the AC would contact the Forward Air Controller on-site to talk about what they needed us to do when we arrived.

While engaging the enemy the AC had the controls because he was aiming the aircraft to fire rockets while the co-pilot in the front used the minigun and thumper in the turret. So, when things were calm on the way to and from, the co-pilot would get some flying time. Flying from a snake's front seat was a bit of fun. The front cockpit was tight, and the controls were on a ledge on either side, with pads to cushion our forearm as we held the pistol-grip cyclic on the right and collective on the left. These controls were at chest level and very short, unlike the controls in the back seat which extended up from the cockpit floor. The different ratio of control movement made the front seat controls very sensitive, making both flying and hovering more of a challenge. When we swapped who was in control, we couldn't glance over to see what the other pilot was doing as we could in a Huey, so we added a third step to the normal two-step process; if I was flying in the front seat and the AC was ready to take the controls he would put his hands lightly on the controls and say, "I've got it;" I would respond, "You've got it," then take my hands off the controls and raise my hands for a second so he could see from the back seat that I had turned loose of the controls, just to make sure in any confusion that *somebody* was in control of the aircraft! Safety first in aviation.

We flew gun cover for Dustoff missions to take our wounded out of the field and jungle. The missions in which Dustoff landed at an LZ were tense because if the enemy did open up and use the red cross for a target we had friendlies to worry about as we returned fire. Hoist missions required the

Dustoff pilot to hover over jungle, very still while lowering the hoist as much as 250 feet then raising the patient up through the trees. Even when the enemy had not been in contact for a while, a hovering Dustoff committed to a stationary position while raising a patient might prove too tempting a target to resist. Our job was to blow the enemy away if they revealed their hidden position by firing on Dustoff.

Terry Garlock with enemy AK-47, foot on elephant skull. No, I did not kill the elephant!
photo courtesy of Terry Garlock, all rights reserved

We covered combat assaults, firing on either side of the LZ or PZ to keep the enemy's head down while our Slicks landed to insert or extract ground troops, or engaged the enemy if they had contact.

People-sniffer missions were very boring, flying cover for and guiding the turns of a low-flying Slick in a back-and-forth pattern over a map grid so the machine they carried could detect ammonia traces, meaning there were probably people, frogs or monkeys in the area. This imperfect process helped identify the movement of enemy units concealed in the jungle.

Our hunter-killer missions were a pair of snakes covering a "Loach," an OH-6 LOH Scout (Light Observation Helicopter) snooping around to find enemy campfires, trails or weapon caches. Scouts flew low and slow to snoop and to draw enemy fire to expose their position. On hunter-killer, the *low bird* snake would fly low in the kill zone, about 500 feet, constantly shifting and turning to be close and in position to cover the Loach quickly when the enemy opened fire and exposed their position. The *high bird* stayed at a safer altitude of about 1,500 feet, just out of small arms range, flying a slow and lazy pattern farther out, in position to roll into a long rocket run to cover both low birds when the shit hit the fan.

One of my first missions in Vietnam was hunter-killer, co-pilot in the high bird, doing nothing but *being ready* and keeping track of our position on a map, which wasn't hard since we didn't move much, while Bob Lugo, the AC in the back seat, did the flying, slow turns to keep us in position, always keeping the low birds out front, far enough away for a rocket run. There was a mirror in the cockpit in front of the co-pilot/gunner so the two pilots could see each other's eyes, and too bad I didn't have my sun visor down because these were the days before air conditioning in Cobras and it was not only boring but hot. My eyes grew heavy, and I dozed off. Bob saw my closed eyes in the mirror. He turned off my radios and then told the low birds to get some altitude for safety for a few minutes so he could teach this brand new snake pilot a lesson. Bob slowly climbed to about 3,000 feet, turned on my radios, nosed it over into a vertical dive, and when we were falling out of the sky like a locomotive he screamed in my radio about enemy 51 cal fire. I nearly dirtied my shorts. But I never dozed off on a mission again.

Gun cover missions for LRRPS were often tense. LRRPs were small teams inserted into hostile territory to sneak around in extreme camouflage for a few days or longer to complete their missions, usually close observations of the enemy, trying to avoid detection or contact. LRRPs were very effective making the enemy's life miserable, as evidenced by their standing bounty on the death of any LRRP. Covering LRRPs always made me edgy because they were astonishingly bold, often whispering radio instructions to guide our fire because they were "danger-close" to the enemy and I feared I might hit them. Some of those guys were only 18 years old and they scared the hell out of me.

We were eating lunch at Bear Cat one day, having covered the insertion of a LRRP team and now waiting to see when they needed gunship cover again. Suddenly the team radioed for emergency extraction, out of breath, compromised and running from the enemy. We scrambled along with the slick that would pull them out and flew to their position as fast as we could. A couple miles out we could see different colors of smoke commingled and filtering through the thick jungle trees as they ran and popped one smoke after another trying to mark their position, but the smoke blended together by the time it drifted and emerged from the trees and we couldn't pinpoint their position or risk firing to help them. There was no time for them to go to the planned PZ, the enemy was closing in and the PZ was too far, so we found a tiny clearing where the slick dropped a rope with a rig on the end while we started firing well to the rear of the team, trying to slow the enemy and make them go to cover. As usual we were just shooting at trees and guessing where they were based on radio contact with the LRRP team below. When the slick pilot pulled up with all five LRRP team members "on the strings," each man secured to the line with carabiners, he gained altitude and headed back to Bear Cat. I thought he might pick a spot to land to bring the team into the aircraft, but soon the LRRP team was swinging back and forth in wider and wider arcs under the helicopter. I called the slick pilot to ask him what the hell they were doing and he said they were just playing with his mind, trying to screw up his flying with their antics. Those guys were nuts, or as we said in Vietnamese slang, *dinky-dau*.

In the area of Nui Ba Dinh, the Vietnamese name for the Black Virgin Mountain near Tay Ninh, we occasionally flew gun cover missions for Special Forces. My best memory of those missions were the flights south to return to our Bien Hoa base when the mission was over at the end of the day. Our flight path parallel to the highway took us near an ARVN artillery firebase that was built by the US and turned over to our ally after training their troops. We would sometimes receive fire from the ARVNs in that firebase as we flew by, just a harassing tracer or two out in front in our flight path, perhaps as rough play but more likely a reminder that not all of our South Vietnamese allies loved us.

One evening there was a lot of chatter back at the 334th in Bien Hoa about one of our fire teams due back from the Tay Ninh mission. It seems one of our pilots grew tired of the insult from the ARVNs and, using the defense of the rules of engagement that said we could return fire if we could see where the fire was coming from, he rolled in and punched off a pair of rockets into that *friendly* firebase. Our CO was waiting for him to chew his ass when he landed, but apparently nobody was killed because after the chewing he was still on flight duty. Our pilot told the CO he couldn't see it was a friendly firebase in the dark, but he didn't even try that excuse with the rest of us because we all knew where it was. Besides, we couldn't wait to shake our fellow pilot's hand for putting the ARVN smartass on the gun in his place, and we bought his drinks for a while.

When life is tough you have to take your fun where you can find it, and we did. We were enthusiastic young men living a dangerous adventure, typically flying above 1,500 feet, out of small arms range, or low-level just above the deck so we flew by the treetops too fast for the enemy to aim and shoot. Flying low level at 150 MPH was a kick, dodging and weaving to stay low as we followed the contour of hills, sometimes clipping trees with the rotor or skids. One of our guys got too close to a large tree limb one day and knocked off one side of the skids. When he came back to Bien Hoa he had to hover over to the maintenance hangar and then hover in place while a crew built up a sandbag pile to hold up one side of the aircraft so he could set it down; hovering soon gets tiring, especially when you are trying to hold the aircraft very still. They made it before he exhausted fuel, which is a good thing because otherwise it would have been a crapshoot whether the disintegrating blades would take his head off when it flipped over on its side. Money changed hands on the bets whether his fuel would last. When you do something stupid, like knocking half your skids off, you have no room to complain when the guys watching make bets on whether you make it or not. That's life in a war zone.

There is a river north of Bien Hoa, maybe the Saigon River, with very high trees on either side, maybe 150-200 feet tall. Now and then on the way back from Tay Ninh we flew a few feet off that river surface as fast as it would go, gently laying the aircraft on its side and gaining a little altitude to let the turn slide us up the tree line, following the river's meandering course, more of a kick than any theme park ride back home.

When I was not busy flying one of these varied missions, I carried out my duties as unit Civic Actions Officer by helping the local orphanage in Bien Hoa. On payday I hung out with the pay officer and asked fellow pilots to contribute so I could buy food and clothes for the orphanage kids. Some contributed, some didn't. Our help went a long way because the locals wouldn't lift a finger to help these mixed-race kids, the offspring of American men and Vietnamese women. For me, this work was not particularly saintly; American humanitarian aid was common in Vietnam.

WHATEVER MISSION WE HAD for the day, we usually flew in light fire teams of two Cobras, rarely in heavy fire teams of three. When we were doing what the snake was designed for, firing on known or suspected enemy positions, we typically used rocket runs, shallow dives to stay on target longer than steep dives so we could fire, adjust and fire again. Rockets fired in pairs, one from each side to keep the aircraft balanced. We had a dial that allowed us to fire as many pair of rockets at a time as we wished with one button push, but we usually fired just one pair at a time, adjusting aim to fire again. Firing multiple pair at one time was for desperate circumstances.

We usually carried twin 19-rocket pods inboard and lighter 7-rocket pods outboard on the stubby wings since weight was already enough of a problem that we couldn't carry a full load of fuel. These were 2.75" FFAR - folding fin aerial rockets - with a solid fuel motor. When the rocket motor was fired its folded fins sprung into place after leaving the tube to help it fly straight.

The rocket motor and warhead were crated separately, and when we landed for rearm we would tell the ground crew what we wanted inboard and outboard, 10 lb or 17 lb HE (high explosive) warheads, with standard or *proximity* fuses, or *nails*. The standard warhead was 10 lb HE with fuses that detonated on impact. The 17 lb warheads were of course more powerful, useful against bunkers and equipment. If our objective was anti-personnel, the standard fuse might still do the job since they usually detonated in the trees and scattered shrapnel down on the enemy. If the terrain was more open, we might ask for proximity fuse warheads that detonated above the ground so the shrapnel would scatter in a downward cone-shaped kill zone. An even more effective anti-personnel weapon was Flechette rockets, or "nails" as we called them. The Flechette warhead was like a shotgun shell with a charge behind a container of thousands of pointed little nails with fins crimped in the tail so they would fly pointed-end first. When we fired *nails* the rocket would fly half-way to the target then the charge blew out the nails high above the target, marking the discharge point with a puff of red smoke and scattering lethal nails toward the enemy. A frequent configuration was 17 lb HE inboard, *nails* outboard. When we told them what we wanted, the ground crew would manually change fuses on the warheads if necessary, screw the threaded warheads into the rocket motor, then slide them into the tubes.

If you are squeamish about these weapons, try to remember the objective was to kill the enemy as quickly as possible. War ain't beanbag.

We didn't have smart weapons back then that would fly faithfully to the target. A rocket run was basically aiming the aircraft as we flew it down the slope of air in trim so the rocket would fly straight . . . well, sort of straight; they did corkscrew a little on their way, but we were pretty accurate when we could get relatively close. We were vulnerable at the bottom of the run as we pulled out and slowed down to climb around for another run, and the co-pilot-gunner in the front seat would lay down minigun or 40mm fire from the chin turret if he had a shot to *cover the break*. Our sister ship would be at altitude, also covering our break with minigun fire just before starting their own rocket run while we climbed to do the same. We did this in racetrack patterns, varying our approach direction to keep the enemy off balance, pulling out well short of the enemy so we weren't overhead to present an easy shot when most vulnerable. Some hot situations called for innovation instead of this standard tactic.

Some missions were uneventful without a shot fired; in some we had firefights with the enemy. When you haven't been in a war you might think of it as constant battle, but it was not that way at all. Many missions were routine patrols, searching for the invading enemy who eluded contact until the time and place of their choosing. There was a lot of boredom with scattered moments of fright, but at least my boredom was in the comfort of flying and not the misery of the grunts in the jungle. Helping those guys, our brothers, was our purpose and our daily motivation.

SOON AFTER I ARRIVED in Vietnam I was introduced to *basketballs* and the *pucker factor* on a night mission.

One of our ground units was in a tight spot in a firefight with the enemy and our fire team of two Cobras was trying to help them. Flying a gunship at night in a third world country like Vietnam in 1969 was tough because night was very dark, and in a remote area of jungle there are no ground lights whatever to help orient a pilot; we see our instruments inside and nothingness outside, or with moonlight maybe irregular dark shapes scattered on a black velvet blanket that seems to cover the world.

I don't remember if we had moonlight that night, but I do remember struggling to locate friendlies and enemies in close proximity while all we could see was an occasional flash from the firefight peeking through jungle trees and an occasional tracer bounce. At last we spotted a strobe light flashed by American troops in a small clearing and with their radio guidance of direction and distance from the strobe to the enemy's position, we started to lay down rocket and minigun fire to ruin the enemy's night.

Firing rockets at night had its own problems. First was the hazard of getting fixated on the target while aiming in a rocket dive, without the daytime benefit of peripheral vision, and flying into the ground. Second was losing night vision from the bright flash of the rocket motors when we fired them. So, one of our adaptations was to set up our racetrack rocket attack pattern, and when we started our shallow dive for the rocket run, while the AC in the back seat lined up on the target with the *pipper* in his fixed rocket sight, in the front seat I would lower the tinted sun visor built into my helmet to protect my night vision. That meant I couldn't see a damn thing except the light from my instruments, and I watched the altimeter to make sure we stayed well above the ground, but after the AC punched off rockets he would then say, "You've got it," whereupon I would flip up my sun visor, grab the controls, say, "I've got it," and fly us out of the dive and around the pattern for another rocket run while the AC recovered his night vision.

I guess the enemy didn't much like the rockets we sent them that night because they fired .51s at us, a helicopter pilot's worst nightmare, .51 caliber anti-aircraft fire, rounds so big and heavy they would destroy anything they hit in our aircraft, especially people. At night the .51 green tracers that reached up to us glowed so big we called them basketballs, and the tracers always looked like they might hit us right between the eyes no matter how far they missed. The *pucker factor* was our own measure of the feeling of our heart trying to leap outside our chest, our mouth going instantly dry, our skin flashing cold even while we sweat and our butt-cheeks clenching the seat because for an instant we were scared as hell. Sometimes when we needed to land to rearm and refuel, another snake fire team would relieve us and we'd tell them on a scale of 1 to 10 how high the pucker factor had been, just a heads-up.

That night we helped our ground unit force the enemy to withdraw, and I learned the flash of real fear because the pucker factor was high for a little while even though we never took a hit. I was just beginning to learn that night that courage isn't the absence of fear, courage is staying focused on your job while you're scared as hell.

BEAUTY CONTRASTED AGAINST MISERY in Vietnam was striking. Some parts of the country were stunningly beautiful. When I saw a picturesque small farming village from the air, in a valley near mountains, surrounded by lush green palm trees and rice paddies, I knew it was a good bet, at some point in the endless war the villagers had known for generations, some of them had probably died in direct attack or crossfire, a family maybe watching a loved one die.

One of the prettiest sights I have ever seen is a white phosphorous (WP or "willy pete") artillery airburst against a clear blue sky, a small intensely dense white cloud with drooping trails, sometimes at eye level from the cockpit, fired by artillery miles away to mark the first or last shot, depending on unit practices. One of the most gruesome sights I have seen was flying low-level cover for ground troops when a soldier triggered a *willy pete* booby trap. That pretty stuff burns right through flesh.

With engine and rotor noise, helmet and the shroud of a Cobra canopy we rarely could hear gunfire on the ground unless we were flying very low. The first time I saw a group of pretty twinkles like stars in the daytime from a tree-line, it was curious to look at for a moment before I realized those were muzzle flashes of someone shooting at us.

A firefight in the distance at night can be mesmerizing. Big gun red tracers from friendlies seem to float a bit slow on their way to the enemy, bright green tracers coming back at the friendlies, careening wildly when they would ricochet then burn out in mid-flight. Scenic flashes of artillery might illuminate a glimpse of trees cartwheeling in the air while rockets or other munitions with showering spark trails send delayed and muffled thumps amidst tiny rapid pops of gunfire. It sounds harmless so far away, almost a soothing scene, but the paradox was that while I watched the dazzling show pierce the darkness I was really watching young men on either side fighting and dying. Sometimes it was me pushing the button to fire into the show. I have never watched a fireworks display since then without these thoughts, and though I can't explain it, there is something almost reverent in how I think of those men, no matter which side they were on, the ones enduring hardship and risking everything while their countrymen enjoyed the comforts of home, and that makes mock explosions of battle to entertain crowds seem to me wrong somehow. It makes pretending to be in combat, like paintball games, childish and obscene to me. I might be the only one on earth who has an irrationally emotional disapproval of fireworks to entertain the masses, and maybe I'm the only one who has ever wondered if the crowd would "oooh" and "aaah" in delight at the brilliant bursts if they had the slightest clue what the display signifies. That is unfair, of course, applying my own personal hangups to people enjoying what is meant to entertain them while I foolishly try not to like an impressive fireworks show. I learned to hide it so my kids can have fun.

SUNDAY, DECEMBER 17, WOULD be my most memorable day in Vietnam. I awoke for the second time that day disoriented, seeing the world sideways from inside my front seat cockpit, the snake lying on its left side, my canopy door pinned to the ground, my lower back and guts hurting severely deep inside. The turbine was still running and I could smell fuel leaking from the crash.

I had to break out! My worst nightmare was being burned into what we callously called a *crispy critter*. There was a breakout knife for just this purpose, a round knurled steel handle with a heavy steel point to break the Plexiglas. I was able to grab the knife out of its cradle but I was weak and could only make scratches or dents in the canopy. I tried to reposition for leverage but I couldn't feel or move my legs.

I was trapped. My chances of getting out of this were not looking good, and if the enemy got to me first they wouldn't keep me alive long with legs that didn't work. I was in *deep shit*.

How the hell did I get stuck in the cockpit of a broken Cobra lying on its side on the ground in the middle of nowhere? I didn't remember until later, my body was in shock.

Just an hour or so before, we had been on emergency standby at the 334th in Bien Hoa, sitting around the ping-pong table, hanging around close to the emergency phone. When it buzzed we scrambled, John Synowsky and Graham Stevens in the fire team lead ship, Ron Hefner and me in the wing ship. We were the Dragon platoon and Ron flew with the Raider platoon, so he and I had never flown together. We were short a pilot and Ron offered to fly with us on his day off.

The enemy had ambushed an allied convoy near Lai Khe in III Corps not far from Cu Chi in a region replete with enemy tunnels. This was a notorious enemy infiltration route from the Ho Chi Minh Trail in Cambodia to Saigon, the capital of South Vietnam. We were on the way to help.

When we arrived at the ambush site we engaged the enemy with rockets, 40mm grenades and minigun fire, and they hit us many times with small arms and automatic weapons. As we pulled out of a rocket run we took more hits and lost our tail rotor, very big trouble.

One option was autorotation to keep it straight, if only we had a safe place to land. As fire team leader, John urged us by radio to pick up airspeed to keep it flying to possibly return to the base for a runway sliding landing. But it didn't work.

Sometimes you learn new things at the most inopportune moment. I knew if we didn't reduce power and descend, without the tail rotor we would spin to the right, in the opposite direction as the rotor blade spin. But even in a deadly situation I was fascinated to see the slipstream holding against the large tail fin and the narrow body of the Cobra preventing the spin, our nose cockeyed to the right and straining against the wind to flip around, and as airspeed bled off it finally snapped around to the right so fast in a 360 degree spin that, being far forward of the center of gravity, I was thrown out of my seat against the shoulder straps until the slipstream caught it again, straining to the right again until it snapped around again and again and again, each time around the turbine surging as the load varied, spinning out of control with the ground rushing toward us. At a time like this I guess a lot of people pray, but I was not inclined to begging. I do recall having plenty of time to think it was a sadly stupid way to die so young, so much opportunity lost, and I frankly hoped it wouldn't hurt too much, hoped it would be quick. I knew my family would be devastated to learn of my death, and it didn't seem noble or worthwhile, it just seemed stupid. Isn't it strange what little things are burned forever into our memory?

Impact and the rotors breaking to pieces must have been a sight, but I don't remember it. Ron said he saw in the pilot mirror that I was already unconscious before impact; my speculation is my lights went out from more centrifugal force of the spin in the forward seat, or maybe just from fear, I don't know. At impact, with my front seat being so far forward I was lucky the downward flexing rotor blades didn't take off my head before they broke apart and flew off to parts unknown.

Even though his right-side canopy door opened straight up with the Cobra lying on its left side, Ron's door was jammed, and he used his pistol to shoot several holes so he could break the Plexiglas to climb out; I was still out and didn't hear the shots, which must have been deafening in the enclosed cockpit. Ron's effort must have been extraordinary, because at impact his chest protector - chicken plate – bounced up to hit him in the throat, crushing his larynx, breaking his jaw and cracking his neck vertebrae. His neck was cut and his tendons were hanging out, but I didn't see that at the time and had the impression Ron had only minor injuries because I later saw him walking in the LZ, probably pumped on adrenaline.

When I awoke in an unexpected position, Ron had climbed his way out of the cockpit and was dazed, probably in shock as well, trying to figure out a way to get me out while I tried to break out with no success.

John and Graham, in the fire team lead aircraft, broke contact with the enemy before we hit the ground, radioed Mayday emergency calls, ignored the rules about keeping their aircraft in the air and boldly landed nearby. We were all at risk of being killed or captured by a determined enemy with a deep hatred of pilots of *whispering death,* their name for the Cobra, and our wreckage was apt to explode any second.

Even in a tense situation like this there is sometimes comic relief. When John and Graham landed near our broken snake, John put their Cobra in idle mode while Graham jumped out and ran to us. I was weak and making no headway with the breakout knife, just scratching the canopy, and I don't think Ron was able to break it, either. Graham had just one thing he could use to break open the canopy, his .38 pistol. So he took his .38 out of his holster, swung it and when he hit the canopy he accidently fired a round through the cockpit. It missed me. And the leaking fuel didn't ignite. Graham knew he had to hammer quickly so, just to make sure it didn't fire again, even though the hammer was now on a spent cartridge, he aimed at the ground to the side and fired the remaining rounds quickly to empty the pistol – pow-pow-pow-pow. When John heard the gunfire as he was stepping to the ground his pucker factor went through the roof because he thought the VC had arrived and Graham was shooting at them with that pathetic little pistol.

Graham hammered through the canopy while John helped Ron, who was injured but able to stand. John shut down the turbine while Graham hammered so hard on the canopy with his .38 he knocked the wooden pistol grips off both sides and later had to exchange it for another pistol.

Though I was bigger than Graham, he dragged me by my collar out of the cockpit and away from the wreckage in case it blew, and I remember as I lay in that field waiting for the Dustoff helicopter, unable to move my legs and my back hurting like hell, I sang, "I'll be home for Christmas." I guess surviving after being certain you will die is a powerful narcotic. I was alive and knew I was hurt badly enough to go home.

*Cpt. John Synowsky (L), WO Graham Stevens (R)
Photo courtesy of John Synowsky, all rights reserved*

Instead of getting airborne ASAP for their own safety when the Dustoff helicopter arrived, John and Graham stayed to help load us on stretchers and into the helicopter.

Standing guard over their injured men in the middle of a hostile area, in the open with no cover, just pistols for weapons and a dangerous enemy nearby - that's how I remember John and Graham that day. Graham was 21, John a few years older.

Pete Atack, callsign Rebel 51, flew one of the two C-model gunships covering the medevac that picked us up. Pete says they frequently took fire in this area and that the enemy had a .51 nearby. He also told me he took 22 hits keeping the VC away from us on the ground, that the VC rushed our broken snake as soon as the medevac departed but his gunship fire held them off until he got radio clearance to destroy the downed snake to keep the radios, weapons and munitions out of enemy hands. They hit it with miniguns first trying to fire it up, then did the job with rockets.

The Dustoff helicopter took Ron and I to a field hospital at Lai Khe, but I don't remember much of the flight since I was floating on clouds with the morphine they gave me in the LZ. I still remember the rush of euphoria! No wonder drugs are a problem. At Lai Khe they quickly determined I needed neurosurgery so I was flown to the larger Evac hospital complex at Long Binh. Meanwhile, Ron and I got separated, but since he was able to walk he tried to find me to check on my condition; we didn't find each other again until 2009, 40 years later. As I was prepped for surgery a doctor gently told me he didn't know if I would walk again but he would do his best. His best was superb. I wish I knew his name so I could shake his hand.

I had a serious compression fracture of my L1 and L2 vertebrae; the surgery removed bone splinters and a disk and inserted metal pins to hold it all together. I would later find that I had lost one inch of height by compression, that one of my lumbar vertebrae was seriously deformed as it was crushed toward the front, and that I was very lucky, indeed, not to have been paralyzed for life. The big unknown was how much of the nerve damage would heal and whether my legs, bowels and bladder would work.

Meanwhile, back at the 334th, after the Commanding Officer chewed a chunk out of John's ass since he was the fire team leader and AC who put a working snake on the ground in an unsecure area, John and Graham were awarded the *Soldier's Medal* for heroism saving lives that day. Both of them earned a number of decorations over two tours in Vietnam, but they both consider the Soldier's Medal to be special. Perhaps that's because amidst all the killing and dying, they were able to make sure two guys stayed alive.

MY LEGS QUICKLY REGAINED feeling and function after surgery, and over some weeks the bowel and bladder control I had lost slowly returned. I was on a specially designed "striker" bed, mounted on a swivel at the head and foot, so that a removable part of the bedding could be secured on top of me, kind of a "Terry sandwich" between two mattresses screwed tight together, and an orderly could pull a couple pins to flip me over then replace the pins and remove the bedding now on top. That was to immobilize my spine and turn me over every few hours to prevent bed sores.

Since I came out of surgery about a week before Christmas I had hopes of being home for the holidays. I missed it and was flown on a hospital C-141 aircraft on a stretcher to a hospital in Japan where casualties were treated after being stabilized in Vietnam. On the flight at one point my nurse sat with me, held my hand and talked to me for a little while to distract me, because the vibrations were killing my back and I hurt so bad I shed tears, and she couldn't give me a pain shot for another hour. When they moved us off the aircraft to the outdoors for a few minutes with just sheets as cover, the cold January air in Japan stuck in my throat and I could barely breathe for a moment; what a surprise after the warmth of Vietnam!

At the hospital in Japan I graduated to a bed mounted between two big steel circular tubes, kind of like a hamster wheel, with a remote control for tilting the bed up or down, and they sandwiched me in to flip me over in much the same way. With my bowel problems and immobility, I was constipated when I arrived. A redheaded nurse who saw the problem in my chart asked me about bowel movements, and she told me, "I'll take care of that!" She brought her magic potion masquerading as bad orange juice, but it had no immediate effect. An hour later, when an orderly pressed me into a tight bed sandwich to flip me over, when I was vertical and squeezed between two layers, I pooped all over myself, the bed, the orderly and the floor. I may have hit the ceiling. If you think you have been embarrassed, try having an orderly strip you naked to clean your poop while your fellow patients try to pretend you haven't made the ward smell like a dirty diaper.

The evening of the first day of my two weeks in Japan I slipped into a funk, feeling a bit sorry for myself, frustrated at being helpless instead of with my fellow pilots where I belonged, tired of the constant pain between shots of narcotics, aggravated at being immobile, anxious to get home.

As I wallowed in self-pity, I was startled when Jim screamed, and during my stay in Japan I would learn from him an important lesson about small stuff. After a pause Jim screamed again, and a young guy with no legs below the knees zipped past me in a wheelchair yelling, "Hang on Jim, I'm coming," and that's how I knew his name.

The nurses were showing a John Wayne movie projected on the wall, but pretty soon Jim's screams drowned out a lot of the dialogue. Nobody complained.

I called to a nurse and asked about the guy screaming. She told me his back was ripped to shreds by a grenade, and when they changed the dressing on his wound, or when his pain medication wore off before they were permitted to give him more, he hurt so bad he couldn't help but scream.

I heard guys talking to Jim, many from other beds so they had to shout. "Go ahead and scream, Jim." "Give-em hell, Jim." "Dammit, nurse, can't you give Jim more of that magic needle?" One of the hustlers in the ward hobbled by on crutches yelling, "Hey, Jim, I got the cards, want to win some of your money back?" Others tried to distract him from his pain by asking questions like, "Jim, where you from, Chicago?"

All around me young men, black and white and Hispanic and Asian, officers and enlisted men, missing limbs or eyes, with all manner of injury from violence, set aside their own problems as small stuff and joined together to help one of their stranger-brothers get through unbearable pain.

Pretty soon the nurse announced she was preparing Jim's pain shot, and the clamor arose. "Yo, my man, help is on the way!" "Hang in there, Jim, it'll be better soon." "Nurse Nancy gonna make you feel better, Jim." And soon after Jim's shot he slept, at least for a while, because he went through this about five times a day.

"What fine men," I thought to myself, even though some of them were still too young to buy a beer. I felt proud to be with them, to be one of them, and knew I was a damn fool for having the blues. I had no business whining about the small stuff. I knew better now; I had learned what is important and what is not.

My theory is that combat changes you by forcing you to focus on things that matter most. When I think of one common trait among all those I know who have been in combat, it is how they instinctively distinguish the trivial from the important. I think that is one reason we are bound so tightly together.

In a setting where young men are doing dangerous work and dying in violence, what can be more important than each other? Maybe when you learn how much some things matter, it helps us recognize the small things, the trivia, that don't.

Since that day long ago, when I see someone frantic with worry about a missed appointment, angry at a flaw in the paint job on a new car, a delay in the delivery of furniture or other minor inconveniences, I think about a guy named Jim and the small stuff that eats up way too much in our lives.

WHILE I SPENT A few months recovering in hospitals, recuperating at home and then returning to non-flight duty, the war continued just fine without me.

I had a roommate for a while in Vietnam named Pete Parnell from Lee's Summit, MO. Pete and I never flew together because he was with the Raider platoon while I flew with the Dragons, and we roomed together because the bunks were available and we were all in the same small area anyway.

Pete drove me nuts talking about his wife and the baby they were expecting. We all talked about our wives or girlfriends but Pete never stopped.

The day after I was shot down Pete received a telegram informing him he had a son named Thad. Somehow he found a way to order roses for his wife.

Four days later, on Dec 22, Pete was the Cobra front seat when his Raider fire team was supporting the 3rd Mobile Strike Force (Green Berets) at the village of Bu Dop, west of Song Be and about five miles from the Cambodian border. There had been enemy .51 fire in the area and Pete's fire team of two snakes was called for help. We don't have details of how it occurred, but Pete and his back seat Harry Zalesny from Plymouth, MI died when their aircraft went into trees, 150 to 200 feet high, at a high rate of speed. The aircraft burned high in the thick jungle trees.

Three Green Berets immediately volunteered to rappel down from a hovering helicopter to recover the bodies, and while they were able to recover Pete's body they were not able to recover Harry's because the heat of the fire was intense and the rockets and grenades and 7.62 ammo began to cook off. The Green Berets willingly sustained burns in their effort to recover the bodies of their brothers.

The Parnell family received word of Pete's death on Christmas Eve. The flowers Pete had sent arrived days later. Whenever I am reminded of the nasty reception war protestors gave our troops coming home from Vietnam through California airports, I think of Pete's family.

One day in early 1970 a friendly ground unit was losing a firefight with a superior enemy force and feared being overrun, meaning they would all likely be killed or captured, which could be worse. They radioed an emergency call for gunship support and John Synowsky scrambled his light fire team of two Cobras. The firefight was west of Cu Chi, near the Cambodian border, not far from the Ho Chi Minh Trail.

With little information on the enemy force, John was caught in a *helicopter trap* where the enemy places .51 caliber anti-aircraft guns at the three points of a triangle surrounding their position, then when we fire on them and pull out of a rocket run, at least one of the big guns will have an easy broadside shot.

That day John took .51s through the cockpit. One round bounced around the cockpit before it lodged in his shoulder and burned him real good because rounds are hot when fired. John was lucky it didn't go through him, and he's lucky it was hot because it burned the wound enough to slow the bleeding. John's co-pilot was hit too, but they ignored their wounds, the aircraft held together and they continued the mission, attacking the enemy again and again through .51 fire because if they didn't our brothers on the ground would die.

Their attack took out the anti-aircraft guns and forced the enemy to withdraw. The families of those men on the ground would never know their loved one lived that day because John stayed with the job under heavy fire. For that mission John received the Silver Star *for gallantry*.

In the March 1970 incursion into Cambodia to finally cut the Ho Chi Minh Trail, Graham Stevens led a fire team helping an American ground unit in trouble. The enemy was *everywhere* and they came under fire from all directions. His front seat was Johnny Allmer, and his wingman, Larry Pucci was just 19 years old. Larry's front seat, Wayne Hedeman from Hawaii, was an *old man* at 25. When they took fire from an NVA anti-aircraft gun, Wayne was hit with a round through the neck and even though Larry flew his Cobra as fast as it would go to the hospital at Tay Ninh, it was too far; Wayne bled to death on the way.

Graham spotted the anti-aircraft gun and set up a rocket run. While he was lining up his target he took rounds through his engine vitals and while warning lights flashed and alarms sounded Graham had to land quickly while he could still fly it down even though the enemy was everywhere. Once on the ground, the dash Graham and Johnny made in boots and Nomex flight suit, running to the cover of a friendly unit, was gold medal material. Graham received the Distinguished Flying Cross *for heroism* that day.

Do you wonder that I admire these men?

AFTER SOME CONVALESCENT LEAVE, I had recovered sufficiently to return to duty as a TAC officer in the advanced flight school at Ft. Rucker, but I was not on flight status. My strength gradually increased and, even though my back continued to be weak for a long time, I managed to convince the flight surgeon to put me back on flight status. I missed flying and at least I could fly enough to meet my monthly minimums.

The vertical vibrations that I never before noticed were painful and distracting. I tried it for several months but I knew my flying days were over.

A little over a year after I returned from Vietnam the US Army offered all Warrant Officers a choice – either sign up for an indefinite enlistment or leave the Army. I knew indefinite status meant a return to Vietnam. I chose to leave, and I left behind not only the Army but flying and the memories of the Vietnam War. It wasn't that I was particularly troubled by the war, I just wanted to get on with life and that part was over. I didn't seek out those I knew in Vietnam because I wanted to forget it instead of continually reliving it.

Forgetting worked far better than I had planned.

I worked on Wall Street in New York City for two years, then took advantage of the GI Bill to complete my Bachelors degree and MBA. After college I was working in Dallas, Texas, when the phone rang in 1977, seven years after I returned from Vietnam, and the caller said he was "Ski" as if I should know him. I had no idea who he was. He reminded me his name was John Synowsky from the 334th in Vietnam, and I pretended to remember but didn't, and we agreed to meet for dinner at his home in Ft. Worth. I couldn't rely on photos because when I was medevaced out of Vietnam all my photos disappeared. When I saw John's face at his front door it came back in a flash, and I was mortified that I had forgotten my platoon leader, the guy who taught me a lot and one of the two who rescued me. I didn't know many post-impact details of our crash in 1969 since my body was in shock at the time, but that night John and I got drunker than skunks and he filled me in on a few things I didn't remember, like the part where Graham nearly shot me. But Graham was rescuing me and he meant well!

I talked to John once in a while over the years but I didn't know how to find Graham Stevens. I knew him in Vietnam as "Steve," the fun magnet, because if grabass was going on he was always in the middle of it. With a common last name and no first name I had no luck finding him, and I never knew how to find Ron Hefner. A few years ago John and I collaborated over the phone, used internet resources and found Graham in Williamsburg, Virginia. In our first telephone conversation in 30 years Graham told me now that he is a civilian he wears a lapel pin version of just one medal, the Soldier's Medal he received for rescuing me, and so he thought of me every day for years when he dressed for work. Imagine that.

FOR OVER 30 YEARS I never attended any veteran gatherings, and though I visited Washington, DC many times and always passed by the Vietnam Veterans Memorial, I didn't stop. Some co-workers who knew I was in Vietnam asked sometimes when I returned from a DC trip if I had visited the memorial. When I answered "No!" I could see in their eyes the wonder if I was like the Hollywood stereotype of Vietnam vets, fragile, suffering from guilt of misdeeds in the war, a victim of horrifying flashbacks, poised to snap into violence or break down from overwhelming sorrow. But no matter what they believed, I never felt the need to go to a special place to honor the memory of good men who died young, and they never knew I avoided the memorial for the same reason I avoided veteran gatherings. I didn't like what I saw driving by the memorial, long-haired and bearded guys hanging around in uniform remnants, selling trinkets, handing out pamphlets about the war, perpetuating the myth Vietnam veterans are losers and victims. I wanted no part of that and I didn't owe anyone an explanation.

But one day I did stop to look up a few names and discovered the genius in the memorial's design. To find a name among the 58,260 on the wall you have to look them up on an alphabetic index to find the panel. Then you have to search for their panel and on that panel the names are in order of date of death, but the dates of death are not on the wall so you have to look through all the names. As I searched for the names I first wondered why in the world they didn't alphabetize them to make them easier to find because there were so many names, and I began to be overwhelmed because there were so freaking many names, and by the time I found the names I searched for I realized something emotionally I had previously only known intellectually, that there are so many, and each one has a story. Each one is someone's son, husband, father, brother, uncle. The numbers just don't tell the story. The memorial is a fine one; I'm glad I overcame my reluctance.

FROM MY VIETNAM EXPERIENCE came my understanding of some of life's greatest lessons. One of them is the common virtue of troops fighting a war.

When John Synowsky and Graham Stevens visited me in the hospital and I thanked them for risking their neck to rescue me, they brushed it off, saying "Any of the other guys would have done the same thing." I realized that, of course, they were right. That doesn't diminish the fact they risked their neck for me, but that's just the way we were. That's the way our troops were in any combat unit. We gave ourselves a tall and unspoken order, it seems: *bring one another home alive.*

I didn't realize it then; it took some years of maturing and reflecting, but that was one of the great lessons of my life. Adm. Chester Nimitz summed it up well in 1945 as he marveled at the brutal punishment taken by US Marines fighting for control of Iwo Jima. He said, "Uncommon valor was a common virtue."

One of the men in the famous photo of the flag-raising on Iwo Jima was Navy corpsman John "Doc" Bradley. When his son, James, was researching his book, *Flags of our Fathers*, he interviewed the decorated heroes from Iwo Jima to try to understand what makes a man do things described as "uncommon valor." James Bradley became frustrated because they all told him the same thing in various words that meant, "I didn't do anything special, I just did my job like the other guys." Finally, Bradley came to understand he should have focused on the other part of the admiral's phrase. It wasn't the decorated few that mattered nearly so much as the *common virtue* of all of them, struggling and risking their lives for each other.

America loves to celebrate heroes, to elevate a select few over all the others as special, and I think in doing so they miss the point entirely. It is the *common virtue* of all the GIs doing their best to protect each other in combat that is most noteworthy, like it was on Iwo Jima, like in Vietnam, like today in Iraq and Afghanistan. All of them, not just a select few.

The public makes the mistake of treating some vets as special because of impressive medals. Medals are nice recognition and sometimes the deserving actually get them, but medals are not what matters. I received the Purple Heart, Bronze Star and Distinguished Flying Cross and I appreciate the recognition. But I also know I was only doing the things the other pilots did every day and

I know other things matter far more than medals. I have been invited repeatedly to join the Distinguished Flying Cross Society, an invitation I have declined for two reasons. First, others deserve the medal so much more than me. I served less than half of my tour before I was wounded and medevaced. I did my duty, but all the vets I know flew two, three or four times as many combat missions, some who deserved but did not receive the DFC. Second, clubs like the DFC Society serve to divide us, to set some of us apart as special, and I won't do that. We were in it together. If anyone ever treats me as special because of a medal, I will think them naïve while I try to think of something polite to say. The highest compliment anyone can pay to me is to remember that I am *one of the guys* who fought with my brothers and sisters in the Vietnam War.

IN 2009, JOHN SYNOWSKY called to let me know he found Ron Hefner in Newton, Alabama, 40 years after our crash. I contacted Ron, then drove to visit him a week later and discovered his injuries were not minimal as I mistakenly thought since he was walking around in the LZ in 1969 while I was on the ground. Before the crash Ron was a tall 6"4", but after the crash was 6'2" as a result of compression fractures of his vertebrae, and he was fortunate to have been able to walk. He spent nine months in Walter Reed hospital and has had back and neck surgeries. But he continued to fly in the Army until he retired and undertook a business career, now retired from both and trying to wear out the golf course at Ft. Rucker.

Ron also told me our commanding officer, Major Dunsford, told him there were over 70 bullet holes in our aircraft. Since it was destroyed in place, I don't know who could have told him that other than John or Graham, and I don't think either of them took the time to count holes. They could see holes on only one side of the Cobra, anyway, since it was left side down. Graham did tell me when he visited in the hospital in Vietnam that one panel under my cockpit had 14 holes and I never forgot that number because of my good fortune in not being hit. All we can say for certain is the aircraft took some hits while neither pilot was hit.

Having found Ron, it was time the four of us gathered for a reunion. We put it together in October of 2010 at John's ranch in Texas, just south of Weatherford, not far from where we all started flight training at Ft. Wolters. We four and four other Cobra pilots gathered for two days of fellowship, during which John showed me the .38 pistol I was carrying when shot down. I didn't know he had it. John told me when he helped load me on the medevac I was struggling to get my pistol out of the armpit holster I wore. Not knowing what I was doing, maybe something crazy under the influence of morphine, John said he tried to calm me with reassurance all would be fine, that Ron and I were on our way to the hospital. I don't remember any of this myself, but John said I handed him my pistol, telling him he could write it off. That meant he had an unaccounted for pistol to give to someone who needed it, to carry an extra weapon or trade for something else. He kept it and brought it home well hidden in the interior of one of the big stereo speakers he bought at the PX in Vietnam.

LOOKING BACK, I HAVE dealt with back pain nearly every day since my crash in 1969; I long ago accepted it as part of my life. I bowled for a few years in a social league, just to be with friends, and tried to hide the pain. I played golf for years, and many times the pain forced me to sit out the last few holes. I drank too much for many years to relieve pain at night so I could sleep. Routine tasks like mowing the lawn, raking leaves, moving furniture or just standing at the sink to wash dishes often requires a break to sit for a few minutes to let the pain subside.

Even after all this time, damaged nerves can deliver peculiar results. I became very familiar with referred pain; when my back is cramped, like sitting too long, my right leg and foot get an escalating ache that eventually becomes unbearable and I must move. That right leg goes numb sometimes and I have to walk it off. Certain spots on my buttocks have been numb since that day in 1969, and don't tell anybody but there are a few permanently sensitive spots down there that, if touched just so, make me suddenly pee a little, an awkward trigger if it happens in public when shifting clothes do the touching.

Doctors told me long ago I would have problems when I was older as arthritis formed on the injured areas of my spine. They were right; I now have Degenerative Disk Disease. Getting fat hasn't helped.

But don't mistake any of that for a complaint; those are just facts of my life, and I tell you these details to emphasize how the many who are wounded in action may have permanent complications as part of the price they paid.

As for me, I consider myself one lucky guy. I'm lucky to have been born in the USA. I was lucky to have lived the day I was shot down, because I didn't burn and the enemy didn't get to me, and because I wasn't hit by any of the rounds that hit our aircraft. I am lucky to be walking after the way my back broke. I am lucky all to hell and back to have served with such fine men, the ones I think of as "The last of the cowboys," and lucky to have served my country as a snake pilot. I'm proud of my role and the pinheads who wear blinders and say it was an immoral war can kiss my ass.

Now that I no longer avoid vet gatherings, I am in frequent touch with some guys who also fought in that war, and some will be friends as long as we live. There's something comforting about their company. Maybe it's because we understand things about each other nobody else can.

I NEVER KNEW WHEN I killed the ox what an important reminder he would be to me of the cost of stupidity. Some will always see my killing the ox as part of American evil in Vietnam, more concerned about the ox than our national purpose and the hard choices involved. That's one kind of stupidity. We permitted, by our own president's order, the sanctuary and safe passage of our enemy's supply lines through North Vietnam's Haiphong harbor, then on the Ho Chi Minh Trail through Laos and Cambodia. You don't win a war by protecting your enemy's assets, which is worse than stupid. Some are more concerned that we called our enemy *gooks*, *dinks* or other insulting shorthand names, or that we casually used profanity, than that we were killing each other with vigor, and that's pretty stupid. Most Americans can't be bothered to discover how they were misled about the Vietnam War, that the things they still believe are not quite true, and that is really stupid since realizing how the news gets twisted by politics makes for more skeptical and informed citizens. The popular belief that an inept US military lost the war is an egregious type of stupidity on two counts: first, because it fails to recognize our fighting forces did a fine job and never lost a major battle, and second because it lets the real culprits off the hook, the dishonorable politicians whose foolish policies got thousands of America's sons killed, gave away the war, betrayed our ally and turned their backs as the communists turned murder into mass production.

Here's a different flavor of stupid that countless Vietnam vets have experienced. In 2005 I was in the Jenco golf cart store in Jenkinsburg, Georgia, discussing plans for the stretch golf cart they would build for my family. Another customer, an older man, was telling a story from his exploits as a fighter pilot in WWII. I complimented him on his role in that war and told him I felt we had something in common because as a Cobra gunship pilot in Vietnam I flew the same kind of close air support missions for ground troops as he was describing in WWII. The man looked at me like I was something on the bottom of his shoe and walked away, as if his service in WWII was honorable, but mine in Vietnam was not. That's stupid.

MY WIFE JULIE AND I married in 1993, neither having had children in prior marriages. In 1998 we traveled to China to adopt Melanie at 12 months old from a Chinese orphanage in the city of Hefei, in Anhui province just west of Shanghai, in the middle north part of the country. In 2003 we returned to China to adopt Kristen, also at 12 months old, from a Chinese orphanage in Yangjiang City, in Guangdong province on the southern China coast near Hong Kong. Having children brought new thoughts about the Vietnam War. What brand of revisionist history will they read in their schoolbooks? How would inaccurate conventional wisdom lead their thoughts about my brothers who fought that war? That is part of my motivation for writing this book.

At this writing Melanie is 13 years old and Kristen is eight. The progressing aches and pains from my Vietnam adventure interfere with my ability to do the things a dad wants to do, like teaching your kid the right way to swing a bat, and exacerbating my tendency to be impatient and grumpy. I take meds for the pain and Mom steps in for me. The first time Melanie saw me walking with a cane when I came to her school to pick her up from elementary school, she yelled "Dad!" and ran to me. I was touched that she was so concerned that my back hurt and I needed to lean on something but she whispered, "Dad, people will think you're OLD!" I guess it's a kid's job to be embarrassed by her parents.

Having kids with Chinese roots has led me to become familiar with some aspects of Chinese culture, like the Chinese Zodiac, the cycle of 12 years with each year named after one particular animal. People born in the year of an animal's sign are considered by Chinese people to have the traits of that animal. Melanie was born in the year of the ox. Kristen was born in the year of the snake.

Whether reminded by my children or by my veteran friends, it seems the snake and the ox are with me to stay.

Photo courtesy of Terry Garlock, all rights reserved

Terry Garlock is a CERTIFIED FINANCIAL PLANNER^{TM} professional. He lives with his family in Peachtree City, Georgia.

Postcard to America

FROM 1968 UNTIL 1972, the 7/1 Cav aided an orphanage in Vinh Long operated by Catholic Nuns. Most of the orphans were mixed race, fathered by white, black and other American troops with Vietnamese girls. You can't stop sex with rules.

The Vietnamese people, being very ethno-centric, would not lift a finger to keep these kids alive.

We realized the US would inevitably withdraw, South Vietnam might fall to the communists and many of these orphans would have short lives on the streets. We worked to facilitate adoption of these kids by American soldiers and their families in the US, and a group of Army and Air America pilots began working to transport these adopted children to their new families.

As I was returning from my tour, I tried to shepherd four of these children to their new homes in America. My four little charges dwindled to two. One child suffered from diarrhea so severe that she could not get a visa to enter the US. Vietnamese red tape blocked a second child.

I left Vietnam with two beautiful children, both in diapers. We boarded a 747 at Tan Son Nhut Airport bound for San Francisco. The Pan Am stewardesses moved us to first class where we shared an almost empty upper cabin with a few embassy types who suffered with me through countless diaper changes and endless crying.

When we finally arrived in San Francisco, the local USO and a group of Army wives met me. The wives provided me with fresh diapers and formula, since the two babies were too small for solid food, and cared for them while I out-processed. The USO staff and the Army wives warned me to change into civilian clothes to avoid a hostile reception by protesters. I changed clothes and gathered the two kids for our flight to their waiting adoptive families. When I went back through the San Francisco airport, my close-cut hair, duffel bag and pilot's helmet bag gave me away. Three young women berated me, called me a baby killer, and accused me and my comrades of killing the children's parents and siblings in Vietnam.

It was a preposterous accusation, but I have never forgotten that insult. I had watched the kids in that orphanage from 1968 to 1971, knowing without our help those selfless Nuns could never have managed. Without adoption by American military families many of those children would have starved on the streets when Saigon fell. Not until the media focus on the crash of the C-5A three years later did most of America understand the fate of children fathered by American soldiers.

My scars from the bullet and shrapnel have long healed and are almost invisible. The stab deep in my heart from my own countrymen will never heal.

Dan Britt, Powder Springs, Georgia
C/&/1st Cav, Vin Long, Vietnam 1968 & 1971

The Burden of Angry Skipper 6
William Neal
Hague, Virginia

THE CHAPTER ABOUT NORM McDonald, titled *Grunt Melody,* accurately describes the life of a drafted grunt in the jungles of Vietnam. I can attest to that because Norm was one of my machine gunners when I was Company Commander of Delta Company, 2/8 Cav operating in the jungles of eastern III Corps in 1971. My call sign was Angry Skipper 6.

To Norm and countless other grunts in Vietnam, the war was a lousy experience to be endured, an unpleasant, physically trying and dangerous duty, something to get through, to finish, to put behind them to get back to the "world." However, for many it was also the defining experience of their lives. Some admit it, some don't.

A career Regular Army officer, though, looked at the Vietnam War through a completely different lens. Vietnam was a chance for combat duty. The war was a unique opportunity to practice and hone our fighting skills and tactics, to test ourselves and what we had learned, to prove our competence in the military profession under real pressure, to lead men and do our work to accomplish an objective under fire, something that cannot be simulated.

This war changed the pattern of combat leadership.

Small unit leaders like squad leaders, platoon leaders, and company commanders had unprecedented latitude on how they fought their battles. There were few open field engagements between large units like in much of WWII and Korea. Most battles were short, under jungle cover of varying density, often surprise contact at very close quarters, and exceptionally violent with both sides firing their weapons at full automatic from 30 meters or closer.

To their utter frustration, battalion commanders, brigade commanders, and division commanders were relegated to observing battles from above in their C&C (command and control) choppers where they could often see nothing but smoke from a firefight drifting above the trees. In many ways it was unprecedented closeness, yet at the same time it was unprecedented distance and disengagement from the real action.

As small unit commanders we walked the same jungle trails with our grunts, but we tried to ignore physical discomforts and concentrate on the job assigned to us, the mission to get done through our men and the resources and tactics available to us. Up through the chain of command all the way to the generals, officers received orders from the level above them and had responsibility equal to their rank. One thing was constant at all levels; when they moved arrows around on maps and gave us our orders, it had a direct impact on the misery our men endured and the injuries they suffered. When we planned our actions and personally led them in jungle combat, our decisions sometimes determined who would live and who would die. We could never predict casualties, of course, which is a good thing because COs are human, too.

Sometimes, long after a battle a CO second guesses himself with 20-20 hindsight, and so it has been with me for thirty something years as my mind flips back to one day in April, 1971, when we found ourselves in a fight for our lives with a tough NVA unit, and I often want to turn back the pages of time and make different decisions in the first moments leading up to that battle. If I could make that magic happen, maybe some of the good men who died that day would be with us now to celebrate their lives and accomplishments. If only . . .

VIETNAM TOUR 1, 1967-68

WHEN MY PATH CROSSED with Norm McDonald's in March, 1971, I resisted taking command of another rifle company in the bush. I had already served my time leading units in the jungles of Vietnam and I felt I was due for a battalion or brigade staff job that would help move my Army career along. Perhaps the story would make better sense if I start at the beginning.

I graduated from Drexel Institute of Technology in June 1966 as a Distinguished Military Graduate and commander of the ROTC detachment there with a degree in Commerce and Engineering. That entitled me to receive a Regular Army commission. But the Army had no regular army engineering slots, so my choice was to take a reserve commission in the Corps of Engineers or a Regular Army commission in the Infantry. I accepted the Infantry commission and went through the Infantry Officer's Basic Course and airborne training at Ft. Benning.

My first duty assignment was platoon leader in C Company, 1/506th Parachute Infantry Regiment, 101st Airborne Division at Ft. Campbell. One brigade of the 101st was already on the way to Vietnam, and the other two brigades were training to go. In the spring of 1967 all my fellow junior officers were getting deployment orders to move out with the Division, but I didn't. No one at Division Headquarters seemed to know why. So I ended up calling Army Personnel Operations to inquire and discovered I was slated to go to Germany because of my engineering background. I asked, "How do I get out of that chicken-shit assignment and get to Vietnam?" Personnel said that the only way I could get out of the assignment was to volunteer for Special Forces or advisory training, so I volunteered for Special Forces. But the Army decided differently and sent me to advisor school at Ft. Bragg, then to Vietnamese language school at Ft. Bliss. While I was being prepared to be an advisor to ARVN (Army of the Republic of Vietnam) units, I was totally pissed about not being able to deploy with my Division.

FINALLY, I ARRIVED IN Vietnam in October, 1967, and was assigned to Advisory Team 87 with the 18th ARVN Division headquartered in Xuan Loc, a provincial capital east of Saigon in III Corps. From there I was assigned as the 2nd officer to a four-man advisory team with the 1st Battalion, 43rd ARVN

Infantry Regiment. The battalion was commanded by an ARVN Lt. Colonel and normally had around 500 Vietnamese officers and enlisted men organized into four rifle companies, a recon platoon and a heavy weapons platoon.

We trained the ARVN battalion for high mobility operations. Heretofore, most ARVN battalions were tied up in defense of key villages and population centers, locally and throughout all of Vietnam. Up until that time, the only high mobility units the ARVN's had were a few airborne and ranger battalions.

Vietnamese troops in new mobility battalions received extra pay for motivation and to compensate for separation from their families. We had a mix of some professional soldiers along with the regular ARVN grunts. We also had a platoon of Montagnards ('yards), a few Hmoung tribesmen, and a few Cambodians. The 'yards and Vietnamese hated each other by tradition, but the 'yards were so good in the jungle one of them could walk up and tap me on the shoulder before I ever knew he was there. Some of our NCO's were former Viet Minh and I was never certain of their loyalty, but they were good at their job.

Locally we called these units *Strike Forces* or *Strike Battalions*, not to be confused with the CIDG strike units formed by Special Forces, which were entirely different with a different mission. We trained them for several weeks on movement and tactics and especially heliborne operations, and then we embarked on our first operation, a combat assault by helicopters into a remote area south of Xuan Loc.

Just as our choppers were landing we immediately got into a firefight with a local VC unit, my first firefight. The ARVN's did well in fighting with courage, tactically sound decisions and execution under fire. As the 2nd officer, the junior NCO and I would normally be up front with the lead unit. We would be in radio contact with the senior advisor who was typically busy working with the Battalion Commander and S3, coordinating artillery, bomber jets, helicopter gunships and anything we could get out of the 11th Armored Cav, the primary US unit in our AO. Sometimes we found ourselves out of range of normal 105 and 155 artillery support, so I ended up being one of the few people who would call in 175mm artillery - the big stuff - for close support, which can be risky. They were in fixed positions in Xuan Loc. I made sure I knew the gun target line and that our position was parallel to that line, so a long or short round would miss us. 175s were far more accurate laterally than they were on distance.

I would learn that when Vietnamese troops were properly trained and had good leadership and air support as American troops did, they fought well, too, but their weak suit was land navigation. They couldn't find shit on a map. I would often walk up front behind the point team to keep them in the right direction. When I was up front, before I became senior advisor, when we got in contact I found myself in the middle of a firefight, and there were a lot of them.

Our mission up until just before Christmas was to be a screening force for the 9th Division, which was pushing straight out Route 1 east to Phan Thiet on the coast. The VC and NVA had closed that entire road from Xuan Loc to

Phan Thiet with ambushes and had blown all the bridges and wrecked the railroad tracks running roughly parallel to and north of the road. The 9th Division moved on the road's axis, clearing vegetation using Rome Plows out about 300 meters on both sides to minimize the risk of ambush, and re-built the bridges as they went. Our job was to provide a screening force to their south, in many areas where allied forces had never penetrated before. We operated in those jungles continuously from late October. We got out of the field just before Christmas, and moved back to our battalion headquarters in Xuan Loc. During this operation we had several contacts with the D440 VC Battalion, and they were good. They hurt us; we lost some good men to them. I know, because most of the medevac helicopters were US and I would run the medevacs from the ground.

We lived in the jungles, searching for the enemy who was infiltrating, establishing base camps, and generally terrorizing the locals.

We lived on ARVN rations, which consisted of lots of rice, canned sardines, and canned pork that somehow escaped being diverted to the black market. One of the staples of the Vietnamese diet is Nuoc mam, a pungent fish sauce poured on rice and almost everything else. I learned to love it. Nuoc mam is made by catching batches of a small fish, like a sardine or anchovy, in nets and layering them in a barrel between layers of salt, then letting them ferment in the sun for several months after which the liquefied result is ready to eat with rice or anything else.

We foraged for other food and this is where I learned a lot from the Montagnards and Hmong. They could survive in the jungle on their own and we all benefited from their knowledge and hunting skills. They taught me to dine on big porcupines, peacock, jungle chickens, and snakes. Now and then the 'yards would catch a big python, *numba one chop-chop!* We also learned to sniff out VC and NVA supply depots using their own trails. So, on occasion, we dined on some fine canned salmon from Japan, canned vegetables and meats from all over Eastern Europe, and even French brandy on one occasion.

When we found a case of Courvoisier I was surprised to be completely drunk after just two drinks. I wonder if it had anything to do with my body adjusting to the new diet and physical regimen. By that time my weight had gone from 215 to 165. I liked what I was eating and was in the best condition of my life.

WHEN AN ABC NEWS crew came to the field inquiring about the combat performance of ARVNs, SSG Zipper showed his wisdom by dragging me away from them on the pretext of checking the perimeter. Our senior advisor and senior NCO told ABC the ARVN performed well, but the prominent news piece reported just the opposite, causing quite a stir. The senior advisor and senior NCO were relieved of their duties, leaving me and SSG Zipper to run the show. That isn't how I wanted to advance.

We had been in the jungle for two months, moving continuously and only stood down one day, next to a 9th Division firebase for a Thanksgiving truce. SSG Zipper managed to swipe a whole turkey from a conex cooler at the firebase, and we made a gift of it to the Vietnamese Battalion Commander, describing it as a very big chicken because there was no word in the Vietnamese language for turkey, and describing how the "big chicken" was a traditional meal at an American Thanksgiving. Later that afternoon, he graciously invited us to join him and his staff to eat the Thanksgiving turkey, and I was looking forward to a traditional roasted bird, though I knew the trimmings would not be there. To my utter surprise, the Vietnamese cook had diced the entire turkey then cooked it in oil with sesame and vegetables and herbs. It was not what we expected, but it was good.

In the middle of January we got a new senior advisor and new senior NCO, so SSG Zipper and I reverted to "junior" status again, which was just fine with us.

AT THE END OF January our Vietnamese allies prepared for their biggest holiday, Tet, or lunar new year. We received instructions to put combat operations on hold because of the supposed cease-fire, and advised to keep alert because our enemy was decidedly losing the war and there was suspicion they might use the cease fire for a desperate surprise attack. We pulled back into our compound in the city of Xuan Loc. We were positioned on the south end of the airfield behind a berm across that area, thankful for a reprieve from combat operations, and writing off the intelligence as just another one of those overblown reports from some armchair staffer in the rear. My battalion was manning this position on the eve of Tet when the VC attacked in treacherous violation of the cease-fire they had negotiated.

Xuan Loc was one of the first targets of the Tet attack because it straddled Route 1 into Bien Hoa and Siagon. The NVA and VC wanted to cut off the 11th Armored Cav, based at Black Horse south of Xuan Loc, to prevent them from moving up to help defend Bien Hoa Air Force Base just northeast of Saigon. There was a hell of a fight for Xuan Loc. The initial attack came from the south right at our positions. We beat them back with the help of a Spooky gunship. Then they attacked from the other three directions, all around the city.

By pure happenstance, elements of the 3/5th Armored Cav was laagering in Xuan Loc that night for the Tet truce and was assigned to cover the west berm around the city, right along Route 1 with their A-Cav 113s. They absolutely devastated the massed VC that attacked from that direction out of the adjacent rubber plantation. I led one of my companies to clear the area the next morning, and there were enemy bodies strewn all over the place, many hit by the .50 cal machine guns from the A-Cavs. Nasty!

During the initial firefight to the south, I was ordered to move a unit over to open the large gates of Xuan Loc to let elements of the 11th Armored Cav Regiment roll through to head west on Highway 1 toward Bien Hoa to defend the huge Air Force Base. They got hit by numerous ambushes the VC and NVA set up at various points on the route to Bien Hoa, but they rolled over those ambushes and arrived before dawn to defend key areas in Bien Hoa. The next morning enemy attacks on the Bien Hoa base were so fierce, with sappers coming through the concertina wire, the US was firing 105mm canister artillery rounds right down the airstrip.

All over South Vietnam, in more than 100 locations the enemy attacked at a time they had negotiated a truce, at a time they knew civilian defenses would be at a low, when ARVN troops on leave would be at a peak, when families were gathered for the holiday. The enemy had been substantially weakened by continuous losses on the battlefield, and the Tet of 1968 attacks were a move of desperation, with high hopes the civilian population would rise up to join them against the central government and the Americans. But the popular uprising never appeared. Instead, the South Vietnamese were generally appalled by the VC and NVA treachery. And they were further appalled by the wanton murder of civilians in each of the places where the VC and NVA got an initial foothold. In Xuan Loc alone, where the VC units broke through the northern defensive perimeter and got into the central city, it was reported to me that they systematically executed nearly 100 civilians.

The attacks of Tet were fierce. I was in 39 firefights my first tour, at least 10 of them defending Xuan Loc and pursuing the enemy during those few days of Tet in 1968. All over the country the VC and NVA were defeated by US and ARVN counter-attacks. The VC losses were so severe, losing 50,000 dead and 100,000 more wounded, that the Viet Cong were never again an effective fighting force. They lost so many men, especially their political cadre, that the NVA essentially took over their guerilla campaign throughout the South.

After the smoke cleared from Tet, the enemy was soundly defeated but the American media came to their rescue. Tet was a screaming victory for the US and ARVN armed forces, but the media reported it as a dismal failure. Reporters had a narrow view of the events of Tet as they gathered in protected clusters in large population centers, reporting on the same small events repeatedly instead of getting the larger story correct from a multitude of battles. They focused on a few isolated images repeated endlessly on TV news reports. Certainly no reporters showed up in Xuan Loc to report on how the ARVN and US forces had utterly defeated the enemy on that battlefield.

At Xuan Loc we decisively won the battles, but there were some very nasty fights, especially in the built-up areas inside the town. I never want to do that again – urban fighting is a bitch. Many of our ARVN troops died bravely defending their city, their people.

On 5 February, 3/52nd ARVN, another Vietnamese Strike Battalion, got chewed up outside the provincial capital of Ba Ria, just north of Vung Tau. The entire American advisory team and the entire battalion command was killed or wounded. Part of the battalion made it back to Ba Ria and part was still holed up in an old French fort. SSG Zipper and I were pulled out of Xuan Loc and sent by helicopter to get the ARVN Strike Battalion back into Ba Ria and coordinate an airlift back into Xuan Loc. We landed in an Australian defensive perimeter and the Aussies, who were heavily engaged with the enemy, took us to the old French fort in their armored vehicles so we could get them organized. They had been through a hell of a battle and lost half their men, killed, missing or wounded. We organized the troops and force marched the five or six klicks to Ba Ria. At that point, I became Senior Advisor to 3/52nd ARVN.

Tet was a wild and wooly time and somehow I made it through without a scratch. When things settled down we took the 3/52nd back into training mode since we had absorbed so many replacements – many of them fresh out of Vietnamese jails for various non-capital crimes. As training progressed we resumed jungle patrols and cordon and search operations.

Bill Neal at the Xuan Loc airstrip just before in-country R&R
Photo courtesy of William Neal, all rights reserved

TIME PASSED, MY REPLACEMENT showed up, and I took an R&R to Hawaii to spend a week with my wife. When I returned the Senior Regimental Advisor was taking leave and left me in command of the 52th ARVN Regimental base camp of Gia Ray on the east side of Signal Mountain east of Xuan Loc. In early August, when I had just 59 days left in country, we were watching a football game on a small TV. The game ended around midnight and as soon as we turned the TV off I heard the whump-whump-whump-whump of incoming mortar fire. I figured it was probably just harassing mortar fire, but instead of typical 60mm mortars they turned out to be the big 82mm, far more deadly, and I knew we were in trouble.

Next came rockets and recoilless rifle fire into our compound and I knew we were in a real fight with a large enemy unit. I had two brand new Lieutenants with me, waiting to be assigned out. We all pulled into the command bunker and tried to use the radio to connect to the Vietnamese and American command net. The radio didn't work.

Our hooch was on fire. I timed the incoming mortar fire and when it seemed I would have a few seconds clear I ran to the burning advisor's hooch, grabbed a radio, a couple M-16s and some bandoliers of ammo and ran back toward the bunker, not knowing the dinks had already come through the wire and were all over the compound. I realized the shit was all over the fan when I saw one dink aiming a B-40 rocket at me when he fired. He missed me but hit a tank of propane gas right behind me. When it blew up, the blast and fire ball hurled me toward the bunker and burned my arms, legs and the back of my body. A sergeant darted out of the bunker and pulled me inside with the radio.

When pain from my burns hit me I was no longer functional. The NVA hit us when the weather was bad; the low ceiling prevented any close air support. Another captain was stuck there with me because the bad weather grounded his helicopter flight, and he took command when I was hit. God, I wish I could remember his name because he saved all our asses that night.

The NVA overran the ARVN artillery compound nearby. The ARVN NCO training camp held and our guys put up a hell of a fight from the central bunkers and held it all night. A medevac helicopter tried to get in to pick me up and was shot down. They tried again at first light. I was out of it with morphine but the medevac brought with him six gunships and several A1 Skyhawks shooting the place up while the NVA were also getting beat back with artillery fire. The medevac set down right on top of our bunker and some wounded ARVNs handed my heavily bandaged body out of the bunker and up into the chopper. As we lifted off and flew out the helicopter took a bunch of hits and one round lodged in my arm. It was nearly spent because it didn't even break the bone. I don't know if it was slowed down by passing through someone else before it hit me, or it was a ricochet.

That was my 39th and last firefight of my first tour in Vietnam, which was over, at least my part in the field. The medevac helicopter took me to the 24th Evacuation Hospital in Long Binh where they diagnosed flash burns, mostly 2nd degree but some 3rd degree, over 37% of my body. They found several pieces of shrapnel in my back, hidden amidst the burns. I was there several days until stabilized, then flown on a hospital transport C-141 to a hospital in Japan for about 10 weeks.

Being treated for severe burns is a miserable, painful process, and I don't remember everything because I was sedated so much of the time. They would give me a shot of morphine then put me in an immersion bath then take me out and put me on a long steel table and use surgical instruments to scrape the dead skin off. Then they would apply Sulfamylon anti-infective cream to treat the burned skin and to prompt new skin growth, but it felt like being burned all over again. As the new skin would grow they would use long surgical tweezers to pull and stretch the new skin to cover more area, all very painful. We did this twice a day, shot up on morphine.

Finally, they flew me to Walter Reed Army Medical Center in Washington, DC, near my home in Maryland. Because burn patients are so susceptible to other infections, they put me and one other guy who had been severely burned, a helicopter pilot, on the generals' ward in relative isolation with just me and him and a few retired generals. My Dad and uncles would come visit a few times a week and bring a big cooler of beer, so my treatment there for burns over a period of weeks was as enjoyable as it could be. I returned to duty in early November and was assigned to the 24th Mechanized Infantry Division at Ft. Riley.

While I was at Walter Reed I started reading the Washington Post newspaper. I was shocked! What was being reported about the war was not even remotely comparable to what I had observed or experienced. I couldn't believe that they got it so wrong. I have never trusted the reporting in an American newspaper since.

VIETNAM TOUR 2, 1970-71

THE ARMY WOULD HAVE sent me back to Vietnam after 12 months stateside, but they tended to let guys who were wounded seriously have extra time to recuperate. After I got out of the hospital and two years of stateside duty, I got orders for my 2nd Vietnam tour.

Many things had changed since my first tour, including me. Anti-war sentiment had spread and deepened, nurtured by news reports with an agenda, strengthened on college campuses where students with a personal desire to avoid serving seemed to suddenly sprout youthful indignation at the war. Even sensible people became weary of a war that went on year after year with no end in sight.

These drifts in our society seeped into the Army through the draft, and squeezing such thoughts out of draftees became more difficult. Pot use found its way to Vietnam and I heard rumors of growing heroin use in the rear units.

These changes were subtle but even among career officers attitudes had changed. In my first tour I was energized and enthusiastic about our role helping the Vietnamese repel the insurgents and invaders from the north trying to take control of their country. By 1970 I had already seen the handwriting on the wall that our pullout was politically inevitable. That makes a big difference in the morale and motivation from top to bottom. It also makes a big difference in the lives an officer in the field wants to put at risk. These were the thoughts I brought with me on my 2nd tour in Vietnam.

GENERAL ABRAMS HAD TAKEN over from Westmoreland and he initiated the incursion into Cambodia in the late spring of 1970. There were reports of heavy fighting in III Corps around the Iron Triangle, Tay Ninh and War Zones C and D – areas I knew about from my first tour.

When I arrived in country in early September of 1970 I was assigned to the First Cavalry Division's 2nd Brigade, intended to be the S2 Intelligence Officer. Here's where that fits in the hierarchy - in the Army we used a simple numbering system that represented the staff function we served in our unit at the battalion or higher level: S1 for personnel, S2 for intelligence, S3 for operations, S4 for supply, 5 for executive officer and 6 for the Commanding Officer. That meant when we were talking to someone on the radio, whether we knew them or not, when their call sign included the number six, we knew we were speaking to the Commanding Officer of that unit. I was now a captain and that S2 staff assignment would be a good next step in my career. I already had more than enough combat experience from my first tour.

When the Brigade Commander interviewed me, he told me before I could step into an S2 role in the Cav I needed to spend a little time in the bush to see for myself how the Cav operates. I caught a chopper ride out to Firebase Audie west of Song Be in III Corps, near the Cambodian border, and hooked up with a rifle company, Delta Company, 2/12 Cav. This was a relatively flat area mostly with low ridges covered in jungle.

There had been electronic intelligence of enemy movement and the area was bombed. Delta Company was tasked for bomb damage assessment. The CO was a 1LT, and though I outranked him I tried to keep my mouth shut, watch how they operate and let him run his company. We did a combat assault of about 125 men on a flight of helicopters into an LZ only big enough for two helicopters at a time. When we hit the ground and took up positions I saw wire on the ground and I knew the enemy were most likely still there; they never left their commo wire behind, rolling it up to take with them was one of the last things they did before they moved out. I also knew it was a sizable VC unit to use wire communications.

I told the LT the VC were still here and showed him the wire, but he said "So what?" and ignored my concern. As we were departing the LZ, the concealed enemy opened fire with 60 mm mortars and small arms.

The enemy broke contact after a brief firefight. We medevaced two wounded soldiers and we moved out on our BDA mission but soon had contact again and more wounded, then again in what seemed a running firefight. I tried to give the LT advice without telling him what to do, telling him, "You know, you're not fighting NVA, you're fighting VC. While they are retreating they like to drop off snipers and small ambushes; they don't stand and fight like NVA. But when the snipers and ambushes make you stop, they come back around and drop 60mm mortars on you. You should plan accordingly." He didn't appreciate the advice and ignored it. But that's exactly what the VC did.

Three days into the mission one of our firefights was serious and we lost some guys. The enemy dragged their wounded into a base camp and abandoned them as the unit escaped. They knew their wounded would slow us down as we had to wait in place for medevac helicopters to take the wounded prisoners to US hospitals. When that was done the LT and his CP crew and one of the platoons headed out following an enemy blood trail and I felt they were going to walk into a trap.

I hustled up to the front and said, "El-Tee, this is not a good idea. The VC bought some time, left a trail for you to follow right into their ambush and this goes up into a ravine." The LT ignored me, took the bait and they continued up the ravine. I went with them, thinking I might help. Then the VC sprung their trap - they had planted a claymore on the back side of a tree waiting for the lead element to walk by it. Fortunately, the point element and the tracker dog they had with them alerted on the claymore. The VC blew the claymore then opened up from both sides of the ravine. The FO – forward observer to call artillery adjustments – was hit right next to me, the RTO right in front of him was also hit. I grabbed the RTO and pulled him behind a log as a medic started working on him. Rounds were chewing up the foliage all around us. Then the medic got hit. From the screams and yells forward, I know we were taking heavy casualties. Then I heard somebody yell that the LT had been hit. I took the FO's radio to try to vector in the inbound gunship support.

We were split in half and I assumed command of the rear part of the unit and immediately set up a defensive corridor to get the lead element back. We got them all out and medevaced. Incredibly, none were killed but 11 were seriously wounded. We thought the LT was dead with a bloody head wound, but he woke up as they were getting him ready for the body bag. His helmet and liner had done their job, diverting a round that ricocheted around and cut his scalp bad enough to make him bleed like a stuck pig and knocked him out. Initially the LT refused to be medevaced, but was ordered to be extracted.

One of the RTO's walked over to me with a PRC-77 and handed it to me. The Battalion Commander told me to take command.

I received the Bronze Star with V – for valor - for my actions that day, and I ran Delta Company in the bush for the next two months.

IN THE MIDDLE OF November I was reassigned as CO of Headquarters Company, 2nd Brigade. That really pissed me off. I asked the brigade XO what I had done wrong to deserve being relieved from my command. I had already been a Headquarters Company Commander at Ft. Riley, my first assignment when I left the hospital, and this was going to do my career no good at all – it's a thankless job and most of the battalion staff officers outrank you. I protested all the way up to the Brigade CO, Col. Louisell, but I was told I had no choice and I would find out why I got that assignment soon enough.

Well, I did find out. 2nd Brigade was standing down, going back to the states. Not only was my promised S2 staff job disappearing, but I had to get everything cleaned up including a property book that was a disaster, lots of things missing and many items in our possession not on the books. Since I had previously been a Headquarters Company Commander, the Brigade Commander figured that I knew how to solve problems like that.

I brought in a warrant officer supply specialist and we became probably the biggest horse-traders in III Corps for about two months. When a helicopter was shot down, the paperwork might say it had four AM radios and a dozen mattresses on board and magically the problem of the missing radios and mattresses was solved. I was short a number of "field kitchens" but I had two radio towers we didn't need and not on the books, so I just had to find the right trading partner. I found some signal guys in Bien Hoa who lusted after my radio towers and somehow they came up with the field kitchens I needed plus the toolkits my aviation platoon was supposed to have but didn't.

I guess fighting creative paperwork is better than dodging bullets for a little while, and when the dust settled our write-offs were minimal and the problem was solved. I was more than ready to find a new staff job opportunity, but things were changing so I went down to 1/7 Cav, rear headquarters which was staying in-country with the 3rd Brigade to see what jobs were available. I was a valuable commodity, a senior captain with lots of combat experience. S4

– supply and logistics – was open and we agreed I would take that job just as soon as I returned from an R&R trip back to the states to see my family.

When I returned to Vietnam in early March, 1971, I reported to 1/7 Cav and told the young captain I encountered I was the new S4, and he said I was to go see Colonel Louisell right away because, as he said sheepishly, "I'm the new S4!" Oh, crap, here we go again.

The colonel told me, "I've got a rifle company out there in trouble; the CO was medevaced with bee stings of all things. We're running it with the platoon leaders but they need a CO with combat experience and you're it." I said, "Colonel, I spent my entire first tour in the jungle, I spent the first three months of this tour in the jungle, and I really would prefer a staff job. I need some battalion or brigade staff time, more combat time isn't going to help me."

"Captain – you've got your orders." "Yes sir." End of discussion.

THAT'S HOW I BECAME Angry Skipper 6, CO of Delta Company, Second Battalion, 8th CAV, 3rd Brigade (Independent), 1st Cavalry Division. Back in 1967, when unit designations or call signs were frequently changing, Delta company of the 2/8 Cav was operating in a different region and acquired the call sign of Angry Skipper simply because the 1st Sgt. nicknamed his perpetually pissed-off boss *the Angry Skipper*. When call signs were frozen in place to reduce confusion, Angry Skipper stayed with the unit. When I was assigned as CO I became Angry Skipper 6 in the bush, which was appropriate because I was, indeed, pissed off much of the time.

Pissed off or not, like any good officer I did my job. We patrolled the jungle, much in the fashion described by Norm McDonald, and we encountered the enemy now and then.

Out of the 10 months I spent in the bush in my first tour, and the eight months I spent in the bush my second tour, there was one day that stands out in my memory as one I wish I could relive to make a few adjustments. That day as Angry Skipper 6 was April 20, 1971, when Norm was on leave in the states.

This is my best memory of that day. I invite others who were there to clarify with their own recollection and their own perspective and improve on any errors I might make in this account.

I freely admit what has always troubled me about this day is a tactical mistake I made. A commander should never split his forces in the face of an unknown enemy deployment. I did that and four men died; many more were wounded.

I was no stranger then to losing men, wounded or killed. I suppose that day weighs heavily on me because these were American young men. In my first tour I was in many firefights and we lost many Vietnamese troops and several advisors wounded and killed. There is a vast difference to me, but not based on any discounting the lives of Vietnamese, who are like us and every other people, wanting to live free, wanting a better life for themselves and their children. I was helping them fight to defend their own country and proud to do it. The American grunts under my command in my second tour were mostly serving where they did not wish to be, late in the war with America's commitment to win a distant memory and withdrawal on the horizon. Their lives were dear, and I lost some that should not have been lost.

Here is how it happened.

FIRE SUPPORT BASE FANNING was in the last stages of completion. It was located just north of Route 1 to the east of Gia Rai Mountain in Long Khanh Providence. Our battalion, Second Battalion, 8th Cav and an attached battery of 105 howitzers, were its sole occupants.

On the morning of April 20 I was ordered to take Delta Company to patrol due north to see what was out there and report any signs of enemy activity. Intel had only vague reports of possible enemy activity in the area.

Delta company had three maneuver platoons - 1st (White Skull), 2nd (Rifle Range), and 3rd (Wild Cat). There was no operational weapons platoon at this time. Each platoon's strength in the bush was typically 25 to 30 men. On this day, Skull Platoon had been peeled off for another mission nearby. As the two remaining platoons moved into the tree line we went into *trail* formation with Range in front, followed by Cat.

The terrain in this area was relatively flat with double canopy jungle. There were also some small open areas - both natural, and those made by native tree cutters. We moved rather easily through the area. There were no real signs of enemy activity although it was obvious that woodcutters had been in the area in the recent past. By early afternoon we were five klicks out when lead elements of Range reported a very heavy east-west *trotter* running on the south side of a fairly deep and wide stream.

Range put out security to the flanks. I took the CP forward to take a look for myself. The CP usually moved behind the lead platoon with the Company RTO in front of me, the Battalion RTO behind me, followed by the FO and his RTO.

The trotter was definitely a major trail, almost two meters wide with recent heavy use. I moved Cat up and ordered Range to cross over to the north side of the creek to see what was there. They reported a similar trotter on that side. We called it in to Battalion HQ.

At this point I was really curious about these trails. I had patrolled this area extensively in 1967 and 1968 with ARVN Strike Forces and I had never seen a supply trail this big anywhere around Gia Ria Mountain. But, the provincial capital of Xuan Loc was about 30 klicks to the west and Gia Ria village was about 10 klicks to the northwest. So I figured this was probably a supply trail between Xuan Loc and some NVA supply base out east of Gia Ria.

I ordered Range to send a patrol west about 500 meters, cross over the stream to the south and patrol back to Cat's position. Likewise, I ordered Cat to send a patrol east 500 meters, cross over the stream and patrol back to Range's position. Both patrols were to call in before crossing over the stream so we would have only one crossing going on at a time. Everyone else was to take up a hasty defensive position to hold tight.

At about 400 meters to the east, Cat's patrol spotted dinks bathing in the stream in front of them, their uniforms and AK's on the bank. I ordered them to lay low unless the dinks started moving out, then fire 'em up.

I then ordered Range to get its patrol back ASAP and Range Six to get ready to deploy east, parallel to the trail on the north side of the creek. The remainder of Cat and the CP were to deploy east on the south side of the creek. I wanted to move in parallel, so we had to wait until Range recovered their patrol. Cat's patrol was feeling like they were really hanging out with backup over 400 meters away through thick jungle - I could hear it in the RTO's voice.

At this point I strongly suspected we had caught a VC supply unit simply taking an early afternoon break, which was typical, and we were going to make them pay for carelessness. A problem was that I didn't know which side of the stream they were going to end up on.

The FO informed me there was a Pink Team nearby, a cobra helicopter gunship covering a scout helicopter that would fly low to spot the enemy's trail. I told him to get the low bird over the stream and vector him in to where the VC were bathing, and see if he could see anything, or draw some fire.

The low bird got over the area just as Range's patrol was coming in. I ordered everybody to immediately move out to the east parallel to the stream, but off the trotter. Cat's patrol reported the dinks were scrambling from the stream and I told them to fire 'em up. Cat's patrol was reluctant to initiate a firefight until they knew we were closer in to back them up, and they held their fire. I remember grabbing the radio mike and growling at them to fire up the dinks.

The low bird reported he didn't see anything, but it was really thick along the streambed. He didn't draw any fire either. Range, on the north side of the stream and Cat (main) and the CP on the south side of the stream were moving approximately parallel to the east. We were both moving outside the trotters, so visual confirmation was impossible. We judged our progress by how far we had moved from Range's original crossover point.

At this point I still felt we had caught a VC supply unit taking a break. I wanted to hit them aggressively, but I didn't know which side of the stream their main unit was on.

Cat's patrol then opened up. I heard just one or two M-60 rounds go off, then a long pause before a few M-16's started firing. The M-60 had jammed. There was no immediate return fire from the enemy.

Then Cat's patrol, the lead elements of Cat (main), and the CP started taking sporadic return AK fire. I figured we had the enemy on the run.

Just then my RTO said Range had spotted a hooch or shithouse. That took a second to sink in, then it hit me like a brick - base camp! That meant a dug in enemy force possibly far larger than I had been thinking. I needed to stop Range before they stumbled into it.

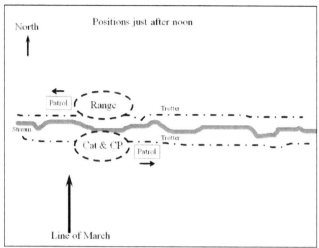

Image by William Neal, all rights reserved

I was just reaching for the radio to stop Range when all hell broke loose on the north side of the stream - heavy AK and M-16 fire. Within seconds I heard at least two Chicom claymores blow, accompanied by the screams of my wounded men. The volume of small arms fire from both sides was really heavy, rock 'n roll all the way.

Intensifying fire from across the creek pinned down Cat and my CP. Only then did I realize that the north bank, occupied by the enemy, was higher than the south bank at this position; the enemy had a distinct advantage. I remember noting our fire was probably going too high and we needed to fix that quick.

Word was shouted that Range Six, LT Bott, was hit.

The firefight on the north side was horrendous. I tried to maneuver the CP element to get across the creek, thinking I should get over there to sort it out, but the crossing in front of us was under fire from the enemy on the north side. The sound of whack-whack-whack told me the tree I was standing behind was taking incoming AK hits, and most of the CP went to ground.

The radio said Range Five was down. SSG Dillon was the most experienced and aggressive platoon sergeant in the company. I knew Range was now in very serious trouble with both their platoon leader and platoon sergeant hit; my people on the north side of the creek were taking a beating.

I ordered Cat's trail squad to move back down the stream to the west, cross over, and come in behind Range, help get them out of there, and hold the crossing site no matter what. The remainder of Cat was to continue firing from the South to keep the dinks' attention.

The battalion RTO told me the battalion log ship (helicopter) was inbound with another basic load of ammo to kick out near our position.

I remember noting that somebody on the north side was raising royal hell with an M-60. I hoped it was one of ours. I remember thinking that maybe it wouldn't be too bad if we still had an M-60 engaged, though the volume of M-16 fire had become really weak.

I ordered Cat to have a fire team back off to the South, find a good kickout point, secure it, and pop smoke to mark a drop spot for the log ship. I planned to fall back into this log site and dig in if I could get Range back across the stream.

Cat's patrol took the kickout and also reported finding an opening big enough for a one-ship LZ further to the south.

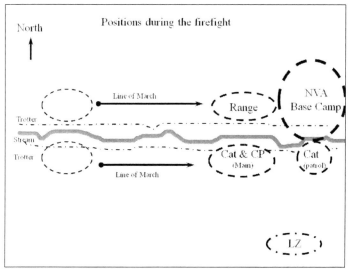

Image by William Neal, all rights reserved

I moved the CP back west to the crossover point. By the time we got there Range's unwounded men were bringing the wounded across and shuttling them back to the perimeter set up by Cat's detached unit. The medics were going crazy trying to stabilize as many as they could. They had a priority one medevac for one of the men, a severe head wound.

About then they brought over SSG Dillon. I walked over to him. He looked up at me and said "Skipper, you gotta get me outta here! I'm hit bad." He had one of the worst shoulder wounds I had ever seen. I could only tell him that that he had to hang on; we had to get a priority one out first. I don't recall seeing LT Bott, but was told he was being medevaced, which later turned out to be untrue. I didn't know it then, and just recently learned that LT Bott, despite his wounds, was continuing to effectively command the platoon from a small depression in front of the enemy bunker complex and direct Range Platoon's withdrawal. He was the last or next to last man to cross over to the south side of the creek.

I kept trying to get an account of who was left on the north side. The last men across said there were three guys up front who they couldn't get to and said they were definitely KIA.

I weighed the possibility of trying to get them out, going back over to the north side of the creek and trying to move up in front of a dug-in enemy force that obviously had their shit in order. I couldn't risk more men under these circumstances to retrieve bodies, and I prayed the reports were right, that they were not still alive and waiting for help. I ordered the pull-back into a perimeter around the small LZ.

Not personally knowing for sure the status of those three men has been the source of great personal anguish since that day.

As best as I could determine, about half of Range was wounded or KIA, including Range 6 and Range 5. The three KIA's were CPL James Melvin Cardwell, CPL Danny George Drinkard, and CPL Joseph Lindsey Hall.

Cardwell and Drinkard had been walking point and caught the full blast of the NVA claymore mines. Hall had initially been further back in Range's formation. When the shit started to hit the fan Hall scrambled all the way up the line past the other troops to help Dillon. He was moving off to Dillon's left to take up a firing position when the Claymores were detonated. How Dillon lived, I will never know.

Gary Collins, who passed away in 1999, was one of the two machine gunners that I heard laying down a base of fire. Apparently he took an M-60 from a wounded machine gunner and continued to lay down a base of fire to cover the withdrawal. Steve Schneck was the other gunner. He was severely wounded in both legs, but also continued to lay down a base of fire that allowed Range to get back across the stream. Steve's comment was, "Well I couldn't walk so I just kept on shooting until I ran out of ammo."

Stan, Joe, Steve, and Gary were real heroes that day and were awarded the Silver Star for their deeds and courage. Joe Hall's Silver Star was, of course, awarded posthumously. There were other heroes. Doug Hilts came all the way from the rear with his M-203 – a 40mm grenade launcher attached to an M-16 - to help suppress the enemy and cover the withdrawal. Medic Harvey Brothers was working on the wounded under extremely heavy fire. Tom Vollmar, who was acting field supply sergeant, went across the stream several times to help retrieve wounded, then moved back to the supply point and ran the medevacs. And there were many others – all working to save their comrades and preserve the unit.

As soon as the medevacs had evacuated our wounded, we radioed for artillery support and fired everything on that base camp the FO and I could get our hands on, including the big 8-inch howitzers.

Then I got an air strike by six F-100's with 500 pound bombs. The first F-100 sortie was short and we took a lot of shrapnel and blast from that bomb; the Forward Air Control officer who flies an observation plane to mark the target and guide in the jets got one royal ass chewing over the radio from this very Angry Skipper. I guess he was the first convenient target for my utter frustration with how badly this battle had turned out. The FAC simply replied "new pilot," prompting another white-hot burst of my abuse. The rest of the sorties seemed to be on target. Years later Tom Vollmar said I threatened to shoot him down right then and there.

I DON'T REMEMBER A lot about the next several hours when we got dug in, redistributed supplies and did other routine things a damaged company does to get itself back in fighting condition. I was trying to figure out how I could have lost the best part of an entire platoon in 10 minutes. COL Louisell, my old Brigade Commander in 2nd BDE and now Deputy Commander of 3rd BDE (Independent) flew in and told me we'd done a good job. It didn't stick with me that day. I knew too well this was no time in the war to be assaulting enemy base camps. US withdrawal seemed inevitable, the political will to win this war was not in the cards, and sending our men home alive to their families became far more important to me, and on this day I certainly had failed.

Skull platoon was choppered in to join us and bring us back up to strength. LT Martin, Skull's platoon leader, took charge of perimeter defense.

Before dusk we moved further east, parallel to and south of the stream, beyond the enemy base camp. We quietly set up a defensive perimeter around an outcropping of rock. I didn't want to be where the bad guys knew we were. The FO and I stayed up all night directing artillery fire on that damn base camp. Eight-inch shell fragments were occasionally bouncing off the rocks above us. We fired a bunch of artillery that night, some of it *danger close*.

The next morning COL Bacon and Command Sergeant Major Cruthers flew in to join us. They wanted to see for themselves what we were up against. I think they also wanted to make sure I was still fit to run the company.

I ordered Skull to lead us over to the north side of the creek to see if anybody was still in the base camp and to recover the bodies of our three KIA's. As they approached the creek they took some sporadic AK fire and I ordered their withdrawal back to the defensive area. I wasn't going to take any more chances on this place. I couldn't believe the bastards were still in that base camp after all the ordinance we had dropped on it.

We pounded the place with more artillery and F-100s the rest of the day and into the night. The next day we tried it again. This time we got in and recovered our KIA bodies, which were exactly where they had fallen when hit at the beginning of the firefight. I got word that our priority one medevac, CPL Gerald Stanton, had died on the 21st. We are not sure whether his head wound was due to shrapnel from the Claymore, or from rifle fire.

When we finally were able to enter the enemy base camp, we found it was extremely well built, and laid out with double interlocking fire zones throughout, and bunkers with six feet and more of mud and log overhead cover. This dink unit knew their stuff. They even left signs behind saying "GI GO HOME" and "NIXON HAS ABANDONED YOU." They were clearly a well trained NVA unit.

When I got back to FSB Fanning, I was informed the Battalion had inserted Alpha and Charlie Companies into expected blocking positions to intercept the withdrawing enemy. Alpha Company hit pay dirt and killed a bunch of them, capturing papers that revealed we had hit the headquarters of an NVA rear service unit and their security detachment and the headquarters for the 33 NVA Regiment. An NVA rear service unit was equivalent to one of our brigade headquarters and the security unit is typically the best and most seasoned fighters who are being given a break from their normal jungle fighting duties.

From Battalion's and Brigade's point of view, this was a very successful engagement – we found and severely disrupted the activities of the main enemy force in our AO. I wish that I, too, could say it was a success, but I can't.

To a CO making decisions in combat, casualties are something to be managed. There's always a trade-off between mission and safety. But no matter how much we compartmentalize to keep our head clear, never far away is the realization that every man killed is someone's son, husband, brother or father; every man wounded takes it very personally and his wounds might radically change his life. How SSG Dillon lived after being shot three times and hit with Claymores, I will never know, but here's how he remembers it, in some of his own words:

> ... as we moved down the edge of the stream, I was 3rd in line followed by my RTO. We were about 30 meters in and moving parallel to the stream when we found a fresh NVA latrine, a hole dug in the ground with bamboo sticks over it and a hole in the middle, just built within the last few days. Uh-oh, we must be right on top of a bunker complex and we had to be real quiet and try to back on out without being seen. I had been in this situation before and knew this was bad news.
>
> I reached over and got the receiver from the RTO and called the latrine in. The jungle was dense and visibility very limited. Then as I handed the receiver back to the RTO the first rounds were fired from the enemy bunker complex hidden from view. One of those first rounds hit under my chin and spun me around and to the ground. The blood from my chin spilled all over me and I remember thinking it felt like a bee sting. I spread the men out as much as I could and placed the machine-gun over to my right and we returned fire. Most of the squad had moved up and were now returning fire. Jim Cardwell was just to the right of me and Dan Drinkard to the front and left of me within a few meters. Joe Hall came all the way up from the rear when he heard the gunfire and asked me, "Where do you want me to go, Six?" Even though I was now Range Five he called me Six because I was in that position before Lt. Bott came on board.
>
> I placed him to the left of me about 10 meters. As I look back on it I think these brave young men would have followed me anywhere. I yelled back for someone to bring up a LAW - light anti-armor weapon - so I could clear out some of the brush blocking our view of the enemy.
>
> Then, a few seconds later I thought I heard the enemy fire an RPG rocket, but I later learned the explosion was a Chinese mine like a Claymore. Sometimes you don't see the one that hits you. I remember a ringing in my ears and feeling a hot flash rush through my body. I yelled out for everyone to get back and reached over to grab Dan Drinkard and looked to my left for Joe Hall. I noticed they were not moving and saw Jim Cardwell motionless also.

The M-60 and the rest of the squad were still firing and most of the squad were trying to withdraw. Range Platoon was terribly shot up but we were still returning the enemy's fire with ferocity. We were fighting for our lives.

I got up and fell flat on my face and rolled over. I couldn't find my arm and I thought I had lost it but found that it was under my back, broken and shattered at the collar bone. I remember after rolling over I pulled my arm out and placed it on my chest and stomach. I tried to get up and walk again but couldn't. Later I found that my knee cap was blown out. I couldn't breathe and could hardly hear, since as I discovered later my lung was punctured and my ear drum had ruptured from the explosion. I just knew that if I didn't get back under some cover I wasn't going to make it. I was lying there struggling when Stan Sargent came by me, holding his head and walking in a daze. I yelled for him to get down but it was as if he couldn't hear me or comprehend.

The medic, Doc Brothers, finally got to me and I remember screaming over the gunfire in pain and holding up my hand and noticing my little finger on my left hand was just hanging by the skin. I told the medic, "I think I've lost my finger, Doc."

He wrapped me up and I can still remember the rush of pain. The men filed by me getting back, most of them wounded.

I was lying there by the stream and one of the men, an Indian from North Carolina, called to a grunt I will leave nameless to help carry me across the creek. The grunt was frozen in a state of shock and could not move. The Indian kept telling him to move and then he said, "Listen, if you don't help me carry Sergeant Dillon across this creek, I'm going to shoot you myself!"

The grunt got up and helped him carry me across the creek. I remember looking down at the rocks searching for a large one to crawl behind if the soldiers carrying me should be shot. I remember saying out loud to my wife, as if she could hear me, "Oh Gail, Gail, Gail, I'm not going to make it!"

We made it to the other side and I told Cpt. Neal, "Skipper, you have to get me out of here, I'm hurt really bad!" He told me I was second in line for medevac, that he had to get Stan Sargent with his head wound out first.

I remember the lift support that they wrapped me up in had the straps missing and one of the guys took off his belt to strap me in. On the way up as the medevac chopper raised the hoist cable to draw me up through the tree branches I remember sliding a bit out of the rig and holding on to the cable with my good hand to keep from falling to the jungle floor. The chopper was taking small arms fire from the enemy, AK rounds singing as they tore through the trees, but it stayed in its still hover, waiting for me. I made it. The chopper was loaded down with wounded guys and lying beside me was Stan Sargent.

They took us to some firebase aid station where they stripped me of all of my clothes and put my arm in a balloon cast and cut off my wedding ring among other things. I guess when they got us stable they were running with us on stretchers back to the chopper where they air-lifted us to the rear hospital.

I'm lucky to have lived through that day.

He is, indeed, and as it turns out, I'm lucky he lived, too. Stan Dillon and I reconnected 27 years later in 1997 at our first Angry Skipper reunion. I learned that he spent over two years in various hospitals, mainly the VA center in Columbus, Ohio. Besides bullet holes he lost one and a half fingers, his spleen, one knee cap, and a lot of his right shoulder. Dillon had 38 operations, over 30 of them on his shoulder. He was eventually discharged after about four years in and out of medical facilities, then worked for the government until he retired. We get together several times a year, sometimes with Bill Bott and Bill Harrington. Stan's wife, Gail, and my wife, Carolyn, are very close. I'm thankful Stan Dillon lived through that day.

SOMETIMES WHEN THINGS GO wrong you want to stop the world for a while, take stock of things and try to get your head right. But the world won't stop. Calendar pages kept flipping and our patrols and firefights with the enemy continued for the remainder of my tour. Out of all those days, all those firefights, all that killing and dying, April 20, 1971 is the day I always come back to, running it over again and again in my head.

Looking back at that day, there have always been good reasons to forgive myself and let this ghost fade away as just one incident among many that I survived while serving my country. My superiors concluded it was a good operation, that Angry Skipper's actions uncovered an important concealed enemy base camp near a vital population center, that we did well in this surprise encounter with an unknown enemy of that size and skill. I also know in my head I should be at peace since I was good at my job and if I had not been there to deal with the situation the result for my men could have been much worse.

Those are sound intellectual arguments that wither in the face of my gut feeling that I knew better than to divide our force with a deep stream at a critical time in the face of an unknown enemy position, and I cannot hide from my gut. My gut thinks for itself and knows me best, and it knows I should have made a different decision. If I had, maybe more of those men who counted on me would have come home to live out their lives.

It is such a small thing, a swift few seconds of decision, just one among so many.

That small thing is my personal ghost. No matter how comfortable my life may be at any moment, no matter how many years pass, always lurking just around a corner is the brief flash of memory and that sinking feeling, that irrational wish that I could turn back time for just one day, for just an hour, for just a moment . . . if only.

Photo courtesy of William Neal, all rights reserved

Bill Neal lives with his wife Carolyn in Hague, Virginia. He left the Army in October, 1977 and recently retired as senior partner in a consulting firm headquartered in Atlanta. He spends as much time as he can chasing fish on the Potomac River near the Chesapeake Bay and the surf near his beach house he named LZ Hatteras, on the Outer Banks of North Carolina.

Postcard to America

IN JULY OF 1969 I flew home from Vietnam with two of the four other guys I trained with in the US to be an advisor to Vietnamese ground units. Our thoughts on that flight were often about the other two, good guys who preceded us on their own trip home in body bags.

We had been warned anti-war demonstrators might try to provoke us, and reminded of our duty as officers to conduct ourselves with restraint at all times. We were also advised to get out of our uniforms quickly to avoid trouble. We did just that, changing into civies in the San Francisco airport men's room, but not before some young lady demonstrators called us murderers when we politely declined their offer of a flower.

Conversely, on the flight to Chicago, a Korean war vet insisted on paying for our drinks.

On leave in a small town in Pennsylvania, I was struck by the TV news I saw for the first time in a long time, pounding their daily drum that the Vietnam War was illegal and wrong. That point of view was a complete surprise, but the people in that small town were ready to face down anyone bold enough to demonstrate against the war. The division in the country was deep, division created by the anti-war crowd and their soulmates in the media.

Ted Reid, Peachtree City, Georgia
Colonel, US Army (Ret)
Advisor to Vietnamese units, 1968-69

Thank You Mrs. Gussye Roughton
Larry Bailey
Chocowinity, North Carolina

I HAVE BEEN ASKED many times what special ingredient qualifies some men to become a US Navy SEAL. You might think I would know the answer to that question, having been a SEAL platoon leader in Vietnam and for three years the commanding officer of Naval Special Warfare Center (NSWC), the SEAL school in Coronado, California. But I have never been able to pinpoint the answer, and I am doubtful of anyone who says they can. The Navy spends a lot of time and money screening SEAL candidates, and the training washout rate is very high. With the incentive to reduce cost and save time, there have been many studies trying to better define what will make one man succeed while another will fail.

For three years while commanding the SEAL school, I searched for the magic ingredient. How could we predict which man had the strength and desire to survive nearly six months of the hell we call Basic Underwater Demolition/SEAL Training (BUD/S)? Strength and conditioning were certainly factors, because so much of the challenge was physical, but beyond a certain level of fitness, being stronger of body didn't seem to make a lot of difference. Sometimes the men perceived to have superhuman strength and endurance were the first to "ring the bell," meaning they quit. The mystery remains.

Of course the training regimen was exceedingly hard, and most reasonable men would quit at some point. But we weren't searching for reasonable men. We were searching for the toughest, most resilient men who had the best chance of surviving the harshest combat, men who would continue a mission long after most reasonable men would give up.

I will always remember three young men who started SEAL training with me in Little Creek, Virginia, in 1963. Besides being good guys, all three were stronger and more athletic than me. They could run and swim circles around the rest of our class and complete three times as many push-ups and pull-ups as we could. But all three quit early in *Hell Week*, and to this day I can't explain why a relative klutz like me completed training and they did not, except that perhaps I wanted it more. Mental commitment is a major factor in SEAL training.

Military missions usually involve difficulty, risk and the application of violent force. When the level of difficulty exceeds the skills and training of ordinary military units, or when secrecy is paramount, the mission goes to specially trained units like Navy SEALs. The ability to deal with harsh conditions, get in and out quickly without detection, gather intelligence, destroy targets, perform rescues and other tasks under extreme time pressure and unsavory odds are the reasons for the existence of special operations units like the SEALs. Since they are expected to do what appears to range between the highly difficult and the impossible on assigned missions, the training to prepare them is sufficiently tough to make most men turn and walk away.

The first weeks of BUD/S training are relatively easy, with intense daily physical exercises, distance running, swimming increasingly longer distances, medical and physical screening and learning basic military values like honor, integrity and discipline. Then things get tough. Trainees learn to launch boats in heavy surf; swim at length with hands and feet bound while retrieving objects on the bottom with their teeth; and keep up 120 hours of continuous extreme physical strain and mental stress with just four hours of sleep. They conduct exercises and perform physical activities such as long-range patrolling, hydrographic reconnaissance, five-mile swims, combat diving under extreme circumstances, weapons and explosives training and practice, simulated capture and severe interrogation stress by enemy forces, helicopter jumps in diving gear, infiltration techniques, close-quarters battle, recon and intelligence methods, drops and pickups in high-speed boats, land navigation, hypothermia conditioning in cold water, fast-rope descent and rappelling, high-altitude and low-altitude parachuting, mountain climbing and running the mother of all obstacle courses with constantly improved times. During all this they run and swim miles every day. For some, the mind games are the worst. For example, the daily physical push beyond prior limits might bring instructor assurance that the next 200 yards are the last, but then they might low-crawl through the surf in full gear another couple of miles.

Interspersed with intense physical regimens are courses in unconventional warfare, counterterrorism, counterinsurgency, special reconnaissance, demolition techniques, underwater demolition, structure penetration, sniper methods, advanced first aid, radio communications and dozens of other disciplines a SEAL has to know.

These are just some of what a Navy SEAL must endure and master. By the end of the program, 75% have quit, leaving the few to earn the right to be called SEALs.

I explain these training elements not to thump my chest, but to help the reader appreciate the commitment of the men I served with - the finest men I know - and to help you understand this: when a SEAL team encounters extreme hazards or obstacles on a tough mission, one thing that helps them persevere is knowing they had already learned in their training how to push themselves to achieve the impossible.

Many of the clandestine combat missions SEALs carried out in Vietnam we cannot talk about, but I can relate an account of a declassified SEAL mission in the Delta region of Vietnam where the heat and humidity are legendary, the water-soaked land is flat and green, and rice, bugs and snakes grow with vigor.

THIRD PLATOON, SEAL TEAM TWO Det Alpha, was excited! An officer from a special POW recovery staff in Saigon, Vietnam, had flown into the airbase at Binh Thuy, just outside of Can Tho in IV Corps, to see if the little band of

SEALs stationed across the road would be interested in rescuing two American POWs.

This was a mission with our names written all over it. It was in mid-May, 1967, and until now the IV Corps SEALs had not had much action due to the reluctance of the American command structure to commit US ground forces in that part of South Vietnam. This mission, however, was important enough to convince the higher-ups that we should make the rescue attempt.

The intelligence underlying this mission was solid. A South Vietnamese Army corporal held prisoner in a Viet Cong camp had escaped and told a story of being held with two American prisoners, one Hispanic and one black. Previous American intelligence reports had indicated that two such POWs were being moved westward through the Mekong Delta to a less-exposed site in Cambodia.

IV Corps, the southernmost military sector of Vietnam, was primarily flat rice-growing terrain broken by intersecting tree lines and canals. It was not an ideal environment for sneak-and-peek operations over long distances, and long distance was very much involved in this operation. The distances were great enough that any rescuers would have to be at least partially transported by helicopter.

Time was of the essence. The Vietnamese soldier had escaped four days earlier, and the VC could figure out that the Americans might launch a rescue attempt; the longer the delay, the greater the chance that the POWs would be moved. We had to get moving fast!

The principal reason that we were so far behind the time curve was that the POW rescue group in Saigon had shopped the mission around before finding the one unit – ours - that was willing to give it a shot. Once the "go" signal was given, our SEAL detachment had the support of everything the Department of Defense and the Vietnamese military could provide.

The first evidence of this support came in a high-priority photo reconnaissance mission flown on Day One by an RB-47 photo-recon aircraft. Within less than 24 hours, we had overhead photography of the camp from which the Vietnamese POW had escaped, and intelligence resources indicated that there was indeed a camp of some sort located there. The escaped POW pointed out the exact spot where he was held, and we then launched a helicopter visual recon of the area.

Once again the escapee correctly identified the area, and it was consistent with his story. We then were confident that he really *had* escaped from a POW camp. In addition, he agreed to go with the rescue team of six SEALs and two Vietnamese attached to the SEALs, thus demonstrating his good faith.

There was a major problem, though. The POW camp was a long, *long* way from our launch point, and we had decided that the mission had to be launched on the night of Day Two if we were to have any chance of finding the Americans. Going cross-country on foot was totally impossible; we *had to* get fairly close to the target by helicopter.

And then there was the problem of navigating across a featureless flat plain without knowing exactly where the helo had dropped us off. We wrestled with this potentially show-stopping problem for several hours before an idea occurred to me. I actually had a sense that Gussye Roughton, my high school geometry teacher, was talking to me. She was telling me to use what she had taught me - that I should use arcs and lines and points.

Voilá! It came to me! In moments I had figured out how we could get to a known point from an unknown point. The solution to that problem lay in figuring out the coordinates of that unknown point. I inquired of our Saigon planners about having several "Firefly" helicopters participate in the operation. They said, "Anything you want, Lieutenant," and they arranged for these helicopters with their gazillion-candlepower mounted searchlights used to illuminate enemy targets at night in support of our mission.

When the Firefly pilots came into the briefing room, I pointed out to them several known landmarks in the area - a Buddhist temple, a confluence of streams, a crossroads, a village - and designated them alphabetically. The temple was Point Alpha; the canal intersection was Point Bravo; and so on. I explained that at certain times I would call on the respective helos to illuminate those points. I would identify a landmark by its code name and give them a "stand by" signal followed by a "mark," at which time the helo called on would paint the landmark with his beacon. I would then get a compass direction and draw an azimuth line on my 1:25,000 map.

I would then do the same thing with subsequent Fireflies, getting compass bearings and drawing azimuth lines on the map. Where the azimuths lines intersected should pinpoint, in theory, where we had been dropped off by the transport helos.

That is exactly what happened. When we were dropped off, we only knew that we were within five to ten kilometers in a southerly direction of our target, or at least that's where we *thought* we were. After exiting the helos and establishing a secure perimeter, I began the geometry problem. It was surprisingly simple. The first three azimuths intersected almost exactly, and the fourth and fifth confirmed it; we knew exactly where we were.

From there it was a simple matter of minding our compass course and counting steps and comparing tree lines with what we had seen on the overhead photography. Moving fast and quietly on foot at night for five or ten miles was no sweat for a SEAL squad, and we arrived at the camp location at sunup, just when the enemy soldiers were making their breakfast. Moving toward their campfire in a classic "L" formation, we took them under fire, killing two. We immediately moved to where the prison cages were supposed to be, but, to our chagrin, there were none.

Interrogation of local peasants indicated that the POWs had been moved the day after the Vietnamese soldier had escaped, which was a prudent move by the VC. US intelligence later confirmed that this was what happened.

The VC we killed turned out to be local force cadre on an order-distribution mission. We found various sets of those orders on the body of the group's leader. I appropriated those documents for intelligence review, but nobody seemed interested in them. I kept the documents, and years later a Vietnamese friend in Baton Rouge, Louisiana, translated the orders contained in the little packet I had kept for thirty-odd years. They instructed some local collaborator in how to teach Marxist principles in his village.

The two POWs we had tried so hard to rescue did, I found out later, return home at the end of US involvement in the Vietnam War. I have often wished I could meet them and tell them of how disappointed we were when we came to visit and found them not at home.

While my SEALs and our attached Vietnamese were disappointed that we had missed carrying out what would have been the only US POW rescue during the entire Vietnam War, we were buoyed by the fact that we had figured out how to do clandestine navigation from an unknown point to a known point. Of course, having the assets of an entire theater of war at our disposal helped!

I'm sure my high school teacher never envisioned such an application of simple geometry. Thank you Ms. Gussye Roughton, for making this mission possible. Sometimes, triangulation is everything.

Photo courtesy of CAPT Larry Bailey, all rights reserved

CAPT Bailey served in the Vietnam War and in various other capacities as a US Navy SEAL. In addition to many other assignments in his Navy career, he was Commanding Officer of Naval Special Warfare Center, the Navy SEAL training center in Coronado, California. He is retired and lives in Chocowinity, North Carolina.

Postcard to America

ON ARRIVAL AT TRAVIS AFB in August 1970 from Vietnam, I was warned to change out of uniform into civilian clothes "for your own protection." That turned out to be good advice since protestors were gathered in the San Francisco airport to harass us, hippies with anti-Nixon and anti-war signs. We got jeers and an anonymous voice yelled "Baby-killers!"

I thought it odd that no one ever said "Welcome home," other than my parents, until around 1985, and then it was from fellow Vietnam vets. In my parent's neighborhood where I grew up, the only person who seemed sympathetic or even curious was a retired British major by the name of Cunningham. Everyone else, even my parents, studiously avoided the subject of Vietnam, as if I had been involved in something quite unsavory and they wanted to pretend I'd just been out of the country on a trip.

Job interviews deteriorated instantly when my recent service in Vietnam was noted. There were no job offers until I finally interviewed with Ralph DiMeglio, a WWII veteran who recalled his own situation in 1945.

I entered graduate school four months after my return and quickly found it was wise to avoid certain professors who had a well-deserved reputation for flunking unrepentant Vietnam vets. My apartment neighbors during graduate school were friendly, but their jokes about my uniform and haircut after I was pulled into the Active Army Reserves in California became a little more insulting if they caught me sneaking out or back in for my drills. I began to think I had a pretty good idea how racial minorities felt about ethnic jokes.

Out of over a hundred men in my Army Reserve unit, I was the only Vietnam vet. Even there, the subject of the ongoing war was taboo, they believed the same nonsense about the Vietnam War as the public believed. I didn't want to talk about it with these idiots, so that worked out fine for me.

Dating one girl several times inevitably opened up personal history, but after my Vietnam service came up there were usually no more dates.

I slowly learned, like countless other vets, just keep it to yourself.

Roger Soiset, Lilburn, Georgia
Rifle Platoon Leader, 199th Light Infantry Brigade, Xuan Loc area, Vietnam 1969-70

Balls of Steel
Nick Donvito
Camillus, New York

ONE THING REMAINS UNFINISHED from my tour in Vietnam. For many years I hoped I would somehow identify the helicopter crew that pulled me out of the jungle when I was shot up, so I could track them down to shake their hand, so I could tell them how I admire their courage and loyalty to grunts like me on the ground. I don't think I will ever find them because there were thousands of air ambulance missions and they had more important things to do than keep detailed records of names, locations, dates and times. Finding them would depend on their memory, and it isn't likely a crew would remember picking me up out of the hundreds of other missions they flew.

I don't know if my rescue helicopter was from a *Dustoff* unit, which were not armed other than the crew's survival weapons, or from a *medevac* unit, which had M-60 machine guns and did shoot back. Those fine points were lost on me and my fellow grunts; we just knew the helicopters with the red cross on the nose and the doors would pull out all the stops to pick us up if we were hit and fly us to a hospital. An hour after being hit in the filth of the jungle miles from nowhere, we would be with doctors in sterile operating rooms. Knowing that was a great comfort. I wonder if the guys in those helicopters know how many of us still tell stories about their *balls of steel* as they came to get us under fire.

I may not be able to find the crew that pulled me out to personally thank them, but I can tell you my story and why I will keep thinking of those guys for the rest of my life. You can decide for yourself what to think of their balls.

FEAR WAS AN UNPREDICTABLE creature, neatly covered and tightly bound by our focus on the job so that most times, even amidst the close gunfire and explosions of a firefight, it seemed noticeably absent but I always knew it was still there, lurking deep in the folds of our gut, poised to make itself known at the time and place of its own choosing much like our enemy. Keeping fear under control comes from training and experience, but occasionally fear would flash through our body in a nanosecond, trying to crush our heart down to the size of a pea. Once in a while fear would slowly settle like heavy air, wrapping us in a cocoon of dread while we tried to ignore it because there was no choice; giving in to fear might make us freeze and get our brothers killed. I was in a lot of ambushes and firefights in Vietnam, I saw a lot of guys shot and ripped apart on both sides, killed some myself, and I learned to manage fear when the shit hit the fan.

Maybe knowing that will help you appreciate what I am saying when I tell you that getting on a helicopter one day was the single hardest thing I did in Vietnam. We had been in the jungle without a rest for over two months and were humping our way back toward a lousy little hole we called LZ Joe.

Helicopters picked us up at an even crummier little LZ hacked out near a rubber tree plantation to give us a ride the last leg; it was about the size of a football field with a tree-line of jungle around all sides. White Skull platoon was lifted out first on the flight of four choppers; each helicopter carried a max of seven armed men. When they came back, my Wildcat platoon loaded and took off while Range platoon waited at the clearing for the return flight to pick them up.

Just as the skids left the ground the enemy opened up. It was a perfect trap. We didn't even know they were there and they seemed to be on all sides. Green and yellow tracers crisscrossed the LZ, scaring the hell out of us. Rounds went zip and whack through the helicopter while we cleared the trees and gained altitude. We were lucky - nobody in our bird was hit, but my buddy Larry Hackett was still on the ground with Range and I worried about him. LZ Joe wasn't far, and when we landed we helped the helicopter crew locate the bullet holes so they could check for damage. There were 17 holes in ours but it was still flyable.

Now the mission had changed. They said the LZ was too hot to go back in to pick up Range platoon, but they were going to take the White Skull and my Wildcat platoon back to reinforce Range, which seemed to be surrounded by a sizeable enemy unit. We barely made it out of there and now they wanted us to go back. The cocoon of dread settled, wrapped me tight and whispered in my head, "I don't want to go. This is going to be a bad one." It didn't make sense that it was too hot to extract Range but not too hot to take us back, but that didn't matter. My brothers were in trouble and I had to go, so I pushed the fear that was trying to strangle me back down in my belly, stepped onto the helicopter and we took off.

There was no landing at the LZ, they just got low and slow and threw us out while the NVA were shooting all over the place. I crawled in the grass lower than I have ever been, clawing my way toward the tree-line opposite of Range, and White Skull platoon took another spot on the tree-line opposite Range, so we could try to secure a perimeter around the LZ. I had never crossed an open field under fire before. The rounds zinging through the grass just above me made me think of cartoons of all things, and I must have become amazingly thin as I hugged the ground trying to merge with the dirt, anxious to pack my whole body under my steel pot with fleeting thoughts of a silent journey home zipped up in a black rubber bag. Scared? Damn right I was scared.

We all made it to our position to form a perimeter and we held off the enemy on all sides. The NVA would probe on one side, then another, looking for a weak spot to get through, but we held them off. It stayed too hot to be extracted and we spent the night with little sleep, the NVA attacks continuing off and on. We expected to end the fight with air and artillery support in short order because battles in Vietnam were brief affairs, most often surprise contact and the enemy disappearing into the jungle when they chose after inflicting some damage.

Not this time.

We were there for a week, surrounded and holding night and day. Our log ship would fly over and drop food, ammo, supplies, a water bag and even the Stars and Stripes newspaper, in which we read that President Johnson was calling a halt to the bombing of North Vietnam as a good will gesture. It was November of 1968, and good will gestures seemed surreal while we were struggling to survive.

One big guy, a real nice guy from deep south moonshine country, got hit in the head and moaned the whole week. That was torture because the LZ was too hot for a medevac.

Every time the enemy tried to get to us, we beat them back, trying to clean weapons in between. My lousy old-model M-16 jammed every time. My buddy Bruce and I would trade off, one on watch while the other cleaned his weapon. Once they hit us while Bruce was cleaning his M-16 and he dropped the cotter pin from the bolt in the dirt, running his finger through the dirt looking for it while he screamed and we took terrible fire.

One night an NVA soldier came strolling into our perimeter to *Chu Hoi*, a term that I'm told meant "open arms" in Vietnamese and was a highly advertised program encouraging enemy soldiers to surrender and be well treated, even rewarded. A surprising number did come over to our side because they were sometimes brutally treated by their own superiors. Incredibly, this young man spoke English and had been a college student in North Vietnam, home visiting his family when he was conscripted. He had a typical buzz haircut in back, but they hadn't given him a uniform yet, he wore a loincloth and t-shirt. Our top sergeant interrogated him and found out our company of 90 men was surrounded by an NVA battalion, about 400 men, but both units had only small arms. Sarge also asked how the man walked so easily past our trip wires and he said he had carefully watched us set them up in the first place and knew where they were; so much for security! Later, when the battle was over, an ARVN officer in a helicopter took the young enemy soldier away. I assume he was just as well treated after he left us. After all he did *Chu Hoi*.

But the battle was not yet over. We had Huey and Cobra gunships, and even jet bombers driving back the NVA, and without them we would have been overrun. One of our choppers with the mail bag was shot down and a pair of Cobras had to destroy the broken helicopter and the bright orange mail bag because of the potential intelligence value; one thing the Army worried about, since the enemy was connected somehow to anti-war radicals in the US, was them finding a source of names and addresses of our soldiers' families back home, and the harassment that might result.

Ultimately, the enemy gave up trying to get to us after a week and moved on. Including that week, we had been in the bush 78 days without a break. They picked us up and flew us to our home at LZ Rita, a short flight of great relief giving me a little time to reflect on what the hell an 18-year-old like me was doing there.

I WAS EAGER TO be a soldier even when I was too young. My prized possession as a kid was a photo of my Dad and uncle Nick, who I was named after, in uniform in WWII. Uncle Nick was an infantry Sgt. in Patton's army, killed in Germany and buried in Holland.

When my buddies joined the Army and the Vietnam War was raging, I was miserable and stuck in a Syracuse, New York dysfunctional high school, much like the old sitcom *Welcome Back, Kotter* but without the humor. I was an unwilling student in a broken high school, dying to join my buddies, a bunch of Italian kids, tough guys that ran with a gang. When they came back in their new uniforms I thought, "I've got to go." Every one of these guys would chicken out when it came time for them to serve in Vietnam; they all went AWOL, were dishonorably discharged and spent time in federal prison.

But I didn't know that then. When I learned I could finish high school in the Army, I was gone and enlisted in no time. My parents reluctantly signed their permission since I was only 17.

My Army recruiting sergeant gave me a signed agreement that I would go to airborne artillery, but he gave me my first Army screwing because at basic training I was told I couldn't get that assignment since I was only 17 years old. Thank you very much. Instead, they made me a grunt, an infantryman, a member of what the Army proudly calls the Queen of Battle since no matter how much we bombed or battered with artillery, conquest required grunts with rifles going in to finish the job.

I spent the early months of 1968 freezing my butt off in Army basic training at Ft. Dix in New Jersey. The snow actually blew through cracks in the old barracks walls.

A lot of people complained that the M-14 rifles they issued us were heavy, but I thought it was a beautiful weapon. I could take that weapon apart and put it back together blindfolded. I loved that rifle. I'd only fired a rifle once or twice before the Army, but I shot second best in the company. I think the top score was 63 and I shot a 62. I would later wish I had the M-14 in Vietnam.

SSgt. Goad, my drill sergeant in basic, was like a Dutch uncle to me. He had been with the 1st Cav Division in Vietnam. He gave us lots of advice about staying alive in Vietnam and said the Cav knew how to train their troops and they knew what they were doing. I decided I wanted to go to the Cav, too, but I knew with so many units over there my chances were slim.

I should have tried harder to enjoy the cold in basic training because my summer would be at Ft. Polk, Louisiana, a miserably hot and humid place for Advanced Infantry Training (AIT), but a great place to prepare me for Vietnam. I was issued a brand new M-16 at Fort Polk and shot expert with it, but I didn't like it.

The M-14 fired 7.62mm rounds, the standard for small arms in NATO countries. The new M-16 fired smaller 5.56mm rounds, an advantage in the number of rounds a soldier can fire at the enemy since he could carry more of the lighter ammo. The Army's theory was the number of rounds fired was more important than a larger, longer distance round, especially in a close-contact war like Vietnam. That may have been true, but the M-14 was a fine weapon, and I missed it.

They had *Tigerland* for jungle training, and a mock Vietnamese village. We were trained on the M-60 machine gun, M-79 grenade launcher, LAW light anti-tank weapon to be used on bunkers and other tough targets and even the .50 caliber machine gun. It was excellent training but very tough in the heat and humidity, up at 3:30 AM to run five miles before breakfast just to get the day started right. AIT put me in the best physical condition of my life, and I credit that and the training for saving my skin more than once. That doesn't mean it was fun. Part of our training was crawling through a pigsty and searching through the pig shit with our bare hands for hidden weapons. I swore I would never do that again.

At the end of AIT everyone else in my company had their orders for Vietnam. No orders for me. Where the hell were *my* orders? Finally, on August 30, 1968, my 18th birthday, I got my orders for Vietnam. It seems the Army held up the process just for me since they couldn't send me to Vietnam before I was 18.

I had a two-week leave during which I married my sweetheart. The Army recruiting sergeant had told me bringing a friend to join would get me an extra week of leave, so while I was on leave I brought a buddy to join. When I reminded him about extending my leave, the recruiting sergeant said he couldn't change my leave since I had orders overseas, but he could write a note requesting a 3-day pass for me in Vietnam. Thank you very much . . . again.

My arrival in Vietnam was through Cam Ranh Bay. I was only there for two days and I remember seeing the Air Force area where they had flush toilets and nice billets while the Army had piss tubes and greasy mess halls. When they called my name and said I was assigned to the 1st Cav Division, I was surprised and pleased. They put me on a C-123 and flew me to An Khe, then two days later to Camp Evans where the 1st Cav Air Assault School would teach me the Cav way of doing things in a two week course.

It was a tough two weeks. This time the combat exercises had live ammo and helicopter rocket fire. Our group sat on one mountain watching choppers fire rockets into a nearby mountain to see what it looks like before they taught us how to call a fire mission for gunships and artillery. They taught us how to clear hooches, disarm 105 shell booby traps and, yes, I crawled through pig shit again. We fired a bunch of weapons and we learned how to load on a slick and jump off in an LZ. They showed us what a Cobra gunship can do with its rockets and miniguns. Did I tell you it was hot? Ft. Polk in the summer had conditioned me to Vietnam's heat and humidity, but a lot of other guys had not been conditioned and they dropped like flies from heat exhaustion.

I went from Camp Evans to my unit in the field, Delta Company, 2nd Batallion, 8th Cavalry Regiment, 1st Air Cavalry Division. We were D/2/8 Cav, call sign Angry Skipper. I met the Angry Skipper CO, Major Buddy Gardner, just promoted from Captain, a big red-headed man who welcomed me with a warm handshake and kind words. The company had three platoons that operated in the field, White Skull, Rifle Range, and Wildcat. There was also a weapons platoon that fired mortars, but they stayed in firebases and gave fire support to us and other units in the field. Each field platoon had two or three squads with a gun team and a point team for operation in the bush. Major Gardner assigned me to Wildcat platoon and introduced me to the platoon Sgt. who turned to yell at the men, "Hey, aren't one of you dickheads from Syracuse?" That's how I met Larry Hackett, my good friend until he died in 2006.

Wildcat had two squads at the time, 1st & 3rd. The platoon Sgt. assigned me to 3rd squad and gave me a choice to hump ammo with the gun team or walk with the point team. I knew point was dangerous, but I watched a little guy named Jerry Hornberger, struggling to drag ammo cans and sweating his balls off. I said I'd walk point. I learned the point team was a crazy bunch of guys.

From the beginning I liked the Cav, they seemed to know what they were doing and had an organized, disciplined approach to their missions. But I was disappointed big-time in how they issued me my field gear and weapon. They told me to go into the supply hut and rig myself up a ruck sack. I had to forage for myself and I couldn't believe I had to scrounge through junk. There were no ruck sacks, I got an old butt pack, a poncho and a poncho liner. After I got to the field, a guy who was leaving gave me his ruck sack.

I must have been issued the oldest M-16 in Vietnam, an old AR-15 with the triple pronged flash suppressor, an aluminum bolt and no bolt forward assist, the old model that was notoriously unreliable. The other guys had the newer improved, far more reliable M-16, but we had the old ball ammo that was really dirty and it was terrible on the rifles; mine would jam most times after firing a clip. It froze up often and I had to clean it every day. It was awful compared to the equipment I had in training. One thing you learn fast in the Army is to adapt, improvise and overcome, so I made the best of my unreliable rifle. Looking back, I had many opportunities to exchange rifles with a grunt who was wounded or killed, but it never seemed right to me at the time to take advantage in that circumstance and I kept my crappy rifle throughout.

We operated out of LZ Sharon and LZ Nancy near Quang Tri. A black guy nicknamed *Queen Bee* from Compton, broke me in. I don't think I knew until we continued our friendship after Vietnam that his name is Linwood Queen. On my first night with my new unit it was still monsoon season and raining. I had just lain down to sleep when my platoon Sgt. shook me awake and said, "C'mon, we got a Charlie Alpha!"

I said, "What's that?"

He said, "Combat assault, newbie. You're airmobile now, get your stuff and get on the goddamn chopper!"

Off we flew into the rainy night for my first insertion into an LZ. I was tense, but it turned out to be nothing. That area near the DMZ in I Corps was hot at other times, but in the month I was there we saw almost no combat action. I think we had just one firefight, but I did see a murder. As a retired police officer, I don't use the word *murder* lightly and thinking about it still makes my blood boil.

In a highly populated area near Quang Tri, a bunch of us, including a few other new guys, were walking on a road near a rice paddy. Two ARVN troops had taken a VC suspect, a girl who looked to be about 16 to me, and were interrogating her at the edge of the rice paddy, full of foul water. Maybe they knew for certain she was VC, I don't know, but they beat the living shit out of her and it happened so quick. They forced her head under the foul water then brought her back up to ask a question, then back down again and again until they kept her down too long, she drowned and didn't respond because she was dead. These little pricks threw her body aside like so much trash and walked away like it was nothing. I and several other guys couldn't believe what just happened, we wanted to kill those ARVNs or at least take them in custody but our sergeants and officers restrained us and said we might create an international incident if we interfered. That was the wrong thing to do, we should have shot the bastards. I was disheartened that I had trained so hard to be part of something disgusting like this. The only good thing I can say about it is that I never saw anything like that again and, even though we killed the enemy with vigor while they did their best to kill us, I never saw an American soldier mistreat civilians, even when we found enemy weapons or food caches in a village in a free-fire zone.

During my month in the Quang Tri area I spent a little time at the Sands, as we called it, the sandy beach on the South China Sea. I had never before even seen an ocean. We dropped our clothes and swam a few times with guards posted.

IN OCTOBER, 1968, ON Halloween, we moved the 1st Cav Division, 20,000 men and equipment and all that entails, several hundred miles south to III Corps to Quan Loi, in the Tay Ninh area, to operate around Nui Ba Dinh, the Black Virgin Mountain, the dominant feature in this area with mountain jungle west to the Cambodian border. We moved in just 24 hours to relieve the 1st Infantry Division; that's air mobility!

Our new AO was called *the Parrot's Beak* because of the border shape. Everything would change for the grunts in this new AO. The terrain was steep mountain jungle and the enemy was well-trained and well-armed, North Vietnamese Army regulars in uniform. I never saw any VC guerillas in this area, just the much tougher NVA.

With the rugged terrain, the practice of prior units in the AO had been to wait for the enemy to reveal themselves, but our Cav unit was aggressive and we would take the fight to the enemy invading South Vietnam. Our patrols in this area were intended to find them and kill them as they infiltrated from the Ho Chi Minh Trail in Cambodia, on their way to Saigon and targets in the surrounding area, preparing for attacks to ultimately overthrow South Vietnam.

The jungle was thick, the going would be slow, and our method would be *ambush* or *search and destroy*. The enemy was invading; our job was to find him and kill him.

The day after our move to Quan Loi we did a Combat Assault into an LZ then humped the rest of the way to our new home, a firebase in miserable condition named LZ Rita. A strong enemy attack against the 1st Infantry Division at LZ Rita had occurred a few days before when they were moving out, including a rare human wave attack. Cobra helicopters had greased the enemy in the wires. There were 15 burned out APCs, about 150 rotting enemy bodies in the wire, and the whole place was a mess with pieces of bone and flesh still underfoot. Now this would be our new home, D/2/8 and a Cav artillery unit with 105 and 155 howitzers, so we got busy to clean it up and turn it into a livable and defendable place.

Each squad was assigned a small area, told to dig a hole and build our bunker the way we wanted; it would be our home when we came in from the bush and we had to defend it. So we built it pretty good, we thought, with high sandbag walls and the remaining walls and roof built with whatever we could scrounge. Our hooch was on the perimeter, next to a 155mm howitzer, because when we were there we were perimeter defense. The tree-line was so close to the perimeter it invited enemy attack so we removed trees and brush to move the tree-line back. It was hard labor to get rid of lots of tanglefoot and barbed wire out there.

They built a mess hall by digging a hole with a bulldozer and covering it with a tarp. It seemed like a trap for a mortar round, so no thanks, we ate C-rations at our hooch. Every time a log bird came in we grabbed a couple cases of C-rats, or C's. At 18 years old I did just fine on C's. I loved meatballs and spaghetti, franks and beans and some others, and when we opened the cans they were fresh no matter how old. I ate all the chocolate I could find in C's for energy walking point.

One night the NVA tried to hit us in LZ Rita with mortars, but they couldn't aim worth a damn. It seems our Cav guys were finding and killing the NVA forward observers who were trained to call in adjusting fire, and their mortars were missing us by a mile while we laughed our ass off, that is until one of the mortars started a grass fire. A few days prior, during the human wave attack the 1st Infantry Division used a Claymore mine to blow a 50-gallon barrel of CS gas, tear gas to slow down the enemy. That heavy gas powder now lay in the grass until stirred up by the grass fire, drifting in on the breeze to gas us, so the blind-assed enemy mortar team got us by accident with our own gas. We couldn't see, we could hardly breathe the stinging air that burned our throat and we had a miserable night sleeping in our gas masks. Our only consolation was the enemy probably didn't know they got us good.

While we were building our hooch the artillery teams were setting up the howitzers. When they were ready and had their first fire mission, the concussion from the big 155 next to our hooch knocked our wall over, so we built it again, shorter and stronger. We stayed in and around our hooch when they were firing; their fire missions were usually a few rounds and we learned to keep our mouth open to let air pressure equalize and not let the concussion wreck our eardrums. We were also told not to fasten the strap of our helmet under our chin lest the concussion rip our head off when our steel pot went flying.

Our weapons platoon, the mortar teams, stayed in place at LZ Rita to provide short-range fire support. The mortar rounds had a built-in charge to fire them downrange when they were dropped in the tube. I never shot mortars but the weapons platoon guys told me they could wrap as many as nine little powder bags around the neck of the mortar round, above the fins, to give it a boost to a max range of 4,000 meters. But I never thought mortars could match the accuracy of artillery.

ONCE WE WERE SET up, our ambush mission began and the enemy, who had been moving freely in the region, got a hard dose of Cav reality. In the mornings our Battalion Commander in his Charlie-Charlie – command and control helicopter – would scout the area, confer with his staff and give each company a box of map coordinates to patrol. If known, he sometimes gave us a specific *trotter*, an enemy trail, to ambush that night. Orders went from Angry Skipper CO to platoon leader to squad leaders to the grunts, and our point team would plan the route by heading and distance. Some days the triple canopy jungle was so thick and miserable we only made a hundred meters.

Our heavy load didn't help in the heat. I quickly learned to always carry extra ammo and water. I carried two magazines taped together in my rifle so I could flip it over when the first one was empty, and I carried two M-14 pouches on the front of my pistol belt with five mags each, plus I carried my bandoleers for the M-60, which I cut in half and tied around my waist. I also loaded up a bunch of extra bandoleers in my pack. I carried a lot of frag grenades, too, especially since my rifle was unreliable. I had two frags on my harness, one on each side of my ammo pouches and a bunch across and inside my pack. The old M26 was an excellent grenade, with a big heavy spring compressed down inside that thing and when it exploded all those little pieces flew everywhere. When they came out with the M33 baseball grenade it wasn't half as effective. You could throw them a lot farther, like a baseball, you didn't have to heave it in an arc, but I'd rather arc it and have it kill something than just make a big noise.

Our company moved slowly though the jungle in three columns, right and left flank, Command Post in the middle. This way when we got hit we could easily form a perimeter around the CP and react quickly to repel an attack. Walking point was most risky because it would be first contact with the enemy or booby traps, but once I learned the art of point, I was good at it and felt most comfortable depending on myself looking for signs of the enemy like broken vegetation, footprints, odors and noises but mostly by spotting their trotters. We reduced our risk by rotating who was point, slack and third in line on the point team. The slack man covered the point man and counted paces, kept track of position so the point man could concentrate on detecting the enemy. While the point man looked carefully to the ground, looking for sign of the enemy or booby trap wires, making sure each step was secure, the slack man just behind point kept his eyes to the front, covering point by seeing ahead as far as the jungle would allow.

When I walked point I never used a machete, never even saw one. We didn't hack the jungle, we usually parted the vegetation and carefully stepped through. A lot of the vegetation was dead from heavy spraying of Agent Orange, attempting to kill the jungle to make concealment harder for the enemy, but this jungle had several levels and killing it all might be impossible.

I used my M-16 bayonet to probe and part the vegetation, hacking once in a while when I had to. The Army took M-16 bayonets away from us because a Geneva Convention ruling said the knife could only be sharpened on one side, and the M-16 bayonet was sharpened about 2½ inches on the 2nd side. The NVA had triple-sided sharpened bayonets on their SKS's and AK's that would penetrate body armor, but they took ours away from us. What utter bullshit! I could shoot a guy, blow him up, strangle him or beat him to death but if my knife was sharpened on two sides instead of one it would be illegal to stab him. How stupid. I traded some stuff for one of the illegal bayonets anyway to use for probing on point, and sometimes I would stumble and catch myself on my rifle as a crutch with my bayonet sticking in the ground. I shot my enemy, blew him apart with grenades, tore him into pieces with Claymores, called in the hell of helicopter gunship rocket and minigun fire to rain on his head, had artillery and napalm and bombs dropped on him to turn him into a cinder but I only stabbed my enemy once; I shot the guy, but he was still fighting and trying to get me with his three-sided bayonet, so I stabbed him in the throat with mine.

During the day we pushed through the jungle on our stealthy mission, slowly probing for signs of the enemy, mostly seeking out their *trotters,* their trails through the triple canopy jungle like tunnels. On point I'd hold the rifle out in front of me to part the bush, alert for snakes and huge, colorful spiders. When I saw an enormous spider web I'd be looking a little frantic to find the spider, to make sure he wasn't on my back crawling up to my neck for a bite. Some of those bad boys were as big as my hand. The spiders were spooky but the snakes were more dangerous. Small, colorful pit vipers would kill us with their bite; we called them *two-steps,* which was an exaggerated estimate of how quick the bite would put a man down. I had a couple of close calls with *two-steps.* Once on point the man walking slack told me to freeze because a *two-step* was close. I asked where and turned my head to see a pretty blazing orange snake as big around as my little finger hanging from a tree branch, his head about three inches from my nose. I slowly backed off and so did he. Another time when we took a break I backed my ruck into a tree and started sliding down to a sitting position as I liked to do when a buddy yelled, "Nick, stop!" I was about to sit on a *two-step.*

I was looking for booby traps, too, but didn't find many in this region. Maybe the NVA were too much on the move in large numbers to screw around with little traps like the VC did.

Some guys had problems with leeches but I didn't. I found extra long bootlaces to tie up my legs, starting to wind tight at the top of my boots and around up my leg several times, tied off at the crotch. When we crossed streams I sometimes had leeches on my face, my neck and my back, it was pretty common, but it was no big deal. With a buddy to help one another we just put lit cigarettes to them to make them drop, or we sprayed bug juice on them and they fell off. Bug juice didn't keep other bugs away but did a job on the leeches.

The going was slow and tedious in the jungle. When we found a trotter with fresh sign we would pick our spot and set up for a night ambush, usually an L-shaped ambush to catch the enemy in crossfire without shooting each other.

WE DUG IN EVERY night, one man awake for every man asleep, listening and watching, straining our eyes to see shadows that moved among shadows, waiting for flares or Claymore mines to be tripped by unsuspecting enemy on the trotters, alert for NVA who detected our position and were sneaking close to ambush us as we waited to ambush them. Shoulder to shoulder, back to back in our hole, we learned to keep each other awake, to rely on each other, to trust each other.

As we helped each other stay alive, we became tight. Once our resupply log ship couldn't get to us for three days while our food and water ran out. We got so thirsty, the foul rainwater in the bottom of bomb craters, *bunker water*, began to look good and we got the word that any man who drank bunker water would be court martialed, because it was certainly full of very bad little critters and diseases. One of our guys, John Crocker, was huge, 6 feet 7 inches tall, so naturally we called him *Lurch*. Lurch came to the field with another new guy, Warren Eskeridge. After three days while our food and water dwindled to nothing, Lurch came up with one can of C-rats ham and eggs. We gathered our squad around as we heated it up, and each man got one spoon of that delicious meal. That one spoon was good, and the brotherhood was better. Maybe that's why these men were the best friends of my life long after we served in Vietnam.

Digging in each night took a little time. Before it got dark, we decided on our fields of fire for the M-60s and spread riflemen in between, put out trip flares and Claymores in the best position to set up our ambush. We had a small piece of wood about eighteen inches long to which we nailed five or six Claymore clackers, the detonators. All we had to do was run our Claymore wires to the detonators and plug them in, and we always set them up the same way so everyone knew how to start pounding the detonators to blow the Claymores when the shit hit the fan.

We also used the large can from C-rat franks and beans for a grenade trip. We'd tie the can to a tree, open side up, put the grenade in the can – it fit so perfectly - with firing spoon on the outside, the pin straightened and loosened so just a small tug would fire the grenade, and run a trip wire attached to the pin across the trotter. When they tripped it, the pin pulled out, the spoon flew off to start the fuse and wham! We used that same technique with trip flares. One rule was we better know where our traps are lest we blow ourselves up.

While some of us set the ambush, other guys were digging and filling sandbags. Each guy carried 10 empty sandbags in his ruck. By digging down and building a sandbag wall up, we didn't have to dig nearly so far, which was a good thing because sometimes the nest of roots in the jungle made digging extremely hard. If we had two guys to a hole, we had 20 sandbags to work with. In the morning when we prepared to move out, we emptied the sandbags into the hole to fill it up. We became sandbag experts.

The men on watch at night tried to make no noise because the jungle in this area was very quiet. There were few animals other than insects and snakes, maybe because of the tons of Agent Orange spread on the area trying to kill the vegetation. It seemed to kill the critters, and it killed some trees, but the jungle was, well, robust. So at night we watched and waited quietly, listening intently, waiting for the footfalls that could only be NVA, waiting for the loud violence and bright flashes of Claymores that would hopefully kill them before they tried to kill us. When I hit the clackers on the board to blow the Claymores I usually kept my head down. I couldn't rely on my rifle, so when the ambush didn't get them all and a firefight followed, I might use my rifle or I might just throw grenades.

The NVA certainly weren't used to opposition in the area; we had contact most nights by ambush and usually killed a few or a lot. The body count game in Vietnam was a bit insane, a twisted way to measure progress, killing far more of them than they killed of us. Our higher-ups were dubious about our unit body count, though, it was so high and consistent at an average of 10 or more enemy dead each night, not counting the dead the enemy managed to drag off with them. Our higher-ups learned we were just unusually effective. Our enemy learned that, too, and documents we captured showed they had a bounty on our heads.

If you go by the book, after an ambush you go check the enemy casualties, but it was a good way to get killed. We'd sit tight, reset ambush trips and wait until morning light to check the dead because trying to do so at night was too risky. One night we ambushed a big bunch of them and found when we checked them the next morning every one of them had a big pack on their back full of brass and links, empty shells from fired 12.7mm anti-aircraft guns, humping them back to Cambodia to be refilled by hand.

We lived in the bush and every month or so we'd get a few days of rest back at LZ Rita. If you were plucked out of the comfort of your home to spend a few days at LZ Rita you might think you had arrived in hell. But for us, after the filth and bugs and snakes and razor-sharp leaves and things trying to kill us in the bush, Rita was a welcome and secure haven of comfortable relaxation. After a few days rest at Rita, it was back to the bush for humping through the jungle by day, ambushes by night.

Our entire AO was a free-fire zone. That meant we could fire based on our own judgment without radio permission from higher-ups, and it meant anyone moving in the jungle at night was presumed to be our enemy. There were few villages in the area, but those villagers knew well that they lived in a free-fire zone and to stay home at night. Since they lived in the path of invading NVA who saw them far more frequently than we did, we suspected them of supporting the enemy and they frequently did. It was hard to blame them; if they didn't do what the enemy wanted there would be immediate and violent consequences, and these people just wanted to be left alone, at least that's what I think. The villagers wore false smiles and tried to appear cooperative while we were there, and I'm sure they were happy to see us leave, but even when we found and destroyed enemy weapons caches, concealed food stores and other materials the enemy would use, I never saw any civilians mistreated when we cleared a village. I'm sure *zippo raids* (burning the village) must have happened during the war, but never in my unit while I was there.

MY FIRST TWO WOUNDS came in the same week. In my upper left arm I took a minor frag wound in one night contact. Our medic, Doc Romero, patched me up. Calling him Doc might conjure up a mature image but Doc was all of 18 years old. In the same week, we set up a big L-shaped ambush on a large trotter and a bunch of NVA walked right into it that night; it was a perfect setup. I hit my clacker to trip my Claymore and blew the arm off an NVA. I had to admire that soldier because with one arm gone he was still throwing grenades at us. Danny Barnett and I jumped in a hole to dodge one of the grenades but Danny got shrapnel in his nose and it got me in the left leg. SSgt Walt McClarnen, my quiet and fatherly squad leader, jumped in our hole with us. Danny and I both threw a grenade back at the NVA soldier and both of us hit trees, the grenades bounced back at us and we franticly threw them again while that very brave one-armed guy continued throwing grenades our way. I pulled a grenade pin and waited a couple seconds before tossing it and killed him with an airburst. If my old-model M-16 wasn't such a piece of shit with a bolt that swelled in the heat . . .

The next morning when we checked enemy dead I found the one-armed guy I killed was an officer, an NVA Lt. Since we always checked for intel from documents I tried to get the pack off his back with one hand on my rifle and one hand pulling on the pack when one of our guys yelled, "Nick look out!" One of the enemy bodies was still alive and trying to shoot me in the back, but my squad leader shot him with a pistol. Our new and very green "El Tee" said to me, "Can't you do anything right?" as he jerked the pack off the dead body, turning it over so that half the head and brains flopped into my hand. What an unnecessary mess that was. Thank you very much.

When we assessed our wounds that morning, Danny Barnett was excited to get wounded enough for a Purple Heart even though Doc Romero said it was just a tiny hole in his nose but said he'd put him in for the Purple Heart. Danny is a great guy, now a preacher in Kansas. He never did get the Purple Heart but he still has the indentation in his nose.

I was to be medevaced, but it was so hot guys were dropping from heat exhaustion, one guy with his eyes rolled up in his head, and they used the medevac to fly those guys out. I had to hump all the way back to LZ Rita with my squad the next day on my wounded leg. Now I want you to imagine a wound in your leg where a hot piece of metal flying faster than you can see tore out a chunk of flesh, then after several hours for it to become very sore and tender to any touch with nerve ends exposed, imagine someone jamming a stick into that wound. That's right, you'd have to scream and want to kill someone, and I just wanted to prepare you for knowing while I humped to Rita I stepped over a log and jammed a stick up into the wound in my leg. Oh boy!

At LZ Rita the requests I had submitted again and again for a replacement pair of glasses caught up to me. It seems the IG was investigating items disappearing to the black market and some genius decided, instead of sending the glasses which failed to reach me twice when they were "shipped," this time they would fly my ass to Quan Loi and place the glasses in my hand. So off I went to get my glasses, and while there I went to the aid station to get care for my wound.

The doctor said he wouldn't touch me until I cleaned up. That shouldn't have been a surprise but I was so used to being filthy in the jungle in layers of sweat and grime, covered in red dirt, I hadn't given it a thought. On those rare occasions we hit a civilized place, we sometimes got a rude reminder when we followed our noses, lusting after a hot meal, and a mess sergeant would throw us out of his mess hall because we were *unclean*.

In the jungle being clean was impossible and soon forgotten. When Queen Bee was breaking me in, he said, "Hell, boy, don't worry 'bout keeping clean and fresh. You're gonna stink so bad, one day soon you'll smell your own ass and yell, 'What the hell is that?!'" We didn't use deodorant or cologne not only because it would do little good, but because we needed to blend in to our surroundings, we needed to smell like the rotting vegetation of the jungle to hide. We wore one shirt and one pair of pants with no underwear to prevent chafing, which could quickly form a rash and turn into something ugly in the sweltering heat. Anything unclean our hands might touch, snot included, was wiped in a new layer on pants or shirt. Our clothes were rotting off our back in the field, our boots were so filthy they turned white from salty sweat. Most importantly we had two pair of socks and changed socks every day. I carried my spare socks in my ruck except when they were wet, then I carried them around my neck to dry out. I never took my boots off at night because we never knew when the enemy would hit us or stumble into us. When I rotated socks the next morning they may not be clean, they may not smell nice, but they better be dry because the guys who didn't take care of their feet ended up being medevaced out with *trench foot*, our term for the terrible fungus that takes hold and eats your feet.

When the doc refused to touch me I had one pair of pants and a shirt that were in the process of rotting off me, two pair of socks, no underwear and that's it. I was able to find replacement jungle fatigues, even extra small pants that finally, at last, fit my 28 inch waist, and I showered in the heavenly luxury of hot water. That was the only change of clothes I had in Vietnam, and other than Cam Rahn Bay and An Khe when I first arrived, that was my only shower in Vietnam. The doc patched my leg and told me to stay a week in Quan Loi to let it heal. Well, OK then!

I ran into a couple of guys I knew and we lived that week on Schlitz beer and slim jims. We had to do something since we were ambulatory so we volunteered for bunker guard duty, which was a complete joke; we were in the 3rd perimeter just under the eight inchers, the biggest, baddest piece of artillery on the ground in Vietnam, and if the NVA ever got that far it was all over anyway.

I never saw anybody smoking pot there or anywhere else; I guess that came later in the war when draftees brought the habit. This was not the rear, it was a firebase, so there would have been hell to pay for doing any kind of drug. Out in the bush, it would have been a far worse infraction. Our CO was on top of things; he walked the perimeter every night tapping helmets to make sure we were awake. My recuperation was over too soon and I was back in the bush to daily humps slugging through thick jungle, nightly ambushes to slow down the NVA.

We made a practice of never staying in same place two nights in a row. We fudged on that practice and stayed two nights in one spot just one time and guess what? They had been watching, they hit us and we got hit bad. Every guy in the other platoon was hit with mortar shrapnel, and when I heard the first incoming mortar round I threw my guys in our hole then there was no room for me! We blind-fired at the jungle but you don't win a firefight that way. I was used to blind-firing anyway because my M-16 would usually jam after firing a clip even though I cleaned it every day, sometimes several times a day, so I would hide behind something or put my rifle over the top of sandbags and fire with my head down in case it blew up in my face. I threw frag grenades a lot but this time the NVA weren't that close.

We were pinned down, they knew where we were and we didn't have their position pinpointed. We called for help by Spooky, an old AC-47 with 40mm cannons and miniguns lined up on one side so the plane would set up a circular pattern and aim the focus of the fire and hold it on a spot with the most devastating firepower you could imagine. Spooky dropped flares, the Range platoon gave him a fix on the NVA because we couldn't pinpoint their position, and Spooky opened up to grease them with a shitstorm of fire that lit up the night sky with tracers every 5th round. Holy shit! When we checked bodies the next morning we accounted for 45 bodies, but they were mostly just pieces, a head here, a leg there, bits and pieces all over, what a freaking mess! Thank God the NVA didn't have a Spooky.

When White Skull went out on a platoon-size patrol from that spot, it was about November 20, it was a place we called Chicken Valley, and NVA snipers shot the first three White Skull guys in the head, killed them.

On Christmas eve, while we were resting at LZ Rita, I got a 42-lb package from my family with all the goodies I had asked for: cookies, coolaid, beefaroni, slim jims, beef jerky and my aunt sent a gol-darned fruit cake. I think half the weight of the package was in that fruit cake. I mean, the big joke when we opened C-rats was finding fruit cake. About the time I opened the package I heard, "Mount up! We're moving out."

"You've got to be shitting me," I thought, but it fit with a grunt's lousy luck. So I split up the goodies among my whole squad, which I would have done anyway but it would have been nice to relax with just a taste on my own. I didn't know it but this would be my last departure from LZ Rita, my last time loading on a helicopter for a combat assault.

Call me crazy but when I got used to combat assaults, I liked the thrill of loading up in one or more helicopters and moving our unit to an LZ where we didn't know if it would be hot or cold, or whether the gunships were going to blast the hell out of the sides of the LZ with rockets and miniguns right outside our helicopter door. The pilots who flew us were with the 227th and 229th aviation units, and they got into it, too. Sometimes our approach to the LZ was low-level, following the contour of the terrain just a few feet above the trees where the enemy didn't have time to take a shot when we flew over them, dodging and weaving with the hills while we sat in the open doorway with our feet on the skids, called *riding the rail*, held in place by centrifugal force and maybe a precarious grip on a steel ring in the floor, and the pilots would look back at us with great big grins when we would scream our delight at the roller-coaster ride, on our way to an uncertain reception in an unknown place. Little slices of pleasure can be found even in dire circumstances.

Carl "Pepe" Pipher (L) and Johnny "Duckbutter" Mays (R) at LZ Carol
Photo courtesy of Nick Donvito, all rights reserved

WE HUMPED IN THE sweltering day, slowly, step by step, and ambushed by night, day after day, night after night. One day ran into the next but I do remember January 27, 1969. My friend, Larry Hackett on the gun team, was breaking in a new gunner on the M-60, Tommy Kunzel. Tommy had just arrived, brand new to combat and firefights.

We found a trotter and we didn't even have time to set up, here came one lonely NVA walking down the trotter like he was out for a Sunday stroll. We stayed out of sight while Larry, in Range platoon nearby, asked the CO on the radio what to do because he didn't want to give away our position, but the CO said take him out. Larry quietly told Tommy to open up on the NVA soldier but Tommy froze, he couldn't shoot. Larry stood up and hit him with all 18 rounds in his M-16, I mean it, he hit the guy with every round and really messed him up, killed him instantly. One of our guys drug the body off the trotter and threw it next to me. Lucky me. The M-16 round makes small holes and the body's intestines were trying to squeeze out some of the holes. My squad leader just had to give the body a kick while I was down there next to the body, making sure he was dead but also squirting fluids we don't like to think about all over me, adding to the rank layers I wore. Thank you very much.

That night our ambush killed a large group of NVA and we knew we were in a hot area.

The next morning, January 28, we got our compass heading and moved out with me walking slack in the point team. Pepe hadn't been feeling well and so I had walked first on point three days in a row, stretching my luck. Warren was too inexperienced to be on point so I walked slack, covering Pepe on point while Warren stayed close behind me.

Warren "Snake" Eskeridge at LZ Rita
Photo courtesy of Nick Donvito, all rights reserved

It was still early, about 07:30 and Pepe was moving too fast. I wasn't close enough to cover him as he topped a small hill and disappeared on his way down, so I stepped faster and power-whispered "Don't get so far ahead, Pepe," and just as I said that a couple of things happened that would change my life.

I topped the hill and looked down because something unusual caught my eye, and just as I heard *Boom!* I saw a corrugated steel roof and under it were huge pots of boiling rice. I knew in a heartbeat, Oh, shit, we stumbled into an enemy base camp with a big mess hall and just as I saw it they blew a Claymore on Pepe. I ran after Pepe through the thick undergrowth while Warren ran after me. When I got to Pepe I bent down to him just as I spotted another American Claymore mine and yelled, "Look out!" to Warren while I instinctively turned my head and that's when it blew and sent one big pellet into my head just in front of my right ear, broke my cheekbone and came out just on the outside corner of my right eye. If I hadn't turned, the pellet would have gone through my head and maybe killed me. The blood on my face and chest came so fast it was like someone hit me with a bucket of hot water and it hurt like hell.

But I was the lucky one.

Warren Eskeridge, the new guy who looked to me for guidance, caught the Claymore's blast full on in the chest and stomach. He was torn up bad and as he lay there dying his last words were "Oooh, Nick!"

They were hitting us from all sides.

Pepe was dead. I wasn't sure if Warren was still alive and I was trying to drag him back out of the shitstorm when an NVA soldier in a tree shot me through the forearm; my ulna bone was now sticking out. They were everywhere, in the trees, in spider holes, everywhere, and we just walked into them, into their ambush on us. I tried again to drag Warren and they shot me in the hip and my leg was suddenly gone, I couldn't walk, couldn't stand up and I couldn't find my leg, then I found it up behind my head. It was weird, my leg was gone but I could wiggle my toes.

The gunfire was deafening but I was in my own zone, time slowed down to a fascinating slow rate. I tried to stay calm. I still had my pack on my back and was sitting up, they were still shooting at me but I guess my training kicked in and I assessed my wounds one by one. I could only feel my face, I couldn't see it, my ulna bone was sticking about three inches out of my arm and my leg was messed up bad. The pain was incredible but I thought I had a chance to live through it. Two medics were hit trying to get to me. Pepe and Warren were dead. The firefight was very heavy, ferocious. Guys were getting hit all over the place.

I didn't know it at the time but my femoral artery was partially cut and pumping blood like crazy, the sun was hot and I was a complete bloody mess. I guess it was the loss of blood that made me woozy and tired, and I must have dozed off because they put me in one of the bunkers we found and covered me with a poncho thinking I was dead.

The medevac was released with a load of less seriously wounded grunts; they didn't realize I was alive.

When they carried Pepe's body wrapped in his own poncho, his hand fell out from under the poncho and I saw the wristband he always wore. That's when it really hit me what a bad day it was. My squad leader, Jerry White, who had 14 days left in country was shot through the knee trying to get to me. Doc McGary got hit. Luckily we were trained to be aggressive and our guys came up to help, including Larry from Range platoon with his M-60.

By the time they realized I was alive, there were other guys freshly wounded but it was too hot for the medevac to get back in to the LZ, so the pilot backed away and dropped a jungle penetrator a little distance away from the firefight but still nearby. My guys humped me over there and hooked me up to the rig first because I was the most serious. They took me up on a Jungle Penetrator with a "Stokes" litter, a covered wire basket shaped for a human body, just attached at the head. The helicopter started hoisting me up.

I remember spinning around on the way up, like an out-of-body experience, watching gunfire smoke from the bushes and wondering lazily whether they were shooting at me or the helicopter. I heard bullets whizzing by my head and thought that helicopter must be a huge target hovering so still overhead. I wasn't so sure I wanted to get on it if it was going to be shot down, and as they pulled me in the chopper I could hear the whacks of the chopper taking hits, which kind of woke me up out of my daze. Now I was anxious to get the hell out of there, but that pilot stayed in a hover to hoist up six more guys while they were shooting at him and punching holes through the chopper. That was when I got really scared. I guess fear hits you when you finally think you're going to make it out and then think again you might not, like finding redemption unexpectedly then losing it again. We sat there hovering in the middle of a roaring battle and they kept pulling guy after guy up. Son of a bitch! That pilot and crew had balls of steel. They got me and six other wounded guys out of there, including my Squad Leader, Jerry White, Art (Doc) McGary, and the last one was Michael McGhie, who pushed himself up on my wounded leg and it felt like he was killing me!

The medevac crew weren't doctors but they worked like a surgical team as they brought each one of us into the helicopter and laid us out on the blood-slick deck, applying bandages and tourniquets, taking vital signs, starting life-saving fluids, giving morphine if not already done by the medics on the ground and telling the pilots what kind of injuries were involved so they could plan their hospital route based on which hospital specialized in head injuries, eyes, etc. We stopped at two aid stations on the way to the evac hospital at Long Binh near Saigon, and at each place they examined us to give immediate attention where it was needed and maybe the helicopter needed fuel. They put a splint on my arm and leg and always asked if I could see because my face was a bloody mess and I kept telling them, "I can see; fix my leg!" When we finally got to the Long Binh hospital they ran us into trauma, surrounded us and a redheaded major looked down at me and said "Don't worry, son, I'm a plastic surgeon, I'll do surgery on your face for ya." I said, "Thanks Doc, but I can see; fix the leg, fix the leg!"

It was very bad for me when they straightened out my leg because the fracture was compound and comminuted, meaning the bullet broke the bone into multiple pieces. In surgery the Doc fixed my face and a few fragmentation wounds I had here and there from all the explosions I was near but the wounds to my leg and arm were too complex. They packed me in cotton, put me in plaster casts and prepared to fly me to Yokohama, Japan for the difficult surgery. When I woke up I could see myself in a mirror and it was a shock.

My head looked like a busted pumpkin and scared the hell out of me. They packed the wounds real tight and put me in a body cast to immobilize my arm and leg pending surgery in Japan. When I later saw my x-rays the first time I couldn't believe my eyes. Not being able to move is a bitch! Any movement I tried was very painful.

The next day they put me on a C141 Starlifter hospital transport outfitted with racks for patients and flew me to the 106th General Hospital in Yokohama, Japan, my home for the next two months. By moving me out of one place and into the airplane, the bouncing in the airplane and moving me out to the other hospital, the cast broke at the hip, right at the gunshot wound, and every time they picked me up the broken bones in my upper leg ground together in excruciating pain and I would involuntarily yell with the stabs of pain. One Spc 6, a medical aide in Yokohama, said to me, "Aw, shut up you big sissy!"

LOOKING BACK ON THAT moment, I think the flip comment by that Spc 6 bothered me far worse than the pain because it was so far out of line, but I was still just 18 years old and I'm not sure I could have put it into words then. I had been at the point of the spear, in the jungle where this man, if removed from his safe job and put in my place, might have cried for his mommy. I had been playing footsie with death and doing without the basic comforts of life we all take for granted, pushing myself physically and doing it willingly with my brothers while watching each other's back because I was a soldier doing my duty for my country. I doubted he would have the physical or character strength to do the same.

That civilians would insult or ignore our service in Vietnam for decades, that was one thing; they didn't really know anything about the war but what they saw on TV or heard from other unreliable sources. But an Army medical specialist mocking a soldier's pain from combat wounds was beyond all bounds. I learned far too young what it means to kill in war. Killing from anger is something else again, but I could have done it right then with my bare hands because my spirit was up and eager in a second, but my body was unable, my broken cast useless and soaked in blood. He was one lucky asshole.

Later while in the Yokohama hospital I encountered a different kind of asshole. Japanese students, protesting the war, broke into the amputee ward in the middle of the night and threw lit strings of firecrackers to harass us. If you think about it, throwing firecrackers in the middle of the night at combat veterans, many immobilized by their wounds, makes the word asshole seem far too kind, especially since I met men in that hospital barely old enough to buy a beer, but men I will always admire.

SURGERY ON MY ARM and leg was done right away, surely a difficult process to reassemble shattered bones that had been rubbing together inside me when I was moved. They didn't use pins or screws, just placed the bones to grow back together and tried to immobilize the leg with a full body cast. I went to the amputee ward. A soldier in the next bed had been a truck driver and had a broken femur; he and I were the only ones who were not amputees, and we were on that ward because we were on watch to be an amputee ourselves.

My arm was in a plaster cast and they put my leg in traction with a Steinmann pin through my shin about three inches below my kneecap, secured to a horseshoe device on the ends of the pin, attached to a cable through a pulley and a weight on the end. I hurt like hell. Every day the medic cleaned the holes they drilled through my leg for the pin, to prevent infection, and he cleaned the surgical drains on both sides of my leg, and on my hip. It hurt so bad when they cleaned the wound. I and my fellow traction patients learned how to clean it ourselves every day and I changed my own drain dressing every day, too.

WHILE IN JAPAN I received word from my buddies in my unit. They told me the day I was hit we had stumbled into a huge NVA base camp just a klick from Nui Bah Dinh, the Black Virgin mountain. They had called in a Cav battalion to clear out the NVA and the battle revealed a huge enemy complex of over 1,000 underground bunkers. The bunkers, some as big as living rooms, were made for barracks, hospitals, officer quarters, not just crude holes in the ground but carefully crafted with walls as flat as in our houses. The Cav cleaned the NVA out of the area. Well, I always knew walking point was dangerous. I guess at least when I got hit it was productive!

THERE WAS ONE KID in the 106th in Japan with no legs and one arm. Somehow he learned how to zip around in his wheelchair to find someone who would play cards with him. We called him Scooter, from New York City, and he was always cheering up the other guys. I asked the nurses why he was in such a good mood because he seemed to have been here a long time. They said he refused to go home, he always came up with some problem they had to treat him for; the nurse said if he goes home he'll be the local side show, someone to feel sorry for, but here he's in his element, among the brothers who understand him.

They brought in a young Marine one day with both feet blown off. He screamed every night with phantom pains in his feet and we all suffered with him.

Benny from New Jersey was in the bed across from me. He got hit with a Light Anti-tank Weapon in his legs. Actually the LAW round hit the log he was hiding behind and took out the shin bone in his legs. The hospital had a bone bank and they were trying to replace Benny's shin bones. They'd attach bone pieces to his leg bone with little pins and had to leave the skin of his leg flayed open.

SOMETHING WAS WRONG.

I could look at the pin in my leg and see it was crooked. I kept telling the doctors on their daily rounds that something didn't feel right. After two months - TWO MONTHS - they took an x-ray and said, "Shoot, the leg's healing wrong!" They had to reset it. Thank you very much.

When they put me under – I counted backwards from 10, saying nine, eight, zzzzzzzz – they laid me out flat and straightened me out and the bone breaks aligned in place with just a little pressure. That's what the doc told me.

They put me in a full two-leg "spica" body cast and decided I was sufficiently stable to go to a stateside hospital. I flew in a hospital aircraft on a stretcher to Walson Army Hospital at Ft. Dix in New Jersey, the closest Army hospital to my upstate, New York home. Benny was in that hospital, having come home before me, so I had an instant friend. Benny's legs were straight out now and he could get around a little in a wheelchair. My Dad got to know Benny, too, even though he nearly fainted at Benny's still-open shins.

After two more months in that hospital the Army doctors let me go home on convalescent leave with the arrangement that my Dad picked me up in a station wagon on an Army issue stretcher, loaded in the "seats-down" back like a canoe, because I was still in the double spica body cast immobilizing both legs. But at least I would be with my family and could stay in my own room at home.

Later, Benny's legs healed enough for him to be ambulatory and he got out of the hospital, too. As it turned out Benny's girlfriend couldn't deal with his wounds. When he was able to drive, he got drunk one night, hit a big tree at Ft. Dix and killed himself. I assume it was an accident. I lost a friend and my Dad took Benny's death hard.

When I first arrived at the Cav in Vietnam they issued me a wallet with zip pockets to keep our stuff dry with a few important documents they wanted every one of us to read and review now and then: the Uniform Code of Military Justice, Code of Conduct and POW rules. I also kept other things like my Ranger Creed card, PX allotment card that allowed me to buy certain amounts of things like beer, as if I was ever close to a PX! In the firefight when I was hit, I lost that wallet in the jungle when they cut my pants off my right leg and I sure did regret losing it, especially my Ranger card. While I was home on convalescent leave a package arrived in the mail containing that wallet and a check for $160. The Cav brought in a battalion to deal with the NVA in the area where we found all the bunkers and a guy I never knew in another company found my wallet in the jungle, with all my stuff still in it, and turned it in. The Army sent me the check for the cash, well, the MPC equivalent, that was in the wallet. God bless him. I still have that wallet.

I would ultimately be a patient at Walson Army Hospital for two years. About six months after I arrived, they finally let me stand up for a moment in the cast, vertical for the first time after eight months in the horizontal. I promptly passed out. So, we tried again and again, my system getting re-accustomed to blood draining from my head as I eagerly worked toward the goal of walking on my own, which would regain several important parts of my life.

First, I was permanently assigned as a patient in a Medical Hold Company. As an NCO I was permitted to live off post when my wife came down from Syracuse, but not until I could walk again when they freed one leg from the cast. I had nineteen year old hormones, I had incentive!

Second, I would be able to get from one point to another, the blessings of which you may not appreciate until you are in a cast in bed for eight freaking months! I would be able to go places, see things, do something with friends, almost like a normal person.

Third, I could find something useful to do because I was going nuts.

So I worked at it. It wasn't easy and, sure, it hurt a lot. But eventually I had a walking spica cast with the left leg cut off so I could walk using my waist, not my hip, to throw my bad leg in the cast out in front of me while supported with my good leg, step forward and do it again to take each step.

I enjoyed the freedom of living off base with my wife.

Getting in and out of a car was miserable, especially with the special crutches I used since my arm was also in a cast, but, damn, it was good to be in my own home, good to be able to stand up and walk and even go places in a car. You don't know what you have until you lose it. It was good to be alive.

My buddies took me fishing and I stepped over a big log in my cast and fell down. They asked each other, "Where the hell did he go?" They found me and pulled me upright and we laughed our ass off. I could prop up against a barstool to have a beer with them. You learn to make do. Adapt, improvise and overcome.

A year after I was hit I finally got out of all the casts, but rehab treatment continued for quite a while. Since I was assigned to the medical company, I had to be at my bunk at 8AM for check-in and at 4PM for check-out and in the ward in between. I was going nuts so I found a way to keep busy. I went to the Hospital SgtMaj. and asked him if there was something I could do to keep me occupied as long as I was going to have to be there so long. He was surprised since this was the first time a soldier ever asked him to go to work instead of taking advantage of their down time.

The SgtMaj took me to the Dependant Medical Records section of the hospital and introduced me to an older SSGT named Ramper. He had bad arthritic hips, it was difficult for him to get around and he was leaving the Army soon. So, I started out doing simple clerical tasks and after a few months I was promoted to NCOIC of the section, an E-7 slot, though I was an E-5. I had a day and night shift of military and civilian employees reporting to me, 15 people.

My new job was a welcome distraction from physical limitations and pain. I met and knew everyone in the hospital, including Col. Eisner, the Hospital Commander, who one day publicly decorated me with my second Purple Heart and second Bronze Star, along with a few other guys.

I was given a P-3 profile, every soldier's dream and mine was permanent. No prolonged standing, no stooping, no bending, no jumping, no running, no strenuous activity. It also made me non-deployable but, the re-up sgts. were after me like crazy. With combat experience and decorated, I was a hot commodity while the war was on.

I took all my entry tests and qualified for Officer Candidate School, but I was heavily recruited by Military Intelligence due to my clean civilian and military records. They did a thorough background check on me that contacted just about every person I ever knew. This got me a nice coveted "Top Secret" security clearance that helped many times in my subsequent civilian career as a police officer. But, when it came down to it, Military Intelligence told me that I would have to voluntarily give up my P-3 profile and I figured as soon my new training was done, I would head straight to Vietnam again. No thanks. I decided to forgo the promotion and get out of the Army at 20 years old.

I already had an apartment and a job lined up at the Syracuse Police Department and went right to work. The war was still raging and what I saw on TV told me most Americans had a twisted view of what our soldiers were doing there.

But nobody bothered me. Many people in Syracuse already knew my character before the Army and actually were proud of me for serving, especially my family and my parents' friends, many of whom served in WWII. Besides, if there were guys who thought of giving me a hard time, maybe they heard I still held the undefeated title in my weight class in boxing at the Boys Club.

Over the next 28 years I would be a police officer and was Chief of Police in two local Departments. I retired in 1999 because the pain was getting to be too much to handle, and getting into and out of patrol cars was an ordeal without strong pain meds.

When I wasn't working I raised a family and rebuilt motorcycles. Larry Hackett, my first buddy in Vietnam, became my close friend for many years. He died of stomach cancer in 2006, one of many vet deaths I suspect were related to Agent Orange but how do you prove something like that?

Linwood Queen and I remained buddies even though he lives in California, and we get together now and then. His daughter calls me "Uncle Nick." I love that.

Some others I haven't mentioned, Steve Atchley, John Lee, Big George Thayler and some others all died of cancers suspected of being related to Agent Orange.

John "Lurch" Crocker had a very bad case of survivor guilt since he came through his tour without a scratch, and died in 2006 of pancreatic cancer. After John died, I am the sole survivor of my squad.

I'm still quite close to Warren's family and Larry's family and these friendships are some of the treasures of my life.

My memory is crystal clear of fine young men doing America's dirty, bloody work, and in particular I will take to my grave the image of Pepe's wristband as his hand fell out from the covered stretcher. I will forever hear Warren's last breath speaking my name.

Other than losing those fine men just as they were becoming adults, I have no regrets and no animosity toward those who were my enemy. They did what their country sent them to do, just as I did. My pain and related health issues are major factors in my life now and I take many strong medications, but in spite of it all I'm proud to have been there when my country needed me and I would do it again.

EVERY NIGHT I PRAY for the souls of my departed friends and I thank God for that helicopter crew. I have tried several times over the years to find them, but I have been unable to and I now understand I probably never will. But I have told you about them and how much I owe them, how much I will always admire them for what they did, what they risked, for their brothers on the ground.

For every one of you who flew Dustoff or Medevac as a pilot or crew member, so long as Vietnam veterans are alive there will be someone, somewhere, every day telling a story about the crazy things you did to save their lives, and they will think about you with gratitude until the day they die.

I am one of those guys. Not a day will pass without me thinking about you and your balls of steel as you pulled my shot-up body out of the jungle. I think of you every day when I remember my buddies, Pepe, Warren, and other faces frozen young in my memory forever.

Photo courtesy of Nick Donvito. All rights reserved.

Nick Donvito lives in Camillus, New York with his wife, married over 40 years. Now retired from a career in law enforcement, his long-time hobby is rebuilding Harley-Davidson motorcycles, like this special bike he rebuilt to commemorate POWs and MIAs.

Postcard to America

AFTER A YEAR OF *flying Dustoff helicopters in Vietnam I came home via Travis AFB where we did our out-processing paperwork then headed for the San Francisco airport to catch a plane home. We were met by a pretty ugly crowd of protesting hippies – ugly as in unkempt, unwashed, and nasty attitudes. They yelled insults and spit at us but we knew if we tangled with them we would buy trouble for ourselves. Besides, we were so anxious to get our asses home we took it and ran the gauntlet.*

My home state of Massachusetts was another anti-war hotbed but I didn't get much outward hostility. It was interesting that the subject of Vietnam never seemed to be mentioned as though it was a forbidden topic, which I guess it was in polite company.

I left the Army but I joined the Guard and with my military haircut I stood out in the days of long and scraggly hair styles. At a 1970 party a half drunken young girl felt she had to berate me because of my military involvement, particularly Vietnam. Someone pulled her away and took her home before I had a chance to retort.

A year or so later, with Vietnam still going on, the Guard sent me to the Instrument Course at Ft. Rucker, AL, where we were told NOT to wear the uniform going to and from post. I became a full-time Guardsman in the same timeframe and was ordered to recruit, which was neither pleasant nor safe in Massachusetts during the war.

After having flown many combat missions, these were relatively minor irritants, just the small things that makes one feel unwelcome and a bit on the miserable side now and then. It gave me some anxiety toward people which I must admit I still harbor today.

Dave Pearsall, Sandwich, Massachusetts
45th Medical Company Air Ambulance, Long Binh & Lai Khe, Vietnam, 1967-68

In the Valley of Death
Tony Armstrong
Heath Springs, South Carolina

Cobra pilot Tony Armstrong
Photo courtesy of Tony Armstrong, all rights reserved

O N THE MORNING OF April 19, 1968, I was descending through a hole in the clouds with many other helicopters, spiraling down for a combat assault in the A Shau Valley. As we descended I looked up from my Cobra cockpit to see the tail boom blown off a D-model Huey helicopter by enemy anti-aircraft fire at 5,000 feet, about 1,000 feet above me. The body of the Huey began to fall and spin, throwing out one of the American grunts who had been sitting with his legs dangling outside. He passed me on his way down a few hundred yards away, gripping his rifle with both hands, arms outstretched, as if that might somehow break his fall. I watched this unknown soldier all the way down to his impact in the jungle far below, and though that was over 40 years ago I still think about him every day, helplessly plunging into the Valley of Death.

The night before, when we were briefed on our mission and the foreboding place we would assault, my platoon leader, who had been to the A Shau before, said, "Fella's, this is gonna' be a bad one."

Like many others, I updated my will that evening and I wrote two letters, in case they should be my last, to my mother and father. I didn't sleep a wink that night and at breakfast I gave the letters to the mess sergeant and quietly told him to mail them if I did not return to retrieve the letters from him. As we waited to launch, I reflected on how a Georgia country boy wound up nervously waiting to fly into a place frequently called the "Aw-Shit!" Valley.

I GREW UP IN the small town of Jessup in south Georgia. When we needed something from the big city we drove to Brunswick. In 1966 I was an unenthusiastic student at Valdosta State College, and to outrun the draft I enlisted in the US Army on January 13, 1967, enticed by the Army's commitment to send me to helicopter flight school. I wanted to fly.

After basic training in a hellhole known as Ft. Polk, Louisiana, I completed helicopter flight school and received orders for Vietnam. While on leave I was diverted to a brief transition school in Savannah, Georgia, to be trained to fly the new Cobra gunship.

I arrived in Vietnam in March 1968, my first venture away from home taking me halfway around the world. Vietnam had just endured the enemy's treacherous Tet Offensive when I arrived. I was 21.

At the 90th Replacement Company in Long Binh I found my assignment on the bulletin board. I was going to the 1st Cavalry Division in Quang Tri, very close to the DMZ. I didn't know it at the time but they needed new pilots badly after losses in the Tet Offensive. We were based near Quang Tri at LZ Sharon, amidst the luscious green rice paddies of the flatlands near the coast, with mountain ranges climbing sharply to the west. The enemy was very active in this area, and I got my wish – I was assigned to fly gunships in D Company of the 229th. The other companies of the 229th were slicks, lift companies used primarily to take our infantry troops to battle in LZs, operations we called *combat assaults*.

At the 229th I had the good fortune of learning under the protective wings of William P. Overmhole and Gary Doyle, who taught me all I know about flying gunships in a war zone. Wherever you are, guys, thanks for keeping me alive while I learned the ropes.

We flew two different types of gunships in the 229th, the older B and C model Huey gunships and the new Cobra. While the unit was building the Cobra inventory, we switched back and forth, flying both type aircraft.

The Huey B and C model gunships were underpowered, so we could only half-fill the fuel tanks. We carried a 7-rocket pod on each side and twin M-60's static-mounted on pylons on each side firing 7.62mm from long belts. To fire the static 60's, we had to aim the aircraft, just as with the rockets. Some of our B and C models were fitted with rocket sights for the right seat, using a red *pipper* of light projected onto an angled slice of glass to sight the target, but we rarely used it. Sometimes simple solutions are best; each of us found our aiming point on the windshield based on our height and seat-height adjustment, and we marked our own aiming X on the windshield with a grease pencil. It worked.

Our crew chief and gunner in the rear had their own M-60 hanging from a bungee cord, and I swear they hit more than we did with the rockets and static-mounted 60s, maybe because they had more range, able to fire ahead, along-side and aft of the aircraft as we flew by a target. They could also pinpoint their shots better than a pilot aiming the aircraft.

LZ Sharon was near sea level, where the air is thicker and more friendly to helicopter rotor blades taking maximum bites of the air. But our B and C model gunships were so underpowered we still had to nurse the aircraft off the ground to a hover, or just get it light on the skids and bounce it along the PSP until moving fast enough to hit translational lift and gain enough lift to get off the ground on a short runway about 100 yards long. Hovering, or just getting it light on the skids so it would move, took a very gentle touch on the cyclic as each tiny unnecessary movement wasted some small part of the cushion of air. Sometimes we had the crew chief and gunner get out and walk along beside the aircraft to lighten the load while we picked up a little momentum, and they knew when to jump in. New pilots had great difficulty getting off the ground, while more experienced pilots had the touch ingrained, and with their calm and steady hand they would take the controls and extract every ounce of lift from their bird. I soon learned to do the same.

Our takeoff could be dangerous if anything obstructed the path. The other price we paid for an underpowered aircraft was short-on-target time since our fuel was typically just half full.

Flying low-level, skimming the treetops, climbing and falling as the elevation changed was our usual method because the enemy had a harder time hitting us when we zipped by close overhead. We clipped a lot of trees and sometimes brought home a branch on the skids.

We also flew Cobras, which were newer, sleeker, faster, more powerful gunships that carried more rockets and ammunition. The Cobra fuselage was just three feet wide, with the aircraft commander sitting in the back and the co-pilot gunner in front and a step below. There was no room for any crew, just lots more firepower and speed.

Whether we were flying B or C models or Cobras, we always flew in light fire teams of two, covering one another as we pulled out of our rocket or gun runs, breaking off short of the enemy position to avoid giving them an easy shot.

The airmobile concept was new to the Army, just a couple years old at the time, and all over Vietnam helicopter pilots like me were learning from guys with just months more experience. We tried and learned new things on our own and passed on our wisdom to the new guys who came in to replace us as our tour ended. Some tricks we learned were focused on how to make ourselves more lethal against the enemy; some were methods that gave us a better chance to survive. We were cowboys, gunslingers, young and adventurous, feeling invincible and having dangerous fun shooting the hell out of the enemy who was trying to shoot us. Now and then we would be reminded of our mortality when one of us was killed, an event that was driven home to me when it was my turn to inventory and pack up the belongings of one of our guys who was going home in a rubber bag.

Our missions were usually escorting slicks on combat assaults and prepping the area with rocket and 7.62mm fire, taking on the enemy when they revealed themselves by firing on us or the ground troops. We also flew gun escort for medevac missions, hunter-killer with one gunship low and one high, the low ship snooping around for trails or evidence of enemy presence and trying to draw their fire. A typically *eventful* mission was flying gun cover for LRRPs; we inserted the team in a diversionary game of helicopter leap-frog at dusk, then while the team was out we were a Ready Reaction Force on standby to be called if and when they were compromised and running from the enemy.

Tony Armstrong at Marble Mountain
Photo courtesy of Tony Armstrong, all rights reserved

SURROUNDING LZ SHARON WERE lush, green rice paddies and farms, beautiful country, scenic villages that looked deceptively peaceful from the air. To the east was the city of Quang Tri, to the southeast the cultural capital of Hue, both cities on the coast of the South China Sea. To the west the terrain gradually became rolling hills, giving way to steep mountains covered in deep, triple-canopy jungle.

About 30 miles southwest from LZ Sharon was the A Shau Valley, a triangular valley between exceptionally rugged mountain ranges with steep cliff faces, running generally north-northwest to south-southeast, about one mile wide and 25 miles long.

The A Shau Valley is one of those eerie places that makes the hair on the back of your neck stand up just by looking at it. The Valley advertises its own mystique by the misty darkness of frequent cloud cover and fog, prompting nervously shifting eyes straining to interpret murky shapes, like kids trying to see the boogeyman. Jungle cover is lush green from altitude, thick and hard to traverse on the ground with steep slopes, leech-infested streams, many varieties of poisonous snakes, unusual insects, so dense that one could not see more than a few feet. The valley floor might look clear from the air but swallowed ground troops in elephant grass nearly 20 feet high and countless bomb craters. There were tigers and elephants, too, but the danger came from the enemy.

Our enemy chose the A Shau as their own because it was naturally protective of their troops, just five miles from the Laotian border and the enemy supply line, the Ho Chi Minh Trail. The A Shau Valley was a distribution point for enemy infiltration into South Vietnam. In fact, the enemy built route 548, a road traversing the valley, as an extension of the Ho Chi Minh Trail, with cleverly hidden gas lines for trucks alongside the road. They forded small streams with PSP to keep trucks out of the mud and built bridges over larger streams and rivers. The road itself was concealed in many ways, with covered rest stops and supply depots, highly unusual infrastructure for an enemy who was infiltrating, constantly on the move and never settled.

Beginning in 1962, US Special Forces established three base camps in the A Shau Valley, but they were closed, one after another, in the face of relentless enemy pressure.

Assaulting the A Shau Valley was always difficult because of terrain and weather, and the entrenched enemy made it deadly. The valley was often crawling with enemy troops, and their defenses were strong with .51 caliber anti-aircraft fire and radar-guided interlocking 37mm guns, some on tracks for retraction into mountainside caves.

The A Shau Valley was the setting for some of the war's most intense and bloody battles. Its name carried with it a stab of fear among US troops. Many called it the Valley of Death.

PERIODICALLY THE A SHAU was assaulted to interrupt enemy operations and infiltration. One of those assaults was Operation Delaware, otherwise known as *Lam Son 216*; Lam Son was the birthplace of a celebrated Vietnamese patriot who defeated an invading Chinese army in 1427. On April 10, 1968, a nine-day campaign to soften up the valley began with B-52 airstrikes, along with Air Force and Marine fighter-bomber jets and artillery barrages pummeling suspected enemy positions.

Our assault was to be on April 19, to establish an artillery fire base on the top of a ridge on the northwest edge of the Valley, to be named LZ Tiger. If successful, Tiger would command the approach into the Valley from Laos.

We were staged near Camp Evans, closer to the Valley than our own LZ. We stood by on April 19, nervously waiting for daylight and the word the lift companies were launching so we could accompany them to provide gun cover. When the word finally came, we discovered that Command & Control forgot to launch the guns when the lift companies launched the slicks. They forgot? Now we were playing catch-up and the lead elements would go into presumably hot LZs without gun cover.

They forgot!

The A Shau was socked in that day with cloud cover topping off at high altitude. We had to fly up to nearly 10,000 feet, the highest altitude I had ever been in a helicopter, to search for a hole in the clouds and then spiral down into the Valley. When you combine an assault into the A Shau with lots of aircraft competing for airspace with limited visibility, you have the makings for a really bad day. Add to that the briefings we had about listening intently to hear the first hum in our FM radio indicating radar-guided 37mm painting our aircraft, a second hum indicating a radar lock-on - the warning to take evasive action to avoid the third hum which would mean we were a half-second away from being dead. The pucker factor was as high as I ever saw it.

Amidst all this tension, we were the guns, and we were late because the brass forgot.

The air was cold as we climbed higher and higher, flying west to the valley. The lead aircraft found a hole in the clouds and we began our descent.

When we broke out below the ceiling and could see how to visually navigate the A Shau, we turned our attention to LZ Tiger, where the slicks had gone in without gunship cover and were shot up badly in a ferocious fight. By the time we got there, boots were on the ground securing the LZ for the next elements, but it could never be really secure. The enemy was everywhere in strength, it seemed, with fire coming at us from all directions. The lead elements landing into LZ Tiger took heavy casualties, aircraft shot down with small arms, machine guns, .51 caliber anti-aircraft guns and 37mm, some aircraft tumbling down the sheer cliffside into a fireball far below. Even huge flying cranes with artillery pieces sling-loaded were shot down. We made one rocket and gun run after another, and I couldn't help thinking all the while that if we had just been there a little sooner some of our dead troops might have survived.

We exhausted our ammo and, low on fuel, we returned to Camp Evans to rearm and refuel. When we returned to the battle it was early afternoon. Once again we climbed high, found a hole in the clouds and corkscrewed down into the A Shau. When we were about 4,000 feet I looked up and saw a Huey far above us, and I noted a line of troops sitting on the edge, as they often did, legs and feet dangling over the sides. All of a sudden the tail boom was shot off, presumably by 37mm, and the Huey started to spin. On the first rotation, one troop was thrown out as if by a catapult, arms outstretched holding his rifle, falling a very long way in full gear with his backpack. I wondered what he was thinking on the way down after all the worrying about the dangers of the A Shau the night before.

We followed the spinning Huey down and the pilots and crew and troops in the back were surely terrified as they contemplated their own death just seconds away. Almost as if to miraculously offset the tragedy of the falling troop, the spinning Huey went down on the side of a mountain, landing right side up. It stopped spinning just before impact and seemed to slow its descent with the apparent cushion of a collective pull at the bottom, the impact breaking off the rotor blades and squashing the skids. On its way down enemy fire reached up to the spinning Huey and to our aircraft as well as all others, seemingly from all directions.

When the spinning Huey hit the ground and came to a stop, the occupants scattered out of it like tiny bugs from our altitude, because the ship was likely to blow up from leaking fuel and would attract enemy fire. These guys were surely shaken and injured, terrified and startled to be alive. We would later learn they had broken backs, compressed vertebrae, broken arms, etc., and that when the AC stepped out of the aircraft he tossed aside his chicken plate just before an enemy soldier shot him in the chest and killed him. We descended as fast as we could, following them down to lay down covering fire to protect them. Enemy fire came from $360°$, as if the A Shau had to live up to its terrifying reputation.

As we traded fire with the enemy and tried to cover the downed men, I witnessed what seemed another miracle.

The commander of our operation ordered his pilot to take the C&C ship down to pick up the downed men from the spinning Huey. It was highly unusual for the C&C to get closely involved in the battle; the commander's role was to stay high, watch the operation and direct actions from a distance. They also carried a helicopter crew of four plus the CO and his staff and a very heavy radio console. The C&C pilot zipped down through enemy fire to hover with one skid on the mountainside and the other hanging in the air while at least seven men scrambled aboard one by one. I knew there was no way he could fly with that load, no way in the world.

As I watched, expecting him to plow into a ball of fire, the C&C pilot gently turned his aircraft and let it fall down the steep side of the mountain, just skimming the trees, letting gravity do the work of buying airspeed while he barely kept the helicopter off the rocks, finally hitting translational lift. We watched in disbelief as he gained airspeed, staying low over the mountainous terrain, dipping into canyons now and then to gain more airspeed, and he found a way over a pass out of the Valley and headed east to deliver his impossible load to the hospital pad, barely clearing each obstacle as if God's own hand gave that helicopter a protective shield. He surely over-torqued the rotor head all to hell and back and probably went over the red-line limits on every instrument on board. I gave him a good luck wish when I saw him dip over the pass and I just knew when he arrived at the Medevac pad he would have to do a running, sliding landing with not even the remote chance of a hover. I thought it likely the aircraft was junked after that flight, used for spare parts.

And then we returned to the fight.

The enemy in the A Shau were not VC, which were often local part-time farmers, part-time soldiers and terrorists. These were uniformed NVA, well-trained, organized, disciplined, well-armed, fierce and determined, a very tough enemy.

We killed them, they killed us and the blood ran deep in the Valley of Death. After a lot of shooting and killing and dying, we established LZ Tiger, and another Tiger at a lower elevation, so they were called LZ Tiger Upper and LZ Tiger Lower. Other LZs or firebases set up in this operation were Signal Hill, Vicky, Pepper, A Luoi, Stallion, Goodman and Cecil.

We went in with the entire 1st Cav Division, we kicked the enemy in the teeth, but we paid a heavy price. We found and confiscated or destroyed massive amounts of war material like weapons caches and food storage in huge quantities as well as large weapons and equipment. Operation Delaware must have set the enemy's plans back substantially. During the operation we set up Ground Controlled Approach (GCA) instruments in the valley so we could descend through the clouds instead of searching for holes in the clouds. That raised my pucker factor even further, and when the operation was done I was very glad, indeed, to cede the remains of the A Shau to the enemy, just my personal decision, you see, because I never wanted to go near that place again.

LZ Pepper after three days of battle
Photo courtesy of Duane Caswell, all rights reserved

I had been in-country just under two months when we went into the A Shau. I thought I had enough combat experience by then, but nothing really prepares you for the A Shau. For my part in that operation I was awarded the Distinguished Flying Cross for heroism, but don't even try to call me a hero! My medal was pinned on by Gen. Tolson, Commanding General of the 1st Cav Division. He was an impressive man who wore an NVA officer's belt and carried an ivory handled .45 revolver. He was a natural leader who cared about his men, and took the time to be part of a small awards ceremony and pin a few medals on some helicopter pilots.

1ST BRIGADE OF THE 1st Cav sent a LRRP team out one day with our gunships flying cover, me as co-pilot and Steve Darrow, a West Point man, as AC. We were southwest of LZ Sharon, up in the mountains at dusk, with the D-model Huey carrying the LRRP team of four, playing leap-frog into multiple LZs to confuse the enemy. After they actually dropped the team in a saddle between two mountaintops, the leapfrog game continued for a few more false LZs to keep up the charade.

As we were leaving the area to return home a whisper came up on the radio, the LRRP leader saying, "We have a problem. We are right in the middle of an NVA communication wire-stringing operation. They're all around us and trying to find us! We've got to get out of here right now!" The NVA felt as though they owned the boonies and they ran wire for communications all over the place.

By the time our two gunships and the slick turned around to hurry back to the saddle, the LRRPs were already in contact with the enemy and fighting for their lives. The NVA were nearly on top of them. The Huey tried several times to go in for a pickup but took heavy fire and had to abort each time. We set up a racetrack pattern for rocket and gun runs and on each run we took heavy fire, too. This continued until we fired the last of our rockets and exhausted our stowed M-60 ammunition, so our crew chief and gunner used their M-60s on bungee cords. The fight was still ongoing when their ammo was gone, too.

Steve said "We've got two M-16s here with about 8-10 clips each" so we fired those until all we had left were our 38 pistols. It was a little like throwing coca-cola bottles at the enemy but we used what we had.

We made gun runs with nothing more than the pop of pistols out the window trying to keep the enemy away from our LRRPs, and finally the Huey zipped in and picked them up. None were killed or even wounded, and we got the hell out of Dodge.

Damn, that was close!

AS ANY HISTORY BUFF knows, the NVA surrounded the Marine base at Khe Sanh and pounded it mercilessly for a very long time, initially as a diversion from their intended treachery of proposing a Tet holiday cease fire then launching an attack on cities all over Vietnam on that very holiday. Apparently they were trying for a US version of Dien Bien Phu, where the French surrendered, as a propaganda victory that might end the war with US withdrawal.

Every time I flew into Khe Sanh I took enemy fire. Every time. The landscape outside the wire looked like the moon, pockmarked by craters from endless pounding by 2,000 pound bombs from B-52s and other air and artillery support, including huge eight-inch guns. It boggles the mind how our enemy withstood the pounding, but I would get a chill looking it over from the air because every day the enemy trenches inched a little closer to the Marines' perimeter. I wondered whether they could hold.

One night we inserted troops on a hilltop in the area for an overnight stay. The next morning from the air we spotted enemy bodies scattered around below the hill crest, where they had been shot or tripped the claymore mines set by our troops.

WEST OF HUE-PHU Bai, in the mountains near the A Shau Valley, we were trimming the top off a mountain one day. No kidding. A huge sky-crane helicopter had a 10,000 pound bomb on a sling load. I was the Cobra AC, flying gun cover for the crane, which was to drop the bomb dead center on a mountaintop to blast a flat LZ.

During the mission I felt the shudder of a faint explosion, which was a surprise because the crane had not dropped the bomb, but my instruments were going crazy, split needles. The rotor RPM needle and the turbine RPM needle are piggy-backed for a reason, because when the engine fails and turbine RPM falls while rotor speed continues, the needles split and the pilot had better lower the collective asap to autorotate to the ground. I lowered the collective and called the fire team leader to tell him I just lost an engine and was going down. Actually it was a compressor failure, but the result is pretty much the same.

Going down in a gunship in hostile territory with lots of jungle and few landing spots will make your butt-cheeks take a hard bite out of the seat, but I got lucky. I set it down in elephant grass about 15 feet high and had my M-16 in my hand and scared eyes wide open when my feet hit the ground – pilots are wussies when it comes to fighting the enemy on their own turf on the ground!

We were picked up quickly, returned to Hue and my Cobra was hoisted out by a Chinook in no-time, preventing the enemy from nabbing the radios and armaments.

The maintenance officer could not reproduce the failure on his first attempt and tried to convince me it was my imagination. On his second test hover it failed for him beautifully and scared the shit out of him, too.

That's the only time I went down in Vietnam, but definitely not my most frightening moment. That moment came when I was flying the left seat of a C-model gunship one day. We flew with the side doors stowed in the open position, pilot door windows open to catch the breeze and cool off in the relentless heat. My frisky crew chief stepped out on the left skid in mid-flight, behind me where I couldn't see him, wearing his monkey strap for safety, scooted up just behind my seat and reached in through the window to grab my arm. He nearly killed me with one jolt of fear. What a laugh.

THE 1ST CAV WAS moving from Quang Tri to Tay Ninh province northwest of Saigon. As part of our move I was flying gun cover for a group of slicks picking up ground troops left to guard fire bases and other places, and ferrying them to Quang Tri for transport to the new AO. These trips required us to stop for refueling every couple of hours.

On November 5, 1968, I was doing a hot refuel on a Cobra at a POL point with many 10,000-gallon fuel bladders. The POL point used gasoline engine pumps with a governor on it to run at a preset RPM and maintain fuel pressure. In the field we always kept the aircraft running while refueling because if the battery was weak we might be unable to restart without APU help. Normally the slick pilot would send their gunner to refuel the Cobra as a courtesy while we stayed in the cockpit because it took a few minutes to unstrap the harness and twist out of the small cockpit. This time the gunner was busy and I left my front seat cockpit to do the refuel myself.

I took the Nomex glove off my right hand so I could use my thumb to pull the release lever on the fuel cap.

The fuel nozzle was about 1½ inches in diameter, 18 inches long, angled slightly. While refueling I got a surge in the line and instinctively withdrew the nozzle from the filling receptacle. The result was that fuel sprayed me pretty good and flew up into the turbine intake where it flashed into flame, setting me on fire in the process.

I remember saying out loud "This cannot be happening to me!"

My Nomex fire-resistant flight suit did not burn, but I did. There was a ditch nearby and I ran 20-30 feet to the ditch and rolled in, but there was no water, just mud and dirt. I was still on fire.

Two huge Jolly Green Giant helicopters had landed for refuel near us, fresh from a mission rescuing a jet pilot who had been shot down in the Gulf of Tonkin. One of the Jolly Green crew saw what happened and rushed over with a blanket. He wrapped me up to put out the fire, then helped me out of my fuel-soaked clothes and gave me the blanket. By this time about half the skin and nerve endings on my body had burned away, so I didn't have much pain. I knew I was hurt bad though I didn't think I would die.

He took me to the Jolly Green so they could fly me to the USS Sanctuary (AH-17), a WWII troop ship converted to a hospital ship, anchored off Da Nang. While we were flying I stood up hanging on to the overhead while they cut the remaining clothes off me, wrapped me up in a sheet, laid me on a stretcher and strapped me down.

I thanked the guy who had helped me, who said as he gazed with lust at my .38 pistol in its holster among my clothes, "Sir, could I have your pistol? Because you're not going to need it any more, I promise you!"

One thing I could never lose, even at a morbid moment, is my sense of humor, so I wasn't offended at all by his remark, or his pistol lust. I told him to take my pistol and I gave him my survival knife as well, thinking maybe he could find a way to skirt the rules and take them home as a reminder of this day.

When we landed on the ship's pad they took me in to the doctor, who asked, "How bad are you burned?"

I opened up my sheet wrapper to show him my naked body, to which he declared, "Holy shit! Corpsman! Take this man to a shower!"

I had mud and debris imbedded in my burnt skin from rolling around in the ditch. I asked the Corpsman if the shower had hot water, a luxury we sorely missed in the field but I wondered if it would hurt, but it didn't matter, nothing hurt because the nerve endings were burnt away on my chest, legs, back, arm and hand. All I felt from the shower was pressure.

The Doc gave me morphine, kept me overnight, and the next day I was flown to Da Nang on a helicopter, then on a C-141 hospital aircraft to a hospital in Yokohama, Japan. It was a long, cold flight with a planeload of patients, most on stretchers. Many of us overlooked our wounds and discomfort of the cold to be entranced with the American nurses!

I would be in Japan just 10 days, and I remember nothing about Japan itself except seeing the sky from my stretcher. I was on constant IV pain medication and my drowsy state makes my memory sketchy except for a lot of moaning and screaming around me now and then. I don't think those sounds came from me! They put me in a steel tank of some kind of solution. My right hand was swollen to the size of a football, looking like a boiled ham. While I was in the tank a doctor suddenly took a scalpel, attached to a hemostat, and split my hand open from my forefinger all the way down to the side of my wrist and into my forearm. I didn't feel anything besides the shock of watching the event. I was immediately sitting in a tub of blood. The doctor's radical step ultimately saved my hand and fingers by relieving the extreme pressure on the skin. I am grateful to him.

They put me on a circle bed, a slab mounted between two circles like a hamster wheel so I could be tilted and even turned over by electric motor. The steel tank soaking treatments were repeated a few times while I was in Japan, then I was sent to Brooke Army Medical Center at Ft. Sam Houston, Texas for intense burn treatment.

I had 3rd degree burns covering 42% of my body. That meant a long series of skin grafts on my legs, chest, back, arm and hand. But first the layers of burnt skin had to come off. Each day the medical staff would scrape off part of my skin with a metal blade, the quantity determined by how much of the painful procedure I could stand at one time.

The treatment of my right hand involved wiring my fingers so that as the skin healed I would be able to have some movement in individual fingers. It was not a sight for the weak or queasy.

Next came temporary skin grafts, presumably from cadavers. In the burn ward I saw grafts of every skin color on anyone, white grafts on black men, black skin on white men, a patchwork mix of white, black, yellow, brown, etc. When the temporary grafts had served their purpose and had to come off, that was another painful procedure to endure. Finally, my own skin was grafted through a series of surgeries.

In the middle of this whole process, a second Distinguished Flying Cross medal arrived for me, from the LRRP mission where we fired our pistols at the enemy and were prepared to throw our boots, too, if necessary. This medal was pinned on me on Jan 3, 1969 in Brooke Army Medical Center, Ft Sam Houston, TX, by Col. William H. Moncrief, CO of the hospital. Dressed in hospital pajamas with my hand on prominent display looking like a bunch of French fries wired together, I was very self-conscious while he pinned on the medal and photos were taken. This little ceremony was reported in the local paper and on TV news that evening. It is important for me to note that my awards presented by General Tolson in Vietnam and Col. Moncrief at Brook Army Medical Center do not set me apart from my brother helicopter pilots, many of whom received well-deserved medals for valor. It was just happenstance that on two occasions my awards were presented by big shots.

*Col. Moncrief pins Distinguished Flying Cross medal for heroism on Tony Armstrong
Photo courtesy of Tony Armstrong, all rights reserved*

DURING THE 2004 CONTROVERSY over John Kerry's three Purple Hearts and how he got them, my friend, Terry Garlock, asked me whether I received the Purple Heart for my burns because he didn't know.

"No!" I answered, a little testy at the very idea. "And if it had been offered I would not have accepted it!"

"Why not?" he asked. He knew why; he just wanted to hear me say it.

"Because the Purple Heart is for those who were wounded or killed by the enemy in combat! My burns were from an accident. It's a matter of honor!" You can decide for yourself about John Kerry.

LITTLE BY LITTLE, MY burns and skin grafts healed and my hand healed as much as it would after several surgeries. I was medically retired from the Army after six months in hospitals.

Life goes on. I married, had children, ultimately divorced and remarried. I had a long career as a photographer. For all that time and even today, I have to be careful because my grafted skin is more tender than tough, easily scratched or torn. It is for that reason, not vanity, that I rarely wear shorts even in Georgia's heat. My hand is frequently nicked with the little bumps of life and I still use a lot of band aids.

Once in a while someone who knows about my burns and my background will ask if I have regrets. I rarely tell them the complete answer because it is hard to find the right words, and I'm not at all sure they would understand. I do have regrets, but not about having served in Vietnam, or even having suffered burns.

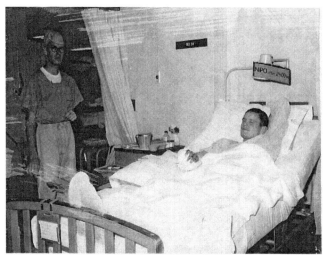

Tony Armstrong at Brooke Army Hospital, Ft. Sam Houston, TX
Photo courtesy of Tony Armstrong, all rights reserved

I would do it all over again in a heartbeat. I am damn proud of having served my country at a time it was not popular, and I wouldn't trade for any treasure the experience of flying into combat with the finest bunch of young cowboys a man could hope to meet, even if being burned is my personal price of admission. I also happen to believe it was a worthy fight, even though our politicians and brass screwed it up and lost the war. But so long as we were fighting the scourge of communism, I'm glad I was there. I couldn't give up my adventurous past by wishing it away, it is too much a part of who I am.

But I do think a little bit every day about April 19, 1968, in the A Shua Valley, and wish the brass had not forgotten to tell us to fire up the gunships on time, wish we had been there to cover the first assault wave as planned. The bond among men that is forged in battle has at its core, I think, a special trust, an unspoken promise to faithfully do our part, even when we are terrified. We didn't talk about it, but now and then we were afraid of dying. I think we feared even more losing our courage, failing to do our job and losing our brothers' trust, or maybe losing their lives. I think we feared that more than anything.

When the first wave of helicopters hit LZ Tiger, those slick pilots trusted us to deliver gunship cover, but we were not there. The mistake by the brass was a simple lapse, the kind of small error that costs lives in combat. When we arrived at the battle a little later we did our job well, we flew into heavy enemy fire to give our brothers cover as they trusted us to do. No matter whose fault it was, after all these years I still feel a twinge of regret that we weren't there to cover the first wave landing in the LZ because our covering fire might have kept more of our brothers alive.

Duane Caswell was piloting Chalk Three in the Yellow One flight that day, the first wave of slicks to land at LZ Tiger. Duane tells me I should let my regrets go, that the enemy was so heavily armed with anti-aircraft weapons that day the gunships would have done little good other than giving the enemy more targets to shoot out of the sky. Duane said out of 60 slicks launched that morning, only 33 could go back in for a second assault, the rest having been shot down or sufficiently shot up they were not flyable for the second wave. Still, I can't help feeling that we might have been able to save just one.

For those who died that day, including the man I watched falling with his arms outstretched holding his rifle, I like to think they rest easy forever in a place of protected comfort, having passed quickly through hell, a place in Vietnam we called the Valley of Death. I like to think that every day.

Photo courtesy of Tony Armstrong, all rights reserved

Tony Armstrong is now retired from his career as a professional photographer. He and his wife, Joan, moved in 2007 from Peachtree City, Georgia to a farm in Heath Springs, South Carolina. Joan developed cancer and passed away in 2010 after more than a year of Tony's loving care.

Postcard to America

ON MY RETURN FROM *Vietnam on July 4, 1969, the girl sitting next to me on the bus from Oakland to the San Francisco airport asked, "Where are you from?" I responded, "I'm from Alabama, but I just got back from Vietnam." She stood up and walked to another seat at the back of the bus.*

I was rather naïve at the time about the depth of anti-war feeling in the country, and I thought she was offended that I was from Alabama! I remember thinking it odd that in 1968 we were required to travel in uniform most of the time, but upon my return in 1969 we were encouraged or required to change into civilian clothes to travel after we arrived in the US.

I had a stateside assignment, so my daily contact was around areas heavily populated with active duty and retired military and those civilians who appreciated our money and kept any anti-military sentiments to themselves. Most of my friends at home in Alabama just expressed sympathy and never brought up the war, unless they were trying to bait me into the "baby killer" or "draft dodger" type of conversation, or political arguments about the war, which I am not good at and tried to avoid. None of them thanked me for my service.

I remember at the time thinking many of the stories about the Vietnam War in the press sounded odd, but I didn't really question them unless they were about someone I knew whose first hand report differed, and then I just chalked it up to shoddy journalism. I have come to believe the effort by the media and academia to distort the truth is a lot more sinister than that.

John Galt, Marietta, Georgia
176 AHC, Chu Lai, Vietnam, 1968-69

Charlie's Elephant Walk
Nick Stevens
Woodstock, VA

AT A TROPICAL LATITUDE between 13 and 14 degrees north in the western Pacific where lush island greens stand out from the surrounding deep blue waters, where the climate varies little from a breezy 75 to 85 degrees with rain every couple of hours, the largest and southernmost of the Mariana Islands is a US territory named Guam. The island covers 209 square miles.

The long, wide runways of Andersen AFB in northern Guam provided an established home for SAC (Strategic Air Command) B-52 aircraft in their role of nuclear deterrence. During the Vietnam War Guam's SAC operation took on a secondary mission of long-range conventional bombing missions in and around Vietnam, and before long those conventional bombing missions overwhelmed the nuclear role at Andersen. I was one of those B-52 pilots.

On our approach to landing, Andersen airfield seemed to leap up 700 feet out of the ocean on steep green, jungle-covered cliffsides, so we called the island "The Rock." Most of our missions from The Rock were about 13 hours round trip, and for relief we periodically rotated to fly shorter eight hour missions out of Kadena AFB in Okinawa and three to four hour missions out of U-Tapao AFB in Thailand.

Every time our mission took us to North Vietnam or the Ho Chi Minh Trail in Laos or Cambodia, we had AAA and SAM missile threats, but since pilots can't easily look underneath them while flying and evading, I only felt the pucker factor of watching a SAM as it fired at us one time. Once is enough, as I was reminded when a friend and his crew took one of the first SAM hits in the war. They came out of it by the skin of their teeth, with some engines dead and others on fire, leaking fuel in small rivers through some of the hundreds of holes, some big enough to jump through untouched, and made a landing at Da Nang that defied all odds. The B-52 was a tough, remarkable airplane.

I'll tell you about the B-52 and crew dogs that still own a big spot in my heart, someone we relied on named Charlie and a striking irony about the B-52's role in the Vietnam War. To properly set the stage, I have to reveal the bumpy start of my rewarding career in aviation.

MY FATHER JOINED THE Army in 1936 and served with the Army Air Corps during WW II. He died in 1969 on active duty in the Air Force after serving for 33 years. I went to school at The Citadel, which strengthened my commitment to serving my country. Most of my fellow cadets would serve on the ground in the Vietnam War while I would fly. I didn't know that at the time but I did have a strong sense of living up to their expectations, and my father's. I joined the Air Force to become a pilot.

The Air Force kindly asked me in 1966, after basic training, where I wanted to go to flying school since there were a number of options, and with either wisdom or good fortune I asked, "Where are the prettiest girls?" They sent me to class 68B at Reese AFB in Lubbock, Texas, where I learned the basics of flying and I met my wife, Margaret.

Competition in flight school for class standing was for priority choice to become a fighter jock, with visions of glory in Vietnam where the USA was taking a stand against the commies. My fire seemed to light more slowly than others, and when I earned my wings at the end of flying school, the hot dogs ahead of me in the class got their jet-jockey dream. Second tier guys like me got bombers and trash haulers and last was the SA-16, Albatross, a flying boat. That's how I ended up in B-52 school at Castle AFB, about 123 miles southeast of San Francisco, a busy place that would feed the voracious appetite of the Vietnam War.

One of my buddies in flying school was "Beachball," a guy named Mike Daly who was 5'4" and about the same width, rather short for the airplane but the program badly needed pilots. Mike was given a 52 slot but didn't want anything to do with it and demonstrated to the authorities that when seated in the cockpit with the seat forward his feet couldn't fully depress the rudder pedals. That was a major issue in a 52 since engine failures and resulting asymmetric thrust problems made rudder control critical.

Beachball just knew that he would be reassigned to C-123s to haul trash or fly SAR (search and rescue) missions. But Beachball didn't get his wish because they gave him a butt-block to scoot him forward in the seat so he could fully depress the 52 pedals. Going to the flight line we had to drag a 50 pound flight bag full of cold/wet weather gear and a small mountain of checklists, and Beachball grumbled a lot because he also had to drag around his butt-block, kind of like a kid's booster seat.

Learning to fly the B-52 wasn't as hard as you might imagine when you think of this huge aircraft. There are really just three things hard to learn about flying the 52, and those are landing, running the crew and AR (aerial or in-flight refueling). The rest is normal flying stuff plus a huge pile of checklists.

That is not to say the 52 was like other aircraft – it was, and remains, unique. Someone once said, "It had the power of fifty locomotives, enough aluminum to make 30,000 garbage cans with over 100 miles of wire and cables, and it flew like fifty locomotives pulling 30,000 garbage cans behind 100 miles of wire and cable." The old bird could be a beast to fly. Most pilots came and went without learning how well she could respond to a gentle touch because the controls often required some strength. Beachball had to be able to not only reach the pedals, but to apply 150 pounds of pressure if lost engines on one side put the airplane in a severe asymmetric situation. Some pilots referred to an AR as "arm exercises" since the aircraft had to be stabilized to take on fuel for half an hour in a four-by-six foot box, which required intense concentration, and strength when bounced around by weather.

As a new student, I didn't know these things yet but over time I would learn that most times you could give the 52 what she wanted with the right trim and power settings to let her fly herself with little input from the pilot.

The first models were built in 1954 with early technology from post-WWII. The story goes that some of the compound curved wing design came from German scientists, and the design was not thoroughly understood by Boeing as they built their first high-speed subsonic jet aircraft, but it worked well and led to advances beyond the 52. The airplane cruised nicely at 0.77 Mach.

The elevator, rudder, and ailerons were mechanically operated by pilot power augmented by hydraulics and mechanical servo tabs. Electronics, fuel pumps, and back-up hydraulic systems were powered by generators or alternators, depending on the model. The generators, hydraulic packs, fuel tank pressurization and air conditioning were driven or pressurized by hot high-pressure air delivered through about a mile of two-inch pneumatic tubing feeding 700 degree 16th stage bleed air from the eight J-57 turbojet engines, hot enough to burn through plate steel. Pneumatics powered other functions like pressurizing the drop tanks to feed fuel; a leak in the tubing at the wrong point could be big trouble.

It took a lot of mechanical knowledge to keep systems like these operational, and with every system there were redundancies and checklists. Fuel tanks were all over the airplane. Keeping the center of gravity under control required constant monitoring and pumping fuel from one tank to another. The Co-pilot had to prepare an elaborate manual fuel sequencing plan before the flight. As part of this he used a specialized slide rule to determine the weight and balance, and he executed the plan during the flight. We didn't have calculators back then in the stone age.

Flying a 52 was a bit like an emergency simulator. There were so many systems with early technology that something was always amiss, sometimes multiples, and we had to deal with engine problems that required shutting one down, pneumatic leaks, hydraulic failures and other surprises, but the systems had amazing redundancy.

During my initial training at Castle, Ron Stevenson and I trained in the cockpit together. Stevens and Stevenson threw a little confusion into the mix now and then. The hardest thing for us was remembering the long list of critical facts being crammed into our heads, and on the controls the hardest was learning the right inputs to land, which took about 20 sorties on the controls before we got it. Our IP (instructor pilot) was "Dad" Porter, a senior instructor in the 328th Bomb Squadron who was the spitting image of Spencer Tracy, including white hair. Flying the airplane was not so hard but keeping up with procedures was another story, and like most young pilots being trained by an experienced man who had seen it all a hundred times, we drove Dad a little crazy just being ourselves.

One day we were doing touch-and-goes, around and around the traffic pattern. I was flying Co-pilot in the right seat which meant I was constantly changing the fuel sequence and updating the fuel log, and my job was to keep the AC (Aircraft Commander) informed. Since the 52 was rather noisy we always wore our helmets and pressed a rocker switch up to talk to each other by interphone, down to transmit to the tower. As I completed one re-sequencing task I informed the AC what I had done, by the book, after which the tower came back to say, "That was really interesting, Bongo Two-Three, but next time try interphone." Dad shook his head, looked at us and said, on interphone of course, "If my dick was as short as your memories I'd have to sit down to take a piss!" While that understandably makes ladies cringe, it is a classic among men – just like Tracy.

We knew what we were training for. The B-52 was specifically designed to deliver nuclear weapons to our cold war enemies in the Soviet Union. The D model could fly about 5,000 miles without air refueling, 6,000 for the G model and 8,000 miles for the H model. With air refueling any target in the world was possible, and so part of our training was flying low level to get under radar to covertly enter Russia, only a few hundred feet off the deck over flat terrain, which was hugging dirt for this huge airplane, bobbing and weaving over hills. Low level was hard on the airplane because of the constant pitching movement and bouncing from low level turbulence, and while the AC was concentrating on flying, the Co-pilot was constantly scanning out the windows, or through a peephole in an actual nuclear mission to protect eyes from a blast flash, monitor the terrain and look for other aircraft. From either pilot window we could see the wings, engines and drop tanks hanging on pylons, bouncing around and seeming to all move in different directions as the ground flashed past at 350 to 425 knots indicated airspeed.

One maneuver we learned was called a *laydown*. Flying in to the nuclear target at low level, we would pop the airplane up to our release altitude by pulling back on the yoke slightly, open the bomb bay doors on the way up and release the weapon which would deploy big chutes to slow its fall to give us time get the hell out of there before it went off. Was there really enough time to escape the shock wave? Thankfully, we never had to test the reassurances we were given.

After learning the basics of flying the B-52, I was assigned as a copilot to the 454th Bomb Wing, Columbus AFB, MS.

MY FATHER WARNED ME to stay out of SAC with its brass-heavy politics and suffocating procedures that were part of operating the network and systems of US nuclear weapons. With Dad's wise counsel bouncing in my head I stepped squarely into SAC, the inevitable destination of B-52 pilots. I would do my part in the cold war, standing by to fly nuclear weapons to our enemies if Armageddon should arrive. Lucky for all of us, I never flew a live nuclear mission, but while a member of a SAC bomb wing I would be dispatched three times on temporary duty to fly bombing missions in and around Vietnam.

The purpose of SAC bases like Columbus AFB was as serious as a heart attack – to make an enemy think long and hard about a nuclear attack on the United States because our retaliation would have three components: land-based intercontinental ballistic missiles, roving and hiding submarines loaded with similar missiles, and B-52's based in many locations that would be scrambled aloft at certain DEFCON levels with directions to hold at strategic spots awaiting coded Emergency War Order radio transmission. When the EWO was transmitted, the crew would open the safe and match the message in classified ways to authenticate orders to carry out a nuclear strike. B-52 EWOs were designed to strike the enemy even after their bases, the U.S. government and the B52 crew's families may have been destroyed, vengeance designed as a deterrence.

Even the flight line layout at a SAC base was designed for EWO, resembling a Christmas tree from overhead with B-52 revetments at a 45 degree angle on either side of a wide taxiway leading to the runway so we could come out from park to sprint our taxi for takeoff as fast as she would go, sometimes taking the runway with airspeed indicators already reading 70 knots, each bird on a solitary mission with MITO (minimum interval for takeoff) just 15 seconds of separation as we would launch as many as possible before incoming might destroy the rest in a real nuclear war. Across from the B-52s, parked in their own revetments also at 45 degree angles to the taxiway, were an equal number of KC-135 tankers, each mated to an EWO bomber. The bombers launched ahead of the tankers if at all possible.

To be prepared for our SAC mission, we drilled constantly, followed detailed procedures designed to safeguard the mission as well as control nuclear weapons, rotated duty on week-long alerts and with mountains of paperwork we certified readiness on every EWO mission our crew might fly. My father was so right!

The B-52 was intended to be heavily maintained on the ground, flown only enough for a high state of readiness for EWO missions. The D model was originally intended to last just 3,800 flight hours. This amazing bird would go on to fly many times that in and out of combat.

A B-52 crew had the Aircraft Commander in the left seat, Co-pilot in the right seat. The most experienced crewmember was often the Radar Navigator (bombardier), considered by some to be second in command although officially the Co-pilot was number two. He occupied the nav compartment and sat to the left of the Navigator behind and below the cockpit, with front-facing seats. The EW (Electronics Warfare Officer), had countermeasures electronic equipment and a rear-facing seat 12 feet aft of the cockpit. These crew members entered the airplane through a belly hatch in the floor of the navigation compartment. In the D-model our tail Gunner, the sole enlisted man in the crew, climbed a ladder up to his compartment aft of the vertical stabilizer through a hatch under the tail. His job was manning a quad-50 to discourage enemy fighters who developed too keen an interest in us.

The Gunner's compartment was cramped, usually either too hot or too cold, and in the rear far behind the center of any turns he had a whipsaw ride during taxi and a backward-facing thrill on takeoff. His rear view was a key part of informing the pilot of the position of following aircraft, clearing for turns and monitoring enemy fighter activity, but the Gunner had to remember his right was the pilot's left and his twelve o'clock was the pilot's six. He could also give us eyes from the rear through his big window on what was happening to our aircraft, like telling the pilot with engine failure problems when an engine was shooting flames out the back. He had a bulkhead door through which he could crawl forward during the flight, nearly 150 feet to the cockpit along an eight-inch wide catwalk on the starboard side through the long bomb bay and past the alternator decks to a bulkhead door leading into the lower crew compartment that housed the navigators.

Co-pilots would eventually be promoted to AC and Navigators to Radar Navigator based on closely evaluated merit and four or five years of experience. The intensity of experience on Vietnam War missions accelerated that transition for many.

On alert, we preflighted the aircraft and put them in the cocked mode so they would be ready to get the engines running two to three minutes from the klaxon horn that would scramble us. Our rotation was seven days on alert, seven days off but still working. While on alert we lived 24/7 in a flight line bunker building mostly underground, though I'm not sure how much good we would be if our building survived a nuclear blast while our aircraft did not. Alerts were new to me, but for old-timers who had done it too many times being sequestered for a week wore on the crew and strained marriages, sometimes to the breaking point. Pressure or boredom was evident in silly pranks and minimal tolerance for edicts from the brass.

The entire SAC base was designed to scramble the 52s and KC-135s, with klaxon horns everywhere and all base personnel drilled in pulling to the side of the road at the klaxon horn so crews on alert could race to their aircraft. But much to the chagrin of Les Hatcher, my first AC at Columbus, there was only so much that could be done for guys like me and the new Nav. We always seemed to be in trouble.

When the klaxon blew we scrambled to launch, never knowing for sure if it was a drill or the real thing, adrenaline pumping, doing everything in rush mode, with the possibility of launching 15 seconds behind other bombers, each carrying the power to destroy cities. If it was ever a valid EWO, it meant our country and loved ones were probably lost forever and the mindset of every crew member I knew was to do maximum damage to the enemy with no hesitation and much vengeance. Because we never knew until takeoff, every klaxon launch was tense.

At one point in world affairs international tensions raised the DEFCON level to the point that the risk of incoming was elevated. We were ordered to line up our aircraft for a straight shot onto the runway, one behind the other with crews in their seats, hours spent waiting tensely for the klaxon to sound off. External APUs didn't have enough power to run all the fuel pumps so when the klaxon sounded we would have to set the fuel panel properly then start the engines by firing explosive cartridges - about the size of a three-pound coffee can. With half a dozen 52s in front, their tanker aircraft behind and all starting up with explosive cartridges, engine start made enough smoke to go IFR on the ground. When the time came, I screwed up the fuel panel, we fired the cartridges and the engines tried to start but then spooled down for lack of fuel. We couldn't move and the aircraft behind us were stuck. I screwed the pooch big time but the planners for this scenario were idiots for not predicting a single problem aircraft could bring the operation to a halt. Luckily it was a drill and everyone learned an important lesson but Les still took a chunk out of my ass, telling me and the Nav to straighten up before we ruined his career.

About a week later on alert the Nav and I talked Les into letting us meet our wives at the base theater for a movie, and he let us take the truck assigned to the crew. As we became engrossed in the movie we didn't realize the theater's klaxon horn wasn't working, and we didn't notice the blue alert light that came on. Soon the usher rushed to tell us about the alert. The Nav and I tore out of the theater, leaving our wives, and while I tripped and fell the Nav jumped in the truck and took off without thinking, leaving me behind. I spotted a tanker crew truck rushing to the flight line and managed to grab the tailgate and swing into the back, hanging on for dear life as traffic pulled over and the truck hit 70 mph racing to the flight line. I yelled into the cab to drop me at my airplane, which they did. Meanwhile when the Nav arrived and climbed aboard, Les asked him, "Where the hell is my Co-pilot?" to which the Nav said, "Holy shit, I forgot him at the theater!" and started to leave to find me before Les told him to stay put, saying, "He'll find a way to get here." I did, and it's a wonder Les didn't kill us. As I climbed in my seat, I finally awoke to the reality I had to get serious.

Practice at SAC included preplanned missions as well as drill launches. Every planned EWO mission ate up the day before with paperwork. The Nav prepared the navigation plan which drove everything else. The AC prepared the flight plan for Air Traffic Control. Our Radar Nav did a target study and set up GPIs (Ground Point Indicators), geographic points that could be identified by radar and were known distances and azimuths from the target to pinpoint the target by triangulation. As Co-pilot I had a lot of slide rule calculations to do for weight and balance control and my fuel log chart spread about 1x2 feet for each leg of the flight. The EW and Gunner briefed on threats. It was a small mountain of pre-mission paperwork just to go fly the airplane, all done manually by the book with SAC standards of zero room for error, quite a pain in the ass for a fun-loving guy like me. Every mission was briefed to the Squadron Commander or Operations Officer.

We certified low level, we certified for each target we trained for, we certified every planned EWO mission to the Wing Commander or his rep, and in a moment of frustration I considered asking if they wanted to certify my dick to make sure I was using it right but was saved by a brief flash of good judgment.

We trained, we flew drill missions, we certified for preparedness to bomb targets like Moscow, Kiev, Leningrad, and others, with plans to recover in places like Tehran; how the world has changed. We were constantly preparing for what made the brass frantic, a surprise ORI (Operational Readiness Inspection), which made it doubly important to do everything by the book. SAC standards and procedures had standards and procedures.

While I was learning the ropes at Columbus AFB, and trying to adjust my fun-loving spirit to the somber mission, the entire wing had been in mourning since I got there. They had lost some of their own in Vietnam in a recent Arc Light deployment, the name of the Air Force's operation with B-52s to support forces in Vietnam. Two crews had gone down in a midair collision. Because the missions from Guam were so long, periodically the lead aircraft would seek a little crew relief by turning over lead to one of the other two in the cell. The cell had been in fingertip formation, a V formation with lead at the center and the other two back on either side, and during the lead change, which was a careful step-by-step procedure, something went terribly wrong. The vertical stabilizer on the lead aircraft was sheared off and everything in the cell went crazy in a heartbeat. George Westbrook was one of the ACs; his crew ejected into the Gulf of Tonkin. George said he was bleeding in the water and just as the SAR helicopter was pulling him up he felt a big shark making a move on him. A number of crewmembers in the two aircraft did not have a chance to eject and were lost including a flag officer in the lead aircraft. As a result of this incident, by policy cells of three 52s went to trail formation one mile behind and 500 feet up for separation.

My father might have enjoyed a laugh if he had lived to watch me squirm in the regulated SAC after his warning, and he may have been wise enough to know my relief would come in combat missions over Vietnam.

AFTER I HAD JUST a few months learning the SAC ropes at Columbus AFB, the entire 454th Bomb Wing moved to Guam with B-52s, KC-135 tankers with their roomy holds full of equipment, crews, staff and a whole lot of brass. Some of our airplanes would be flown back to the US by crews who had been flying combat missions for 179 days.

Crews would be rotated to Guam for 179 days, because in the Air Force's wisdom if they stayed one more day and hit the magic number of 180, they would get credit for an in-country tour and they couldn't be sent back for a year. By sending crews for no more than 179 days, the Air Force could send them home to fly SAC alert for six months and then back to Guam for another 179 day cycle with no end in sight. While on this "temporary" combat duty we collected TDY pay in addition to combat pay. Is this a great country, or what?

Les Hatcher, our AC, had a prior rotation, so he knew what he was doing. Darrel Fisher was our Radar Nav, and I can't remember the EW's name or the Nav who joined me in hot water now and then. Our Gunner was "Scottie" Burns, a Scotsman who rolled his Rs and I can still hear him on the interphone reporting, "Clear-r in the r-rear." For those of us about to get our first taste of combat, it was good to arrive with a known quantity crew.

The Buffs - big ugly fat fuckers - we flew out of Guam were mostly D models from 1966-1972, called *tall-tails* because of the striking black tail over 48 feet up from the ground. While they were designed for a life of 3,800 flight hours with a lot of pampered ground time, they already had over 10,000 hours on them and still going strong. Their bomb bays were modified to carry 84, instead of the original 27 500 pound bombs, or 72, instead of 27 750 pounders. This "big-belly" change allowed the D model to carry a far heavier conventional payload, about 60,000 pounds, than even later models. In addition to the internal load, we could carry either 500 or 750 pounders on MER (multiple ejector rack) pylons under the wings between the inboard engines and the fuselage, 12 on each side. The bombs were M117s designed for air drop, developed in the 1950s for the Korean War.

In our lingo an A-load was 108 500 pounders, 12 under each wing and the other 84 in the bomb bay. B-loads were 72 750 pound bombs in the bomb bay, and an AB-load was 12 under each wing plus 72 750 pounders in the bomb bay.

B-52 G models taking off from Andersen AFB in Guam during the Vietnam War
Public domain photo, no rights claimed

The D models could carry over 270,000 pounds of fuel, but couldn't get off the ground with all that weight, making air refuel early in our mission a necessity. With the weight of about 200,000 pounds of fuel and all these bombs, the D model still needed help to get off the ground. Eight Pratt & Whitney J57 1950s technology turbojet engines were under-slung in pairs on pylons to help keep engine fires away from the wings because an engine fire was survivable but a wing fire was not. Engines one and two were outboard to port, three and four inboard, then on the starboard side were five and six then seven and eight outboard. These engines were designed to inject water during takeoff to cool the compressor and diffuser sections so it could eat more fuel without burning up. The water also increased air density, thereby boosting the normal 7,500 pound thrust on each engine to a takeoff force close to 10,000 pounds. We had a water tank for that purpose just aft of the crew compartment.

Our flight days on The Rock were about 18 hours with four hours of preparation, 13 or so hours in the air then an hour for debrief before beer and hellraising.

Four hours before launch the AC and I would review fuel log charts, weight and balance, aircraft maintenance condition, and other things dealing with general review of the mission, while the Radar and Nav spent an hour on target study and the EW and Gunner were briefed on threats and countermeasures. This was followed by the strike pre-brief for all cell crew members, a dog and pony show provided by the Bomb Wing staff including roll call, aircraft assignments, weather, timing, frequencies and callsigns, targets, threats, etc. Following that, the pilot and Nav team attended a short briefing by the strike ABC (Airborne Commander), usually one of the cell lead ACs, on plans for departure and joining up. By this time flying would be our relief.

With briefings done we would load up the pubs and charts in our flight bags and head to the crew bus; the paperwork equaled the weight of the crew, at least one large flight bag each with the Nav and EW lugging theirs plus the classified. We also had cold weather gear in duffel bags in case we lost the air conditioning (read that "heat"), flack vests or bloopers and God forbid we forgot the flight lunches, water and coffee. The crew bus facilitated this process by stopping en route to the aircraft for each of these items.

A lot of guys added to this weight the personal weapons they carried, just in case. One friend, Waldo Dodd, liked to carry a small arsenal so he would be well armed if he had to bail out, and we joked that if the rescue helicopter dropped him after rescue they may have to declare him as a secondary explosion.

Dropped at our bird, we would start our pre-flight inspection. One of the crew was assigned to keep us posted of the launch time schedule which would continue after airborne to include rendezvous times and all other mission checkpoints until we touched down on The Rock 12 to 13½ hours later. The AC worked one side of the bird while I took the other for a visual inspection for fuel or hydraulic leaks and general condition or anything else unusual. With bomb bay doors open, the Radar and Nav donned goggles to preflight the internal weapons; with 84 M117s in the bay, each with a safety wire protruding, it was not uncommon to get stuck but nobody wanted to lose an eye. The Gunner used a ladder stand to reach the tail section for preflight of his four radar controlled 50 cal machine guns.

The EW was usually the first inside to begin the process of his internal equipment preflight, the rest of us would soon follow up through the belly hatch, into the lower-level two-seat navigation compartment, up the ladder into the upper level cockpit. The AC climbed into his left cockpit seat, the Co-pilot into the right seat. Last to climb in were the Nav and the Radar Nav, and they secured the hatch on the floor on the starboard side aft of their seats. We soon got used to the cramped quarters where nobody could stand up; besides, we were busy wading through pre-flight checklists.

Our seats were explosive ejection types with a built-in parachute, one man raft, beacon and survival gear. The chute was contained in our seat back cushion and the way we strapped in made us prepared for ejection. The seat cushion we sat on was the survival kit containing a CO2-inflatable life raft, emergency supplies, blood chit, map and a 38 revolver among other things. If we ejected the seat frame would separate when the chute opened, the locator beacon would automatically start broadcasting beeps and the survival kit would dangle below on nylon cord.

Both Navigator seats ejected down, the pilots and EW ejected up and the gunner in the rear of a D model bailed out by pulling a jettison handle that triggered the entire gunner enclosure aft of the tail assembly to separate from the aircraft, leaving him lots of room to jump out. In case of seat failure or with additional crewmembers, manual bailout with a strapped-on parachute was out the lower openings left by the navigators or any opening available. As we went through checklists, we checked the seats and chutes and we pulled the ejection seat safety pin so it would be ready to fire if needed.

When we climbed in for a mission, the aircraft was always uncomfortably warm and soon our flight suits were sweaty as we ran checklists and started engines. An experienced crew could run every verbal item by stateside EWO standards in a little less than an hour but an Arc Light crew could do it in less than 20 minutes. By necessity, the books get laid aside in the real world.

The ground Crew Chief assigned to the bird was attached by ground cord to his headset for constant dialog with all crewmembers during the preflight process. The Crew Chief was "ground" in our lingo, our direct line with maintenance that seemed like a living part of the airplane. We thought of ground as another member of the crew and he or she would often fly with us, sometimes going along with toolbox to fix or troubleshoot something while airborne. When a major discrepancy was found it would be reported to the AC and copilot who would coordinate with Charlie and maintenance. As checklists were completed and all systems go, each crew member called in to the AC, and when all clear if our AC was cell lead it was time to check status on two and three then call Charlie.

CHARLIE WAS THE CALL sign of the man who controlled all B-52 operations at Andersen AFB from a small tower situated midfield between runways. The airfield conventional tower monitored Charlie and controlled all other departing and arriving traffic in normal Air Traffic Control fashion. Charlie's name came from the Alpha-Bravo-Charlie pecking order, the Deputy Commander of Operations being third in command in the Wing though he didn't personally fill the role of Charlie in the tower. Charlie was a carefully selected pilot with enough 52 experience to help nervous pilots in unusual situations figure out how best to deal with a problem or emergency. Actually there were several Charlies so they could double-up when needed and cover 24-7.

Here's what Gen. James R. McCarthy wrote in his book, <u>Linebacker II: A View from the Rock</u>: *"The Charlie Tower was without comparison by any standards. Over the years in Southeast Asia the Charlie Tower developed into more of a personality than a place. The original concept must go down as one of the most superb techniques of on the spot controlling of large masses of bombers ever developed, but there was more to it than that. The people in the tower were what made it click, but there was a certain charisma about the Charlie Tower which is difficult to put on paper. The "Charlies" who ran the tower for the DCO were drawn from the most proficient of crews. A base or unit command post has always been envisioned as the nerve center of operations and so it was at Guam and U-Tapao. But, once the scene had shifted to the flight line for the launch and recovery phases, it was Charlie who choreographed one of the greatest shows on earth. The ground movement of dozens of B-52s has been affectionately called a parade of elephants and the elephant walk. Nobody could make them parade like Charlie."*

As a crew member, Charlie was who you trusted, plain and simple. The buck stopped with him every time. No problem was too great or small but you had better be prepared to play the game if your problem served his purpose to amuse and instruct the rest of the crew force. Charlie controlled our launches and was our go-to guy when we had in-flight trouble, and we all had occasion to thank God he was there for us.

Similar to Charlie, Red Ball was a maintenance truck outfitted with a handy selection of the parts and tools needed for frequent repairs, manned by an eager, experienced crew. While changing a 52 tire usually took two hours, Red Ball could do it in 15 minutes. When a bird had a mechanical problem that seemed fixable in short order, we called Red Ball and they dashed to help, circumventing the normal maintenance backlog.

WHEN CHARLIE CLEARED US for taxi to runway 06 Right, the elephant walk began, six tall tails, two cells of three, moving to the takeoff hold line while the spare crew listened intently for signs of an aircraft with problems possibly building to prevent their takeoff. While taxiing we tested our brakes and did a hundred other things on the pre-takeoff checklist. With 20,000 pounds of fuel in the drop tanks outboard of the engines, the wing tip gear were firmly on the ground.

The process of launching our strike, called a "ballgame," involved the choreography of many parties struggling to get all systems working and planes in the air. Maintenance on 52s was intense and many were grounded as malfunctions were found during startup and preparation for takeoff. Aborts on takeoff were also common, and so a spare aircraft and crew were always standing by fully briefed, loaded with bombs, preflighted, started up and ready to fly if needed to get the ballgame launched. If our aircraft didn't make it past checklists, Charlie ordered a "bag-drag" to the spare aircraft. With ground crew help we scrambled out of our grounded bird, threw our bags and equipment into a pickup, jumped on a bus and raced the pickup to the spare, where the spare crew had already de-planed except the AC and Co-pilot who remained in their seats with engines running waiting for their relief. The ground crew threw our bags into the nav compartment and we scrambled aboard, making sure we didn't lose the lunches and coffee in the madness.

When an aircraft aborted a takeoff, the spare was hustled to the takeoff hold line and the spare crew flew the mission in their assigned slot. On spare duty our fingers were crossed the ballgame would launch without us and we would get a day off because we could only fly the mission we had briefed. If the ballgame launched without us, that meant we turned our prepped spare aircraft over to the spare crew briefed for the next mission.

When Charlie was ready to launch the ballgame and cleared number one for takeoff, the AC would advance throttles on all eight and we'd start to roll as I set the crosswind crab a few degrees if required. The 52's wheel carriages, called trucks, were narrow under the fuselage to support a heavy bomb load while most aircraft had a wider wheel stance under the wings, and so the 52 needed some unique help stabilizing against crosswind. The crab function allowed the nose to turn into the wind while the wheel trucks remained true to the runway. With crab set I turned on the starter switches, placed the air conditioning to RAM and the AC placed the steering ratio selector lever to takeoff. We both checked takeoff EPR (engine pressure ratio) and Les hit the water injection switch. The engines screamed at full throttle, noise doubled as water injected and the J-57s wound up to maximum thrust.

At 454,000 pounds, our acceleration was nothing to be excited about but airspeed was building and Les called "70 knots . . . now!" By now the wings were flying and Les stowed the tip gear and was making the lateral inputs to keep the wings level and stayed on centerline with rudder and steering. The Nav started his predetermined time to S-1, our decision speed which was a calculation based on weight, usually 105 knots. If his time expired or we had an engine failure before S1, we had to abort the takeoff in order to stop within the confines of the runway. After S1 we would fly no matter what happened.

As we reached 105 knots the Nav reported "S-1 time . . . now!" We were committed and we knew what to expect from an old tall tail and the hill on 06R. As we started up the grade acceleration stopped slightly above 105 knots and seemed to stay that way longer than comfortable. Finally the runway leveled and speed began to increase.

The B-52 did not rotate nose-up for takeoff like other airplanes. The fore and aft trucks were about 100 feet apart with the long bomb bay in between, and takeoff by design was flat, at about 150 knots. When the wheels lifted off the ground, or "unstuck," the pilot not flying called S-2. D models tended to unstick anywhere between 7,500 and 10,000 feet of runway depending on conditions. Sometimes the aircraft used the entire runway plus the gravel. Many an airplane was saved by the cliff height of 700 feet since they could trade a little altitude for airspeed to get a struggling Buff flying. Not all made it.

At about 155 knots the AC eased her airborne and called for the gear. I'd press the brakes for three seconds and the Buff would shudder as the wheels stopped spinning, necessary to stop the gyroscopic effect before retraction into the wheel wells. I then placed the gear lever up and within a minute he called for retracting the flaps, following an accelerating speed schedule. Halfway through flap retraction, the water ran out and it sounded for a moment like everything quit.

At water run out, I would push the throttles full forward and the Buff would slip away from all the drag of the flaps and gear and smoothly accelerate to just under 350 knots on our departure climb. The next aircraft would be ready to roll with MITO about one minute if we were doing a wave of lots of cells, but with a single or two cells we tended to use a little more separation for safety, about 90 seconds.

With wheels in the well and climbing, we looked for the Russian spy boat that was always hanging out in international waters near our launch pattern, thinly disguised as a fishing vessel, surely counting bombers and radioing warnings to our enemy. The Navy on The Rock had a fishing boat and once while out fishing we pulled alongside the Russian trawler to trade cigarettes and beer with a crew that looked in sore need of some R&R.

Our post-takeoff checklist ended with an oxygen check by each crew member at 12,000 feet. If the ABC reported in the green to Charlie, the ballgame was over until recovery and our strike was turned over to mission directors.

The Radar Nav called the position of lead and the Co-pilot started the usual rendezvous maneuver to place us in trail formation. When number three called into position the Gunner confirmed all birds were in trail a mile behind on centerline, stepped up 500 feet for separation. The Nav announced our planned time, usually three to four hours, to the air refueling control point where we would fill up for our 6,000 miles round trip.

Whether our launch was day or night, whatever the ground weather, when we leveled out on top at 35,000 feet we could usually count on clear skies with sunshine or moon and stars. As I would remember later in life, the beauty in flying big airplanes is the perfect day that always lies above the weather where the air is cold and clear and if you are flying in a B-52 cell, so easy to follow the long contrails of your cell mates.

We had switched from Charlie tower to Andersen mission directors for flight following, would switch over a couple hours later to Clark in the Philippines, then finally to a ship-born controller in the South China Sea who kept tabs on all our operations over Vietnam. The EW, with all his electronic and scrambling gear, was our ground contact in flight to avoid compromising our mission with broadcasts in the clear. From Andersen to Vietnam we were in open airspace over open seas, separated from all other traffic.

The Co-pilot, or Co, was always busy pumping fuel from one of the 12 tanks to another in the fuselage and wings to keep our center of gravity under control. By departure the crew was either comfortable or bitching at the Co, who also controlled the cabin temperature as we climbed to colder levels.

Sometimes one guy was too hot and another too cold, which might mean the Co would be buying his own nickel beer at the end of the mission. Sometimes it took a little while to get comfy as the sweat drying out of our flight suits boosted the cooling effect and added a piquant aroma to the noisy cockpit. We all had a helmet with mic and earphones for comm, but some of us bought our own more pleasing headphones since the helmet seemed to become heavier and more confining as the hours passed. That was not a good idea for the Gunner, who needed to protect his head with his helmet when he was bounced around in his tight confines.

In flight, the B-52 cockpit was like no other place on earth. Boeing clearly gave little priority to human comforts in satisfying the Spartan military specs provided by Curtiss LeMay and SAC. There was no insulation and therefore the minus 50 degree temp and noise outside at 35,000 feet were shielded only by an aluminum skin less than 1/8 inch thick. The air in the crew compartment was very dry with little odor and after six or seven hours our skin began to dry as well as our lungs, especially for smokers. The sound and vibrations of the plane became a part of our body. For someone attuned to the beast and most capable of flying her, it was the comfortable feeling of being attached but few ever went there.

Even I cared little for the Nav compartment. A stairwell about 10 feet behind the pilots led down to the hell hole of the navigators and the myriad of electronics comprising the bombing system. It was a dark place with no windows or any reference to where the earth was except on the big 12 inch radar scope in front of the navigators. I often felt airsick there and no other place. You had to be tough to be a Nav or Radar.

The urinal was several feet behind the Nav to the starboard side, near the entrance hatch, which had a grate over it in case the hatch came off in flight. The joke was that, if standing to take a piss and the hatch came off, you would be strained through the grate. Behind the urinal on the D model was the honey bucket in case a crew dog was in unavoidable need of a bowel movement, but there was no flush and it had to be emptied when we landed hours later. By an unspoken oath none of us would ever foul the Nav compartment with a deposit in the honey bucket, so we took care of business before and after flights, which was a good thing for the Radar and Nav since the hell hole already smelled like urine despite the dry air. Behind the urinal and honey bucket was the pressure bulkhead door that led to the alternator deck just forward of the forward trucks. That led to the eight-inch wide cat walk through the bomb bay and on to the gunner far in the rear. The instructor Nav's seat was attached to the pressure bulkhead door. Later, in sympathy to the Navs, a large scope was placed in front of each Nav and the pilots that was a forerunner to glass in modern aircraft, giving a low light TV and IR display as well as a myriad of other flight information.

Mac put this all in perspective as he talked about being a Gunner on the older B-36. They flew EWO missions carrying the real stuff for over 30 hours with a 24 man crew who not only flew but maintained the aircraft. In the tail were gunners with a galley who could serve a steak with all the trimmings to the AC, traveling through tubes to the forward compartment to deliver meals. They had little trolleys like in the Great Escape to move back and forth. Not many people know these stories.

Anyway, the Buff was not a Holiday Inn but I don't remember too many crew dog complaints except for the Co to turn up the heat. The thing coming to my mind, and one which I still ponder, connecting all these things is the crew element. We see it every day with stories from Afghanistan, Iraq and wherever the American soldier serves, but mostly in combat where all come to a common mission because they are tied together whether in a flying aluminum tube or in a dense jungle. If we must go to war, one thing we are blessed with is this special connection to those who were there with us, a connection that never fades, a connection that can't be explained to others who were not there.

During the cruise to the tanker rendezvous point the Nav took several sextant readings from the EW, who mounted his instrument to a port on the ceiling just aft of his position. B-52 crews were trained to use the same sextant techniques Magellan used in the early 1500s since an EWO mission required us to be self-sufficient in plotting our position, not necessarily reliant on ground navigation systems that might be destroyed in a nuclear war.

The Radar Nav would call the tanker position, one tanker for each bomber, and the cell separated. Each AC set up three miles out, picked an aiming point and slid into position at 255 knots 50 feet behind and slightly below the KC135. When we called pre-contact stabilized, the boom operator cleared us to the contact position a little closer, giving silent directions to move up, down, left, right, forward or aft with a light panel on the aircraft belly. We opened the fueling slot and the boom operator controlled tiny wings on the end of the 45 foot long refueling boom to bring the nozzle to slide into the slot centered on top of the aircraft aft of the pilots, just like sex between airplanes. Hookup to the boom and maintaining the connection in a four-by-six foot flight box was not only difficult but required intense concentration for the half hour required to take on 100,000 pounds, plus talent and strength to stay connected in the bouncing of weather. Disconnects were common. While taking on fuel we had plenty of time to talk to our tanker brothers and listen to what always seemed to be their gripe of the day, and so as guys will do we gave them a name that stuck, "tanker toads," or just toads. But we still loved them.

As each of us finished loading fuel we would back off from the toads in loose formation, forming up again when the last Buff was finished, coming back up to cruise airspeed and onward with the strike mission.

When we left the tanker we still had a ways to go to our target. During the long flight we'd sometimes move around a little even though there was no room to stand. The Nav and Radar Nav would climb up to the cockpit from their electronic cave to look out the windows while they smoked a cigarette and shot the bull. With time to kill we tried to learn the basics of each other's job for cross-training and understanding. The pilots would try sextant shots or sit at the Nav position to get familiar with the systems. Everyone was welcome to sit in the pilots seats to hand fly the plane or play with the controls. Only the gunner was left alone by himself. We had no idea what went on back there and we didn't ask.

As we approached Vietnam, lead would alert the cell that mission ciphers were now in effect. We would now use the mission's code words in our broadcast radio traffic to confuse a listening enemy. And this was the time the lead Nav would begin to sweat.

The Nav had three points to hit and the crews throughout the cell were watching to see how far he would miss. With so many variables, hitting them at the exact planned time and location was nearly impossible. Other Navs in the cell or wave were watching most closely as they did their own navigation to back up lead. Bets were won and lost, as were nicknames and reputations, on the Nav's accuracy so he worked hard at it, calling turns and airspeed variations, correcting for windage, trying for the impossible perfect hit. In the mid-twentieth century, we were not far removed from Columbus and Polynesian navigators, when navigating required far greater skill than reading a GPS.

First came ADIZ, the planned entry point to a country's Air Defense Identification Zone about 200 miles out from its border. When approaching North Vietnam, we were edgy because we knew they had heavy anti-aircraft defenses. Around key assets the North Vietnamese had Anti-Aircraft Artillery, which was not really a factor until they fired the 100mm big boys at us, the only ones that could reach our altitude, and of more concern were batteries of SA-2 Surface-to-Air Missiles. SAM's were called *poles* because they had the look of telephone poles with fins while in reality they were larger. They locked on to us by radar and when fired they reached Mach 3 by the time they reached our altitude, usually 31,000 feet or more over the target, and their 250 pound warhead used a proximity fuse trying to detonate underneath us and scatter upward a cone of deadly ragged and sharp shrapnel. Our EW's job was to detect the initial SAM radar ping, called "*threat radar*," and then call out "*uplink!*" when he next detected SAM radar lock on, at which time the EW would do his work to jam their signal so the SAM would be confused and unable to see us if fired. The pucker factor went up when EW reported "*Missile launch!*" and we hoped that never happened from the IP inbound to the target when we couldn't evade with a diving steep turn. Over North Vietnam we always heard SAM threat radar and uplink from the EW but they did not always fire the SAMs because the EW's jamming seemed to work or they were just watching us and learning.

But whether we were crossing into the ADIZ of North or South Vietnam, Cambodia or Laos, whether there was a threat or not, if we did not hit these points and times close to the plan, a number of timing elements would be upset resulting in possible degradation to mission effectiveness, unacceptable to SAC standards. The Nav worked hard to hit the ADIZ as close as possible to the plan.

If the mission plan included fighter escorts, CAP or Combat Air Patrol, they would join us 20 minutes out from the target and, with their limited fuel budget, keeping to the mission timing was crucial. About the same time, our lead crew would made contact with the ground radar MISCU site and the countdown process began. These sites were the same scoring units used in the states to score bomb runs. They were set up in Vietnam to do the same in reverse because the B-52 bombing system was not sufficiently accurate for the primary target.

The second point for the number one Nav to sweat was the most important, our mission IP (Initial Point). It was a clearly recognizable geographic point, the beginning of our bomb run where our cell made a turn toward the target. The Nav calculated his guts out as he played the competitive game of hitting the IP time on the nose, at which time we turned toward the target and the Radar Nav began to follow instructions from the site to include a countdown with a hack to release. The IP turn marked elevated risk of AAA or SAMs or MIGs if over North Vietnam or the Ho Chi Minh Trail because we had to maintain a steady course to the target and could not take evasive maneuvers.

Before we hit our IP, MISCU initiated the warning on 243.0 UHF guard, "Attention all aircraft, attention all aircraft, heavy artillery warning. Avoid by 20 nautical miles the following area (map coordinates given) until (safe time given). Attention all aircraft . . . " Pilots always monitored guard for emergencies and knew from this announcement a Buff strike was coming which clearly meant – get the hell out of the way!

As we made our IP turn toward the target, the Nav was working on the third critical navigation point, the TOT, trying to hit it precisely at Time On Target as planned for our mission. Not only a matter of pride, this time was the final coordination of a number of events including the critical movement of our own forces on the ground. About one minute from TOT the Radar Nav called for opening the bomb bay doors and called for any course adjustments he needed, "IP inbound to target. Pilot, turn right to 287 degrees, center the FCI and keep it centered." Opening the doors at a precise time just before release was important not only because the more ragged profile and all that iron in the bomb bay made the SAM radar uplink more effective but the simple error of not having the doors open would negate the release. The pilot monitored a light on his instrument panel to confirm the doors were open.

"Twenty seconds," said the Radar Nav as he listened to the MISCU countdown. At the right instant Radar Nav executed and marked the release with "Bombs away!" so Radar Navs in two and three could start their own release countdown based on separation and airspeed. When Navs dreamed good dreams, they were surely visions of hitting ADIZ, IP or TOT on the button, zero-zero.

When 84 internal 750s and 24 500 pounders are released it takes 10 to 12 seconds but there is a time limit to control the size of the target. The Nav watched the time and after a predetermined number of seconds he told the Radar Nav to salvo the rest. While the crew watched an array of release lights to make sure the release was complete, a long string of bombs were falling to rain hell on the enemy, a series of heavy explosions running so quick in succession they could not be counted, with individual shock waves that were visible through the Radar Nav's scope from seven miles up. A few seconds later two and three would release to make the same target bounce again. Our targets were usually either some kind of hard enemy asset or enemy personnel, and bombs killed far beyond the blast and shrapnel that tore flesh apart and burned it to a cinder; the shock wave would burst organs like eyes, brains, livers, hearts and lungs, and if a man was far enough away to live through it he could still lose his eardrums. I didn't think too much about the effects on the ground, I just did my job, and when I did think about the suffering on the ground it was mostly about our guys getting killed by the enemy in the jungle.

We could vary the size of the path of destruction by increasing or decreasing our TOT airspeed and by staggering the two and three aircraft left and right, making the max ground box ½ mile wide and two miles long. Ground or ships radar systems scored our target results to give us proficiency feedback.

As soon as release was complete we had immediate things to do, like reducing the threat by making a steep PTT (post-target turn), closing the doors and starting through our post-target checklists. On occasions the doors were jammed by a "hanger," a bomb that hung up on something and didn't fall as planned. Butts were clenched tight when a Buff landed with a hanger since there was no telling whether it armed and whether it would stay put or bounce all over the airfield and maybe pick a spot to blow. Hangers were common but thank goodness they behaved most of the time.

We always had a secondary target in the event weather or other problem prevented a good run at the primary, and since the MISCU radar targeting assist had to be scheduled, we relied on the Buff's bombing system for the secondary target. Lead Radar would use GPI offsets to triangulate to the target, the ASQ38 radar system would figure time and heading to the target, then number two and three Navs would initiate their times. The Radar would give the pilots precise heading, closure, and distance information to put the aircraft in the right place at the end of the Nav's time and release.

If the mission planners had their heads on straight, our PTT and egress path was a relatively short path to feet-wet or coast-out to reduce the SAM threat as we flew over the South China Sea; if we had serious aircraft damage and had to eject we would be less likely to be captured and more likely to be rescued by SAR helicopters as they followed our emergency beacons. Now that our target was behind us, if we detected a fired SAM we would put the Buff in a 45 degree diving turn to evade, a little extra protection while the EW was jamming.

If we had battle damage but still able to fly it for a while, we could divert to U-Tapao AFB in Thailand, much closer than Guam, and if the situation was more dire we could land at Da Nang air base just south of the DMZ in South Vietnam where the 10,000 foot runway was long enough to stop safely, never mind that the base was not set up to support Buffs.

Under normal conditions we could reduce our stress level once we coasted out and settled in for the long flight home, and many missions in the south didn't involve detectable threats at all. By the time we arrived to land at

The Rock, fuel consumption made us well below the D-model max landing weight. In fact, we were almost always near minimum or bingo fuel.

With front and rear trucks and tip gear lowered with a careful setup to land, we didn't flare like most aircraft, we flew an approach speed that was above the landing speed. Once close to the runway we would flare the aircraft using both stabilizer trim and elevator, gradually bleeding down airspeed to arrive at the touch down point with the aft trucks lower than the front. If the speed was too high and touchdown was on the front gear, a nasty bounce always occurred. Once on the ground we would punch out the 38-foot drag chute to help slow us down as we applied the brakes. After taxi to our revetment and shut-down, we'd climb our worn-out asses down through the belly hatch, dragging our gear to the bus for a ride to mission debrief, which usually took under an hour. Finally, we would claw our way to the O-club for a drink or three or six, usually disappointed to find it filled with senior officers whose feet rarely left the ground. It seemed like our SAC wing was overloaded with colonels trying to outshine each other as they reached for the next rung on the promotion ladder.

The club was a big one, always with several hundred crew members in various stages of eating or drinking. There were several bars and a big dining room where most hung out. The dining room opened onto a large covered patio facing the pool with a table assembly at least 50 feet long where the biggest accumulation of full bulls outside of the Pentagon gathered in perfect pecking order with an occasional flag officer at the head. Most of these guys were there to punch a combat ticket on their way to the next level and few really occupied a position of any importance. Contrasting to state side operations, there was an interesting hands off relationship with the crew dogs occupying the rest of the building, almost as if the brass knew who the club actually belonged to and cared not to pull rank to challenge that.

While we were slugging it out on long missions, the gaggle of high ranking officers on the island seemed to impede, rather than grease, the wheels of getting things done. There were 135 colonels in maintenance alone, and nothing seemed to work quite as well as it should.

Despite the challenges of dealing with all that went with the Rock, our morale was good, an irony none of us spent any time trying to understand. The B-52 was built for and devoted to the SAC nuclear mission, but only in this conventional combat over-use of the aircraft did its strengths and weaknesses become well known. Flying SAC alert in the US, we tried to learn all we could about the Buff, but we wouldn't really know the aircraft to the fullest extent possible until flying over Vietnam again and again drove it into our blood. Only by pushing the airplane to its limit day after day did we know what it could and could not do. Only under the pressures of combat did we slowly become very good at our jobs, knowing the airplane's systems and control responses intimately. When systems failed, as they often did, we learned how to repair them, jury-rig them or how to circumvent them because we had no other choice.

We were pushing ourselves flying Arc Light missions over South Vietnam, North Vietnam, Laos and Cambodia, separated from our families and sometimes in mortal danger, and still our spirits were high. After all, we threw away the procedure book and our pre-mission mountain of paperwork was done for us by staff on The Rock.

Just one out of four Buffs on The Rock was flyable due to breakable systems and maintenance backlogs, and the one bird flyable might have 30 or 40 open writeups that would ground an aircraft in stateside SAC. We flew anyway, making judgments on what flaws we could deal with. Instead of being buried in what sometimes seemed like chickenshit in the US, we were doing what we were meant to do, the intense work of a combat crew and learning a lot about the airplane, about each other and about ourselves. We became crew dogs.

One measure of morale was DNIF. A way to weasel out of a combat mission was to visit the flight surgeon if something was wrong, and hope he would put you on duty not including flying status for a while. As a matter of honor, every self-respecting crew dog on The Rock would fight hard to avoid DNIF. If it hurts, suck it up and fly, Nancy!

ON ONE MISSION THAT first rotation as Les Hatcher's Co-pilot, our Wing Commander was on board the cell lead aircraft as ABC. I didn't know Colonel E.O. Martin, and didn't care, especially since he wasn't on our bird, I had my hands full doing my job and keeping Les off my back. The target was somewhere in South Vietnam.

We hit the IP on time and MISCU started the sequence of heading changes that would lead to our release of an A-load of 108 500 pounders on each bomber. Darrel Fisher, our Radar Nav, was nervous because all he could paint on his radar was the thunderhead sitting right over the target. At the last second, lead aborted the run and broke away from the storm to our secondary target.

That bomb run would be off the lead ship's radar. As we approached the release area thunderstorms were in all quadrants with an especially big one right over our target. At the IP we made our turn and the ABC broadcast "We will release!" We were at 33,000 feet as number one called the countdown to release. We watched lead disappear mid- point into the thunderhead which towered above us at least another 20,000 feet.

Seven seconds later we flew into the storm and released a second later. While bombs were falling all hell broke loose from hail to lightening and turbulence so bad none of us could tell up from down. As the bombs pulsed off for what seemed like a full minute, I feared a mid-air collision. Someone called, "Break right!" which was never a good idea when you don't know who is breaking which way, and Les heaved her over to the right, assuming the other aircraft would do the same. In a moment we emerged into clear calm blue air. The sudden stop of turbulence was like a shock and the old bird shuddered just before we ran mid-point into the next storm as if it had been waiting for us. Since there was no telling where the bombs went, I began to wonder about our leadership as exemplified by our CO in the lead ship, taking a big risk of putting bombs in the wrong place to flex his muscle.

A few days later we were flying cell lead out of Guam and number two had our storm bird. They complained the whole mission that something was wrong with their plane as it would not fly right. We didn't have the heart to tell them it was bent.

OUR PRACTICE WAS TO fly a month out of Guam, then for a change of pace we would rotate to U-Tapao AFB in Thailand for two months then to Kadena AFB in Okinawa for two months. I'll tell you first why Kadena didn't last long after I arrived.

Flying out of Kadena was almost like R&R. It was a civilized place with real restaurants with good food, real ice cream for Pete's sake, even maids who would dig through our clothes to find what needed washing while we were on a mission. We'd return from a mission to find fresh clothes and even socks and underwear folded.

Kadena had a great officer's club with a panoramic view of the runway. One night I sat at the bar nursing a drink as I watched a Buff cell with A loads launching. One AC for some reason aborted too late, after S-1; he almost got it stopped but hit the ditch at the end of the runway. The airplane's back broke, it immediately caught fire and the crew scrambled out, but the Gunner had to drop a rope and on the way down he got tangled up in the drag chute webbing. The EW scurried back to help the gunner untangle, whereupon they ran from the airplane but not fast enough because they were caught in the first blast as the 500 pounders started cooking off. They lived for only a couple of weeks while the other crew dogs were fine.

When the bombs started cooking off the shock wave knocked me off the barstool. We flew a mission the next day, passing over the wreckage on takeoff; there wasn't a piece of that airplane left as big as a piss cutter, it was just a big black smudge on the coral island. The natives were restless anyway about the warmongering US military base on their island and that incident pushed the tension up a notch. We continued to keep our tankers there but no more bombing missions out of Kadena were allowed after the spring of 1968.

We still had the relief of U-Tapao in Thailand, and we liked it quite a lot even though it was constantly 95 degrees and near 100% humidity. To rotate we would fly a mission out of Guam, recover at U-Tapao and stay for two months. Our missions were daily 3-4 hours and a much more relaxed atmosphere. The missions couldn't be any shorter because it took over two hours to get through our checklists, and without the need for a lot of fuel we didn't need to inject water to boost our takeoff thrust. Failed water injection was one more reason to abort at The Rock and Kadena, but was off the table at U-Tapao. We also didn't need air refueling, eliminating a tense part of each mission.

Our U-Tapao schedule was flying six days with the seventh day off. Things got done at that base, giving us the notion we had discovered a cosmic principle: the efficiency and morale of an operation is inversely proportional to the amount of brass involved. There were only six bird colonels and one brigadier general in the entire wing staff at U-Tapao, where nearly every bird was flyable and we ran the same number of daily sorties as The Rock with about half the crews. U-Tapao worked far better than brass-heavy Andersen on The Rock.

On one mission out of U-Tapao, after release we started picking up guard radio traffic of a desperate unit of US Marines needing help in a fire fight near the coast. We started relaying requests for close air support and other things. They were pinned down near the beach with heavy casualties. I spoke with their radio operator on and off until help arrived. Over the years I often wondered how they made out. Only recently I spoke with a friend I have known for some time at Hatteras, NC, about his experience in Vietnam and he related a similar story. We think we may have talked with each other that day, but we don't know for sure.

A few days later we were on a routine run back to the Mu Ghia Pass area to harass the Chinese workers maintaining the Ho Chi Minh Trail after our bomb craters gave them more to do. Somewhere on that mission we lost a hydraulic pack, one of nine operated by pneumatic air. The Co-pilot ran the emergency checklist and we isolated several systems along with the suspected pneumatic leak. This was a fairly common problem, still serious, but in this case there were no further failures.

A day or so after our uneventful air leak problem, a friend launched with his crew and shortly after takeoff everything went to hell in a hand basket on their airplane. First, the flaps would not retract, then they lost generator and hydraulic systems. Many of the navigation team's and the pilot's instruments were erratic or not functioning. Various fire lights were on including one in the bomb bay. The crew knew there was a major problem, but with no idea where to begin they clearly needed to get back on the ground. In a B-52 that was not easy, and of course they were already talking to U-Tapao's Charlie.

Landing a Buff above the maximum landing weight could be a serious mistake. The obvious first step was to jettison the weapons and then burn off fuel until landing weight. They worked their way towards a planned jettison area and managed to get to the minimum altitude of 7,000 feet. After clearing below, they tried numerous times to jettison but the bombs would not release. Charlie and the crew were running out of ideas. They considered ejecting but eventually decided to bring the bird in somewhat above the maximum landing weight. U-Tapao had only one runway so it was a big gamble; if they had a bad landing that blocked the runway, they wouldn't be able to recover other aircraft going bingo on fuel, and a Buff ate about 10,000 pounds of fuel on a go-around.

With the pucker factor high and rising, the AC took it in after one of the shortest B-52 flights on record, about 11 minutes, and made what may have been his best landing ever. The drag chute deployed and the crew stopped the Buff in the gravel off the end of the runway, then with a little help they got the airplane the hell out of the way so operations could continue. With EOD and maintenance all over the plane, a pneumatic leak was found in the bomb bay.

The leaking super-hot air had burned through a M117 750 pound bomb casing and bundles of wires that powered who-knows-what? Tritonal, the explosive mix in these bombs of 80% TNT and 20% aluminium powder that accelerates max pressure of the explosion, was leaking onto the ramp, giving crew dogs the willies about how razor-thin and miraculous had been their escape.

I can remember drinking Pabst Blue Ribbon beer, eating chili dogs and talking over this incident with several other crews at the 105 Hooch while watching the python in the cage with its next meal, a chicken, sitting on his back. Thus, war stories became lessons we passed on to our peers and others who followed.

We flew 55 missions with Les and learned a great deal. With 179 days under our belt, we flew our airplane home to Columbus in January, 1969.

BACK TO SAC ALERTS, certifications and klaxon launches with solitary sorties. In Guam we certified for EWO missions since that had been Andersen's primary SAC purpose, and now back in the US the certification rounds recommenced.

Going back to doing things by the book after flying combat missions was a little irritating, but the EWO mission to protect the country and secure nuclear weapons remained a very serious undertaking and had to be done right at every step. At the time, I didn't quite realize the enormity of what I had learned about the B-52 or how I had matured on those bombing missions. The irony is how much better we were prepared for an EWO mission because of our conventional warfare missions in Vietnam; we were more effective as a crew, better at our individual jobs, and far more capable of nursing the Buff through tricky situations.

In June I was transferred to the 19th Bomb Wing, 28th Squadron at Robins AFB in Warner Robins, GA, where I was upgraded to AC a year later in July 1970 at 27 years old. With just three years flying, I got my first crew: Co-pilot Bill Gottschalk, Radar Navigator Ed Herr, Navigator Jim Garrett, Electronic Warfare Officer Tom Bardsley and Gunner Mac McGuffin. This was a good crew; Ed and Mac had strong experience and Mac was old enough to have been my father. We were flying G models now, with stronger engines and much reduced risk of takeoff engine failure, a stiffer wing, fixed external fuel tanks and a smaller payload since Gs didn't have the big belly modification. Mac's Gunner seat in the G was rear-facing right next to the EW, where he controlled the quad-50 in the tail by electronics and radar.

We had a few months to get settled with each other as a crew before we received orders for a rotation on The Rock.

THERE WOULD BE A sea change my second rotation to Guam. As AC I was responsible for the airplane and the crew on every mission, and that changes your perspective in a hurry. I still had much to learn, as I was reminded on my third Arc Light mission after the first two had come off without a hitch, hitting targets in the delta near Saigon.

This time our D-model Cobalt Cell target was in the Ban Kari pass on the Laotian border with a secondary target just south of the DMZ. Our pre-brief started at 1900 with a planned launch of Cobalt at 2130. After brief, pack-up and bus ride, It was raining as we started the walk around. By the time we were complete outside, the rain had stopped, but no problem, it usually felt good at Guam and crewmembers had a way of doing a walk around while remaining reasonably dry. Rain didn't slow down the ground crew, especially the weapons loaders.

We climbed inside and started our checklists. In less than 15 minutes Ed had found a problem with his bombing system. We had Red Ball on the problem in no time but three or four more things went wrong and by engine start time we reported to Charlie the bird was a no-go. Charlie instructed, "Cobalt two, shut it down and bag-drag to 5070, Spare one. Spare one, taxi to the hold line second position behind Cobalt one and stand by to turn it over to the Cobalt two crew." The ground crew and my crew scrambled ourselves and equipment, racing around the flight line in vehicles to catch up to Spare one, soon to be Cobalt two. This was just one small part of the choreography of the elephant walk as six tall tails moved toward launch and a host of supporting vehicles buzzed among them, nursing them on their way. The spare crew had done their job and by the time we got our gear on board, relieved their AC and Co-pilot and strapped in our ejection seats the engines and all systems were running and ready. Since we all had spare duty occasionally, we fully understood the big smiles of the spare crew because they narrowly escaped this grueling mission.

When we started to roll it was dark even with the ramp lights on. Black tall tails were taxiing or being towed everywhere and they seem to suck the light away. My worst fear was having a taxi accident and having to explain it, so I did the taxi and let Bill do the takeoff. Things were starting to fall into place and I began to have that sense of rhythm or whatever it is when a formation is working together.

The launch was countdown perfect, departure and join-up went without a snag, but there was a typhoon that night in the vicinity of our air refueling tract. We knew that we would be above the heavy stuff but high cirrus might be a problem with visibility. A week before, one crew had experienced some nasty turbulence during refueling and had skipped the boom off the EW's ejection hatch. It scared him so bad he ejected, unfortunately, into the oncoming storm. SAR had been trying to find him for several days unsuccessfully. Everyone felt miserable about that but to make matters worse, the beacon from either his life raft or gear dropped by SAR was still going off and we knew we would have to listen to the sick puppy beeping while trying to refuel.

On the way to the tanker rendezvous, Tom, our EW and the only bachelor on the crew, kept us entertained with all his exploits with women past and future, real and imaginary. When we got to the AR point just short of the Philippines, sure enough, the beacon was still going, and with every beep we felt for our fellow crew dog. Visibility wasn't bad but there was moderate turbulence.

As Cobalt cell approached the tankers, which were out of Kadena, the formation split and each made rendezvous on their respective tanker. We stabilized and connected in unusually bumpy air. Air refueling is tricky even in calm air, though the top pilots at the 52 instructor school could run the limits over 60 degrees of bank while steadily connected to the toads. Mere mortals like me had to work hard to get the gas if we wanted a paycheck. I was still learning every time I flew. We needed 110,000 pounds from the tanker to complete our mission back to Guam and we had only 30 or so minutes to get it. I cannot remember being rougher on the controls than that bumpy refuel but after about 25 minutes and only a couple of disconnects the boomer called offload complete. We backed off, joined up and continued the mission. Nav informed the crew that we were 90 minutes from ADIZ.

The Navs were earning their paycheck here and they backed each other up in the cell to make sure lead was on time. Wendell Weaver was the AC flying lead. We went to flying school together but he was ahead of me by 20 missions and was wave lead qualified. I knew he had a good crew and I had plenty of confidence we would hit the target without scattering 324 weapons in the wrong place. Our ADIZ went well and we started a predetermined route to the target, meeting our CAP of F-4s 20 minutes out. Somewhere inbound, Tom picked up signals from a SAM site that he said was under construction and of no threat. It was a galling thing to know the location of sites under construction for weeks or longer and not be able to hit them until they were fully operational and capable of uplink, but such was the political nature of the game we played.

Our cell hit the IP on time, turned to the target, opened the doors one minute out and released our A loads in what seemed pinpoint fashion, immediately starting our PTT as we closed the doors and started post-target checklists. I remember that we were very near the coast. At 38,000 feet you usually do not see a whole lot on the ground at the target especially since it's behind the aircraft. This time Mac checked in with unusual excitement in his voice, saying, "We had secondaries and big ones! The blast and dust is coming up fast." I couldn't resist rolling steeply back to the right to see a cloud pushing an estimated 14,000 feet above the target area and still climbing. It gave us a clear understanding of why we needed at least 7,000 feet altitude if we ever elected to jettison a full load.

More importantly, I had just made a mistake with the D model at that altitude; my little maneuver had stalled out number four and five engines. We got back behind lead but soon realized we could not keep up with him on six engines. I remembered learning in one of the schools this could happen and the solution was to either maneuver the aircraft or descend. Descending was not good because we knew the enemy had 85 mm AAA up here so I elected to maneuver. That was not a good option either because soon I had another two engines stalled and we were losing altitude. I don't know which was worse, being embarrassed or feeling the pucker factor trying to take a bite out of my seat. We watched as Wendel, his crew and number three grew smaller in the distance, leaving us behind. By the time we got the engines back and regained our altitude, Cobalt one and three were 130 miles ahead. Since those guys left us in the dust, we vowed as a crew to beat them back to Guam. As we coasted out, someone asked the Nav what our arrival time would be and he said something like "don't ask." We beat them home. It took 5,000 pounds more fuel but the tanker had given us a little extra. After 13 1/2 total flight hours we landed ahead of lead and they agreed to buy the beer.

On the bus to debriefing, Mac took me aside and said he wanted me to see something. It was his helmet, split down the top after he had been bounced off the ceiling of his Gunner compartment during my rough refueling. He just smiled and never said a word to anyone about it. It taught me a lesson and over time I got pretty good at taking on the gas.

Debriefing took about an hour this time since they were really interested in the secondary, saying it had to be a sizable stockpile the enemy was staging for an operation down south.

We flew again in two days and maintained that schedule for a month until we deployed to U-Tapao where the EW would be much happier.

I WAS WISE ENOUGH as a new AC to keep my head down, eyes and ears open, but that becomes harder to do as you gather experience. And so came my crew's first distinction of Wave Lead. This gave all of us some sense of accomplishment even though Ed, Mac, and the EW had been there before. On our first mission out as a lead crew the Nav, Jim Garrett, would distinguish himself even more. I'm not sure what happened but the Nav just didn't have any backup from the number two and three Navs when he screwed the pooch.

We missed the ADIZ on our first lead mission by eight minutes and 25 miles. That was as bad as the Radar Nav bombing the wrong target or a pilot screwing up a tanker dip and failing to get one drop of gas. We stumbled through the mission eventually hitting our primary target but it was a somber cell going home. Our crew was looking the other way except Ed, the Radar. He was in the slow process of reducing Jim to the lowest thing on earth, whale shit on the bottom of the ocean. Our crew's reputation as wave lead was stained and Jim had become as untouchable as a leper, left to stew in his own juices and pray for a chance at redemption. Visiting the bar or cafeteria was uncomfortable when he was there, his fellow crew dogs mercilessly pretending he was invisible.

A week later the wing staff was sufficiently desperate for a wave lead crew that they gave us another shot. Yippee. It was also our turn to rotate to U-Tapao.

Launch and refueling were uneventful. Navigation at the time was still in a primitive stage using basic heading and time, celestial navigation with a sextant and occasional semi-accurate updates with the radar when over land.

Between the Philippines and the Vietnam ADIZ was the pit Jim had to deal with, several hundred miles of open ocean.

We hit the ADIZ zero, zero. Not only did he pinpoint the geography, but he nailed the time perfectly, a Nav's dream. Jim wallowed in glorious redemption. We were all happy for him, eager to be rid of the discomfort we all had with his failure while everyone was watching. From that point the mission mood was upbeat and that we would recover and stay at U-Tapao was a bonus. Crew chatter was not directed at Jim but we were all relieved and glad he had survived Ed's torture. It was a long way from Navigator to Radar Navigator on a B-52 crew and that standard had been on stark display to the rest of the crew.

Bill made an excellent landing at U-Tapao, a world apart from Andersen where every full bull in SAC tried to make himself look good by punching another Arc Light ticket. As we climbed down the hatch beneath the crew compartment, the strange fragrance of palm oil, lavender and sewage welcomed us back to Thailand. The EW led us through billeting where most of the young women working there remembered him and fondly called him E-dub and talked about butterflys. Shortly thereafter we checked into our quarters, air conditioned trailers without windows so the time of day didn't matter. Our routine of six days on and one off began without a complaint in sight.

AFTER A WEEK AT U-Tapao the routine was setting in. The Thai people were friendly and pleasant, the women pretty and graceful in their fresh necklaces of sewn flowers. The Thai people were so easy to like and they took care of many of the functions on base very efficiently.

The missions were shorter and seemed to be more directly connected to our ground forces in Vietnam, giving crew dogs the satisfaction of some accomplishment, hopefully of assistance to anyone unfortunate enough to be slugging it out on the ground. Simply put, U-Tapao ops were more interesting. We dropped some unusual ordinance, like some bigger bombs or weapons with long delayed fuses that wouldn't detonate until the gooks dug them out and were halfway to Saigon to make a booby trap out of them. And then there were Blewies.

Now and then we were blessed with a bright idea from one of the staff geniuses who didn't fly, whether on The Rock or elsewhere, whether an Air Force or other brand of genius, a bright guy when it comes to theories he does not himself have to put into practice. One of those ideas was something we called "Blewies." These were small anti-personnel cluster-bomb type weapons designated CPU though I don't recall what that means because we just called them Blewies. They came in batches of 100 in an aluminum box two feet square. When dropped, the box would open up, Blewies would scatter with little fins popping out and when they went off on the ground they caused nasty wounds with little ball bearings. I don't remember whether they went off on contact or if they waited for an unsuspecting passerby in the bad-guy boonies where we dropped them. Some genius figured a fighter can carry two boxes of

Blewies but a B-52 could drop a whole lot of boxes out of the bomb bay. So we had a few runs with Blewies and discovered these little bomblets weren't designed to be dropped in 500 knots of turbulence around the bomb bay of a Buff and with the help of their cute little fins they would fly all over the place. Blewies would end up when we closed the doors rolling around in the bomb bay, as many as hundreds of them, sometimes armed, some went off. So after a Blewie drop we'd land away from other aircraft and sit there while EOD guys would come out and pour plaster of Paris around these things so they could be moved, one more argument for requiring the genius with the idea to stand by ready to personally clean up the mess he caused.

THE MOOD WAS DIFFERENT in Thailand, at least at U-Tapao. All the troops in support of the mission from ground crews to EOD and especially the maintenance folks were happy workers. The 92 degrees and 100 percent humidity did not seem to matter. Maybe it was that whiff of lilac coming out of the jungle that made a difference. Whatever it was, we weren't too surprised when we learned the E-dub was in love.

Our only single crew dog had been smitten with one of the dancers in the club and was now spending all his available time at her hooch, several miles off base in an unsecure area. Intel smelled the bad guys planning some kind of local strike and urged everyone to use caution; we were ordered to restrict personnel from overnight stays off base. That made no difference to Tom while he was dipping his wick, but it caused the rest of the crew a lot of worry, especially me since as AC my ass would be the first to burn if he was caught. We explored creative ways short of a direct order – he was our own crew dog after all - to keep Tom in one piece. Finally, the problem took care of itself.

One night after flying the Ebony cell, Tom left the debriefing around 2300 and headed off base to his gal's hooch. Around two in the morning he showed back up at the trailer visibly shaken. He told us that when he got to her place he knew she would be working for several more hours so he lay down on the bed for a nap. When he woke from a deep sleep it was very dark but he could feel something warm and heavy on his chest. He thought it was her at first but then as his eyes focused, he realized it was a rat the size of a cat inches away from his face. As he started to move the rat lunged forward and bit him on the lower lip. Tom said he was instantly airborne, rat and all, as they fell from the hooch into the khlong below, one of the artificial canals used for transport and discarding sewage. We took him to the doc and a few stitches and shots later, Tom was back flying with us forever cured of overnight stays off base.

That story was worth more than a few beers as it was told a thousand times at the bar, Tom laughing with the rest of us.

OUR TIME AT U-TAPAO was coming to an end and the crew was busily gathering things for our return to The Rock, including goodies they purchased to take home about a month later. Margaret and I still have things in our home from that deployment including one of my favorites, a cistern with dragons along the sides that cost maybe five bots, equivalent then to one U.S. quarter.

The Young Tiger bunch, our tanker toad counterparts who we lived with on SAC alert, the toads we deployed with and talked to occasionally at the bar and on the radio during the refueling, the same toads we helped to locate the air refueling control point, hauled tons of this stuff around with them, but we never depended on them to help get our booty back home even though they essentially flew cargo aircraft. With a special place for toads in my heart, I'm going to turn on them and tell you why.

B-52 crew dogs were used to long hours and rarely complained about being over worked, while tanker toads never passed up a chance to complain. Crew dogs were used to lugging heavy things around so cisterns, hibachi pots, brassware, motorcycles and probably a few Thai women were no problem. You could depend on crew dogs to keep their word and even drop the bombs on target if they could reach it by dry tanks and then eject if that's what it took to do the job. Tanker toads, more attuned to protecting their own hide, always talked about going to Pitcairn Island when the nuclear balloon goes up and I believed them. Besides, we had bomb bays that could haul over 60,000 pounds of weapons and deploy non-stop anywhere in the world, off-limits to Customs inspections, so who needed toads to haul our stuff home?

Part of the accumulation of stuff was intended to please our wives and make coming home a little more like paradise. Lucky for us we had Johnny's Jewelers at U-Tapao, one of a dozen Thai concessions with the BX. Unlike the others, Johnny offered a deal for a crew that spent over $150 each with him during the tour; Johnny would put the crew up for two nights in Bangkok including a 15-course Mandarin dinner. We bought princess rings, sapphires, tiger eyes and opals until we met the requirement. The whole crew except Mac, who was older and may have known better, were off to our Bangkok adventure for the first time. We lost the Co-pilot right off the bat and worried about him for about 30 seconds.

I think we went from one end of town to the other about twelve times. The Thai people were so easily accommodating that even Ed, the tough-guy Radar, was having a good time. At the end of the second day we started worrying about Bill, our Co-pilot. He surfaced at the last minute on wobbly legs and we realized that he had exceeded his own personal limits. But, who cares, we were now back together and the grand finale would be the 15 course Mandarin dinner. I have to admit, after all these years, I still remember that great meal. We started off with drinks, then salads, then prawns the size of small lobsters, followed by duck and on and on for three hours. We had a mission at 1700 the next day, and we would recover after the long trip back to Guam.

We slept on the ride back to U-Tapao which was a testament to how tired or drunk we were since you never took your eye off the road with a Thai driver, and never, ever, said "Lao-Lao!" which meant "Go faster! The Nav was so drunk he chanted "Lao-Lao!" most of the way back. Somehow we made it in one piece and found our way back to the trailer where it was 72 degrees and eternal darkness.

We were up early the next morning because the Co-pilot insisted he needed to see the flight surgeon, Doc Stritchard, who happened to be a friend of the crew after we took him on a mission over the delta. He was much fun and told stories about open heart surgery on Thai dogs to remove heart worms. He also had a crush on our trailer maid and we had baited him on that issue many times. He would play along with our gag; we couldn't help ourselves setting up our Co-pilot to be told by Doc Stritchard that he had a communicable disease since he was only 30 days from home. We must have under-estimated how lousy Bill actually felt because he was not in good humor when he returned from the Doc and we weren't about to ask any questions.

Preparing for the final mission out of U-Tapao that day was a little longer than normal since we needed to out process for some reason. This happened to be right near the BX and while sitting there waiting our turn the Base Commander came by. He stopped and spoke with us, giving platitudes and appreciation for things none of us understood, then he offered to sell us carved Vietnamese elephants of three different sizes. I was the only taker and hope to think that is as close to brown nosing as I ever came. I paid the Base Commander $36 for the carved elephants that were to be delivered to me by tanker to Warner Robins as soon as I got back. I remember thinking, "Good luck on that!"

We went through our briefings, loaded all our gear including our booty with the assistance of maintenance personnel specifically assigned for that purpose. Our personal items, and some for the ground crew, were placed in 4x4 heavy duty cardboard boxes, carefully wrapped and placed in the 47 section of the plane behind the bomb bay. They knew who to trust to get their things home so it was a quiet mutual understanding.

Our mission went without a hitch. Our target was a suspected build up of VC. We released on time in total darkness and made the post-target turn towards our egress point and on to Guam.

Nobody knows for sure which crew dog first felt rumblings in his bowels, but we think it happened shortly after PTT with one hell of a long way to go.

A B-52 crew honors the sanctity of the honey bucket. It was never to be used for its intended purpose as a place to drop personal bombs. You had to be tough to fly the heavies. There was no flush and if you use it, you clean it. That rule kept the honey bucket dusty and unused until the day after Johnny's big dinner treat in Bangkok, the day of a heavy honey bucket workout. In my 6,000 hours on the Buff that was the only time the honey bucket seal was broken.

It hit us all without warning and one by one we took our turn until someone announced that the honey bucket was full and there were no extra plastic bags. Why would there be? It still had in it the original plastic bag when Boeing built the plane in 1954. We then resorted to helmet bags using the Nav's first since he was junior and got us in that mess with the ADIZ. We then went down the chain of command until we discussed whether we should share our predicament with Charlie because we were feeling green and kept the pressure on the honey bucket. Charlie could fix anything, right? Reputations were at stake and we had to think about that for a while.

Knowing Charlie, he would use our misfortune to cajole crews for a month into taking any piece of shit that could fly and do it with humor and zest, at our expense.

We then seriously thought about coming up with some excuse to divert into Clark AFB in the Philippines, but that would get us into even deeper shit because we would probably take out most of the runway lights with our tip gear on the 150 feet wide runway and really piss off all the TAC troops. We decided there was no other choice but to suck it up and find more helmet bags.

Somehow we made it to The Rock. When the crew chief opened the hatch after parking, he got an involuntary whiff from the honey bucket adjacent to the hatch and backed off, yelling something like, "Who the shit flew this thing here?"

I suspect there's a ground Crew Chief around today still telling the story.

After cleaning up the crew compartment and disposing of untouchable helmet bags, we apologized profusely to the Crew Chief and quickly asked the bus driver to take us directly to the flight surgeon.

Mac, the wise old guy who passed on our Bangkok adventure, had been mercifully far away from us in his Gunner's compartment in the D model tail. On our way to the flight surgeon we dropped Mac off at the Arc Light Center laughing so hard he was crying. No telling what he told them but we were never asked to debrief the mission.

With the exception of Mac the crew was DNIF for the better part of a week. To make matters worse, about the time Tom and I were feeling well enough to play some paddle ball, Tom took a wild swing and caught me in the upper lip requiring stitches. Mac would not let us be. With Tom still healing where the rat chewed him on the lower lip and now me with six stitches in the upper lip we avoided going out in public together.

The Co-pilot was in really bad shape. Maybe our gag had gone too far with him because he was sweating bullets. None of us had any idea what the doc told him at U-Tapao and none of us had the guts to level with him for fear of getting a beating; Bill was a pretty big boy. He started keeping to himself and I had a feeling that we all needed to get back in the air to settle things out. We finally flew Sunday night and it was the Gunner's turn.

When we got to the Arc Light Center for strike mission brief, everyone was prepared for the worst. None of us said a thing as we went through our briefings. We thought Mac had served us well until during the dog and pony show slides popped up showing us removing the honey bucket with thirty or so maintenance folks watching. Everyone long since knew what happened and we had no choice but to go along with the fun. They had probably been showing those slides all week. Now we had to face Charlie with the rest of the crew force and the Russian trawler listening. Damn Russians.

As we were doing our walk around, Bill was helping Mac load his gear up the ladder stand into his tail compartment. When Mac opened the hatch, he pulled himself in and then suddenly flew back out and beat Bill off the stand which was a good ten feet high. He started yelling that Tom's rat was in the gunner's compartment. We called Charlie, figuring this might be a good diversion from the honey bucket issue as we knew he would be loaded for us. He replied, "Stand by for rat detail, honey bucket crew!" We were screwed. The rat was only going to make it worse.

Within minutes EOD showed up acting like we had an armed hanger. They went to the tail section and confirmed that we did indeed have a rat, a good 24 incher and that there was no way they could get it out before our scheduled launch so Charlie instructed us with the rest of the world listening to fly with Mac forward and the rat aft, manning the 50 cals. EOD assured us that with the aft cabin depressurized the rat problem would be fixed at 43,000 feet and minus 50 degrees C. Now Mac was like the rest of us, taking the hits as they came.

Even the toads had heard the story and we had to endure their comments all through the refueling tract. But the kicker was a comment from the MISCU radar center about how the rat was doing. Charlie had done his job well and the crews loved it.

Thirteen hours and 45 minutes later we parked on Sugar row and maintenance opened the gunner's hatch. The rat jumped out and ran across the ramp, undoubtedly home to tell his kin about honey buckets and scared gunners.

SO MANY THINGS ARE learned in the pressure of combat missions and enshrined in training programs. One of those is dealing with the loss of the vertical stabilizer in ways a new pilot would never guess.

On each wing are the ailerons, but on the D they were tiny and well inboard on the wing. To cause the huge plane to roll, Boeing also designed the airbrakes to act as spoilers. When you turn the wheel, one aileron goes up, the other down, but a giant bank of spoilers on each wing also raised, much like a picket fence, causing the plane to roll. On a 52 when you turn the wheel - or apply airbrakes - the airplane would also pitch up, a nuisance for which pilots compensated. Spoilers also served as speed brakes by disrupting lift. The book said with the speed brake lever pulled all the way back, you could get 33,000 feet per minute rate of descent at high altitude for tactical descent into a target area. All of this helps me tell you about how knowledge has been preserved among crew dogs over the years.

Not long after the first bombers were built, Chuck Fisher, the Boeing test pilot, lost the vertical stabilizer in severe turbulence over Colorado. To compensate, he lowered the aft trucks, which increased drag on the aft of the airplane. Then he pulled the airbrakes inboard control circuit breaker so only the outboard airbrakes worked. That gave it some lift aft, helped stabilize the airplane and he was able to fly it. As a result of this incident, on D models they reinforced the vertical stabilizer, and on G models they took off the top 12 feet of the fin. They taught Fisher's maneuver in 52 school. In fact, they made a short film of it which is still shown today even to airline pilots.

This was a good thing since one day with Ed White on our wing on a mission out of Guam, the same thing happened, second time ever. Coming off target and feet wet, Ed's Gunner was taking his post-target nap. When he stirred, he caught movement in his peripheral vision and looked up to see the huge vertical stabilizer slowly moving off to one side. He rubbed his eyes to look again and it took him a minute to believe what he was seeing. He said to the AC on interphone, "You're not going to believe this but the vertical fin back here is moving around on its own."

About that time they started having control problems. The vertical stabilizer is held in place with two big pins called coke bottles. One pin worked itself out and the fin was free to pivot. Ed had seen the training film from Fisher's tricks; he lowered the aft truck and pulled the inboard speed brake circuit breaker, raised the speed brakes and stabilized the airplane. When he landed at Kadena, the fin just slowly lay down on the horizontal stabilizer and became a new problem for the maintenance crew.

IT WAS EARLY FEBRUARY, only a week before our deployment back to the states, where we had missed winter. At some point the honey bucket jokes had worn off and we had been in the three-day cycle at Guam long enough to be ready for a break. Home was getting to be our main focus. It was, however, never too late to do something stupid.

All four Charlies had developed some degree of fondness for my crew since we had been the brunt of crew humor for longer than most. We weren't getting any special treatment but at least there was a slight tone of respect in his voice when we discussed a problem with the aircraft. Flying Blue Cell lead one night we could not pressurize the plane on climb out.

Everyone stayed on oxygen while we made a lead change with two, and then leveled somewhere around 20,000 ft.

"Charlie, Blue two"

"Blue two, Charlie, go ahead."

"Rog, we can't pressurize. We ran through the procedure and no go."

"Gottcha Blue two, (without hesitation) open up the panels below the left circuit breaker panels behind your seat and locate the outflow valves. There will be a dozen or so Zeus fasteners holding those panels in place."

"Roger, Charlie what then?"

"Well, what do you think? Look to see if they are open or closed?"

No comment, I know when to shut up. After several minutes; "Charlie, the aft outflow valve is open about two inches; what next?"

"Put a number 12 brogan on it and if that doesn't work get a lightweight flight jacket, soak it down and wrap it around the valve. It will freeze, seal the valve and the other one should regulate cabin pressure."

"Roger that, Charlie." The boot didn't work. We looked around and decided to use the Nav's fight jacket.

After about 5 minutes; "Charlie, Blue two, it looks like we can get about seven pounds of pressure but not much heat to go with it."

"Roger Blue two, if you're happy with that so am I. See you tomorrow."

"Nite Charlie."

For the next eight hours it got colder and colder until about the time we were crossing over the Philippines on the way home we must have been close to ambient. No telling how many times we cussed having to haul around our arctic gear flying just a few degrees off the equator. We now knew the wisdom of that and initially joked about whose German Shepherd they killed to make Tom's collar. We looked like a bunch of Eskimos with helmets, parkas and mukluks.

After a while the jokes faded as we realized we had four more hours to go and not one runway between the Philippines and Guam 1,500 miles away. No way we could go to a lower - warmer - altitude and have the fuel to make it.

After all the harassment we had taken, and after I agreed with Charlie all was copacetic, we were once again in the same boat, seemingly without paddles. Damn Mac, back in the tail was toasty and giving us the usual. The Navigator even talked about crawling back there to join him since we had taken his light weight jacket while the rest of us were wearing that plus our parkas. The crew took turns coming up to my seat or Bill's to warm their hands on the windows which were heated. I wish we had a camera when Tom put his feet up there. Ed the Radar was in a grin and bear it mood but insisted that I would buy the beer for the rest of the tour. I told him that if he screwed up and missed Guam, we would cut him open and warm our hands on his liver before we ditched. There was some discussion of how our obituary would read, "Bomber crew freezes to death before ditching in the equatorial Pacific".

At some point Bill came to life and made some rather amazing calculations. He figured that we could start our descent 400 miles out to a warmer altitude and still have gas to make it back. We knew we had to tell the boss about that maneuver so I asked Tom to give Charlie a call on HF and tell him what was up. I figured they probably had a shift change by now since they worked 12 on and 12 off.

I monitored the call and Charlie responded, "Understand Blue one, sorry you guys had such a cold one. I'm sure you checked the circuit breaker on the left load central right behind the pilot so I'll pass this on to maintenance. If you get short of gas, let us know and we'll get a tanker to you."

Tom responded, "Thanks Charlie, see you on the ground."

Bill and I looked at each other without saying a word. I think we both got out of our seats at the same time leaving George to fly.

We had checked that breaker before we called Charlie the first time 12 hours before. How could we have missed it? There were several hundred breakers on that panel but they were all black and when popped the white ring would stand out. Bill checked the book and it was breaker number 102, cabin temperature control. We looked closely . . . there was no white showing. In fact, there was not even a circuit breaker. It was gone. We looked down at the deck and there in pieces was the circuit breaker. After it popped someone probably brushed against it and broke it off.

There was no way around telling Charlie. We would have to report it anyway and maintenance would then complain to Charlie about clumsy, stupid, crew members, and we would catch it again. We were in UHF range now so I made the call.

Charlie responded, "Got it Blue two, that's a new one and I appreciate knowing that could happen. By the way, take the eraser end of a pencil and insert it in the hole to see if it will reset the circuit.

The eraser worked! Within a few minutes the cabin started warming up. We could hear Mac laughing his ass off 150 feet behind us.

The Honey Bucket Crew
L-R: Rep Wayne N. Aspinall (D, CO) and wife, Nick Stevens, Bill Gottschalks, Ed Herr, Jim Garrett, Tom Bardsley and Mac McGuffin, photo courtesy of Nick Stevens, all rights reserved

The next day I got a call from the Operations Officer, who I didn't know from Adam.

"Nick, it seems Charlie was impressed with the way you handled the AC problem the other day and put in a good word to the boss about your crew.

I responded something like, "Thanks", hoping that was the end of the conversation.

He went on, "The DO would like your crew to escort Congressman Aspinall around a bomber this Friday. We have a substitute crew for your sortie, so that's it and you guys are on your way home. Good ER stuff Nick."

I hung up and thought to myself, "How am I going to tell the crew?"

Friday rolled around quickly enough. The Co was really in a funk and Mac was eating it all up. Ed thought it was a good punch for the ticket and Tom and Jim could not have cared less. This old couple showed up in a staff car at old dog 54105 on Nancy row. I learned later they had come to Guam to visit their son, but we never knew if he was in the service or not. The Congressman looked like he was eighty, never said a word, and I was concerned he might fall and break a hip or something. His wife was asking all the questions. When she got to the bomb bay where there was a full load of 750s, she looked at me and asked, "How do you know you are going to drop these things in the right place?"

I could see the DO and Wing Commander leaning forward to hear my answer. Ed, who was about six foot two, stepped out and gave the most wonderful dissertation about crew training any of us ever heard. The DO and Wing Commander beamed proudly while the congressman's wife never changed the scowl on her face.

We could almost hear Charlie laughing at what he pulled on us. No telling how he would spin this story the next few days but we didn't care, we were on our way home.

DURING MY 2ND ROTATION we had transitioned to a G model crew, and since G models were accumulating on The Rock as the buildup started, there were none available to fly home.

Switching priorities to the G never made sense to me for Vietnam. It is true the performance difference was an advantage with G engines flat-rated at 13,000 pounds thrust per engine with water while the D maxed at about 10.000 pounds. At Guam, almost every G would break ground at 7,500 feet but the Ds were anywhere from 7,500 feet to the gravel on the 12,000 foot runways with engines screaming. It was easier to fly, too, but the Gs didn't have the big belly mod, could carry only 26 750s internal as they never added racks for the externals, less than half the D model payload. But it didn't matter much what I thought, there were none available to fly home, so we needed to hitch a ride with tanker toads.

We loaded our gear and booty on a KC 135 and trusted ourselves to a fellow 19th Bomb Wing toad crew also going home. The airplane behind the cockpit was a large aluminum tube with red canvass seats running along each side. In the center were cargo items including our 4x4 cardboard boxes maintenance had rigged for us. There were also spare parts, ground equipment, and at least one jet engine. At the end of the tube was the boom operator compartment with all the controls for the refueling boom. Outbound to Hawaii we refueled some thirsty chicks also en route to Hickam AFB on Oahu. They would leave us and pick up another tanker out of Hawaii to take them the rest of the way since our tanker needed the gas for its own legs. We played Four-Five-Six and Yatsee to pass the time, watched the refueling, then most of us finally dozed off for a much needed sleep. Somewhere along the line I heard Ed and Jim arguing about something. It seemed they were trying to help the tanker Nav, who was lost.

Pretty soon most of my crew was up in the cockpit. The toads did not seem to mind the assistance, which was the scary part as it told me that we probably were, indeed, lost. How can you miss the Hawaiian Islands? After a while Bill walked into the cockpit, looked over the AC's shoulder, and announced, "I always just look for the cloud build ups that form over the islands." In the distance at least two hundred miles in our two o'clock position we could see a taller group of clouds than elsewhere. The combined bomber and tanker Nav team continued to argue for a good fifteen minutes before agreeing Bill was probably right and made the turn. Pretty soon we had Honolulu center on UHF. In another day we would be back home to a different world with family, EWO certification, ORIs, checkrides with CEG (Combat Evaluation Group) and week-long alert cycles. We would be back at SAC, flying by the book.

OVER THE NEXT YEAR my crew broke up and I was given another one, selected in a careful process by the squadron and wing staff to balance out assets. Bill put in his papers and started building homes. Ed was rewarded with a Bomb-Nav staff job. Mac and Tom stayed with me and we picked up a new Co-pilot, Radar, and Navigator. It was to be my best crew and we would have some great adventures together.

During that year I attended the Central Flight Instructor Course, the SAC instructor school that would in later years become a home to me. They had the best pilots in the command, just seven bomber pilots and seven tanker pilots; these guys could do anything with the airplane. They'd fly and hook up to a tanker that was doing a series of lazy eights. My instructor was a guy named Roy Steigle, and there was John Dean, SAC's version of WC Fields, by looks and style. John would take off 30 seconds behind a tanker, they'd be hooked up in the first turn while John sang something he learned in West Virginia, then for the next 2.5 hrs I would learn everything imaginable about refueling a big airplane, just for starters. I'd look over at John and he had a hole cut in the side of his oxygen mask and a big stogey sticking out, unlit but he was chewing. The confidence factor from that two-week school was enormous, and I'm proud to say in later years I would have a chance to build a new five-week program at CFIC.

But this was still 1971. I returned to Warner Robins as an IP. My first student was John Alward, checking him out as AC, and he did just fine.

When I wasn't giving check rides I was flying EWO training missions. We'd fly at 488,000 pounds in an EWO profile with takeoff and departure, teach the Co-pilot how to air refuel, taking on 30,000 to 100,000 pounds, do a celestial navigation leg with the EW shooting the sextant and the navigator making his comps and course corrections. We would do bomb runs low and high level and once back at the base make as many touch and go landings as fuel would permit. We always trained for emergencies in the B-52, flying engine out approaches and go-arounds whenever we had the opportunity.

The biggest killer was engine failure on takeoff. Engines would blow and lose thrust, with outboards most prone to blow. This would cause severe yaw, taking as much as 150 pounds of rudder to compensate to get it in trim and keep it on the runway. From S1 of 105 knots to S2 or unstick, if you lost an engine you had to do things exactly right. At CFIC the IP would pull the outboard engines between S1 and unstick just for practice. At this point we had lost 20 B-52s because of engine failure on takeoff. Guys would lose the engine, horse the airplane airborne, come out of ground effect, yaw too much which would turn into a roll and spiral dive to a fatal crash. I knew I didn't want to do that.

John Alward was on my crew one day for a training mission. While taking off from Robins AFB, seven and eight blew right above decision speed. The engines were dead and I followed the procedure, got the airplane airborne with a pretty good engine fire. Since there were no suppression bottles to put the fire out, we had to starve the fire of fuel and do things to keep it away from the wing. I increased airspeed to compensate for the asymmetric problem, staying on the deck and letting the airspeed build after flap retraction to close to maximum. That would keep the fire away from the wing and once the throttles and fuel shutoff or T handles were closed according to the procedure hopefully the fire would blow out. We were in a G model with a roaring engine fire, quite a spectacle for ground observers I'm sure. As we approached 325 knots, just as Roy Steigle had taught me, the fire went out.

While I was dealing with the fire John had called the emergency. Atlanta Center replied, said "Big Boy Three Two, understand you've declared an emergency, squawk seven seven hundred, advise souls on board. When would you like to land?" John hesitated a minute, thinking that we couldn't land until we burned off a lot of fuel to get down to landing weight because we had no way to dump fuel. He finally responded, "We'd like to land in 6 ½ hrs." There was a long pause. "Six and a half hours? I thought you had an emergency!" I would have loved to have heard the chatter in that control room. Just about any other airplane you can take off, turn around and land, but not a 52.

I couldn't help thinking about a friend with battle damage trying to land at U-Tapao with two outboards out. The asymmetric problems he was dealing with drove him into a mountain; there were two survivors. Landing with two outboards gone is no picnic, but we eventually had an uneventful landing.

Word spread fast in SAC stateside about an incident on The Rock almost too good to be true. Some crew dogs had 400 missions and were weary of pussyfooting with questionable targets while prime enemy assets around Hanoi and Haiphong harbor were off limits. The overload of brass at Andersen in Guam didn't help the mood. At the long table in the Andersen O-club where colonels sat by pecking order, one day someone said the wrong thing and crew dogs threw every one of the brass at that table into the pool, then all the furniture, including the grand piano, went into the pool, too. Damn, I missed the good stuff!

In early 1972 I attended the Instrument Pilot Instructors School at Randolph AFB, Texas, probably one of the best schools in the Air Force, five weeks learning all you could imagine about instrument flying,

At five in the morning after getting back from IPIS, the phone rang. It was some Colonel calling from U-Tapao in Thailand. I could barely make out what he was saying with the echo you got on international calls those days.

My first thought was that we had screwed something up the previous Arc Light tour and it had taken them this long to figure it out. The guy kept saying something about elephants and that they would be in Robins the next morning. I had completely forgotten about the carved elephants I bought. When I went to pick them up the next morning, the AC told me they had been stored on that tanker for almost a year and had at least 25 combat missions. Gotta' give it to toads, slow but sure.

THAT SAME APRIL DAY in 1972 I got orders to deploy, without my crew, back to the 43rd Bomb Wing at Andersen AFB in Guam. Operation Bullet Shot had kicked off, the big build up of G models, tripling the size and rebuilding the operation that would eventually lead to Linebacker. Missions were continuous day and night.

I had no idea what they wanted me to do. They placed me with another crew to fly a bomber non-stop to Guam. We launched at max weight from Warner Robins, picking up a tanker as we coasted out from California and then another from Hawaii, each time loading over 100,000 pounds. After about 18 hours airborne we found that familiar rock in the equatorial Pacific and landed on runway 06 Left. Nothing seemed to have changed. Charlie urged us to quickly get out of the way of business and not screw up our taxi instructions around the D cell working their way to the number one position. Welcome home.

The next day I was instructed to pay a visit to Bomb Nav for mission orientation. I found Dad Porter, my old 52 school instructor working there. We chatted for a long time before he said, "Well, I guess you can stand up to take a piss now, 'cause they tell me you're going to be one of the Charlies!"

Holy shit! Me, a Charlie? My warm fuzzies still came from knowing Charlie was there to help *me*! How am I going to know all the answers when they get in a tight spot? Filling Charlie's shoes would be a tall order, and all of a sudden I was nervous as hell.

A few days later I reported to the Charlie Tower, a place not familiar to me other than knowing where it was, centrally placed between the two big parallel runways, elevated only about 30 or so feet above the surrounding terrain but almost dead center in the field of action. It had been placed there with some forethought. I felt as close to ridiculous as one could feel after climbing up the metal steps, entering the glass building and having someone introduce me as one of the new G Charlies. I was just a kid wondering how this would possibly play out when a familiar voice spoke above the conversation and further introduced me as the leader of the honey bucket crew. Damn, will that story never die? Somehow everything seemed to fall into place and the next few months would never have a dull moment, some bitter, some sweet.

I soon learned that the time to be a new guy is when things are in the flux of a major change. The NVA had launched a major offensive in March, dubbed the "Easter Offensive" in our lingo, and all branches of the US military were preparing for a significant bombing campaign that would become operation Linebacker I and II. The buildup effort was operation Bullet Shot. Our missions continued while the buildup complicated things. Close to a hundred G models were headed to Guam and there were not even enough revetments to park them much less the logistics to support the planes or provide for the crews. Three other "kid" Charlies showed up and we each fell in line helping to set the stage for one of the biggest bomber buildups in aviation history. The old Charlies were behind us with encouraging humor, providing knowledge and a buffer to insulate us from the normal bullshit from senior wing staff.

And so I settled slowly into my Charlie role, learning quickly to listen to several radios at once, seeing things from a new perspective, like how D models struggled to get airborne on the 06 departure and its uphill slope until past midpoint. People in the Charlie Tower were used to seeing the D's belching smoke and straining up the hill loaded to the gills with bombs and fuel. Some unstuck at the predetermined distance and others would not break ground until well into the gravel at the end. This was the characteristic performance of the D with less thrust and predictability than the G models with their flat-rated engines which always produced the advertised takeoff thrust at any given temp or pressure altitude. With a D cell launch, things got quiet in the Charlie Tower with all eyes focused on the action. The Gs we almost took for granted until one day the number three ship in Tan blew engines seven and eight just starting up the hill.

It was an eye catcher with smoke, fire and parts flying. I didn't see the aircraft even attempt to rotate, which the G required at the higher gross weight of 488,000 pounds and 161 knots at S-2. Instead it simply disappeared over the cliff with gravel and debris flying. No one in Charlie Tower spoke as we expected to see the 750's and 312,000 pounds of fuel cook off. Finally, with enormous relief we saw the aircraft a mile or so out but 30 degrees right of the runway heading, apparently off course due to severe yaw, and no more than 100 feet off the water. The pilots had managed to get the wings level and were slowly nursing airspeed. Gradually the aircraft started to climb and the crew began their flap retraction cycle. At some point the AC checked in and with some obvious pride announced, "Charlie, we don't have to worry about seven and eight any more 'cause they're at the bottom of the Marianas Trench!" The crew recovered nicely back at Andersen after jettisoning weapons and burning down to landing weight. The plane was back in the elephant walk a few days later.

Just like crew dogs, Charlies were learning every day. As a new G-Charlie I was concerned with gaining the confidence of my fellow crew dogs. One day a crew landed and complained that, after taxiing a short distance, the pilot could not steer the plane. Maintenance inspected and found a red cleaning rag in the right forward wheel well jammed in the steering cables. Several weeks later a crew was taxiing out that stopped almost in front of the Charlie tower but several hundred yards away, calling to say they could not steer the airplane. Without hesitation, but as a leader to a joke, I told them that there was a red cleaning rag caught in the steering cables on the right main gear. Maintenance went out and found exactly that. Sometimes there is magic.

Every day new G models were arriving from the states and soon were flying missions in their own cells to the war. None of us knew about the top secret plans for Linebacker but we could smell something big coming and we all wanted to play our part. I remember spending every free moment studying the G, wanting to be ready to help a crew whenever they called in a tight spot. The proper brain cells must have been firing because 40 years later, after having flown almost all the modern Boeings, I cannot remember anywhere near the details of any aircraft like I remember the B-52 and that comes to me as if it were yesterday.

THAT APRIL, 1972, THE NVA managed to hit their first Buff with a SAM. On every mission in the north we detected the SAM threat and uplink, but they didn't always fire because they learned our electronic countermeasures worked quite well defeating the SAM's radar guidance. In fact, I had seen just one being fired since pilots are too busy with other things to try looking straight down under their ass in a steep turn. But one day I did just that, well, sort of.

As Charlie, I would fly a mission when I could, usually with someone I knew, riding the IP seat, catching up on reading and sleep when things were slow, swapping out with the Co-pilot or AC to give them a break. I flew a mission up north with AC Stu McTaggart, a night sortie, and when we came off target the EW called uplink, no big deal, he's jamming as usual. Then the EW yelled "Missile launch!," and since we were post-target and didn't need to keep it steady Stu put the Buff in a 50 degree diving turn to evade, during which I stumbled over to the window, trying to keep my foot off the Co-pilot, and looked down to see the SAM lighting up the night sky as it screamed up at us accelerating to Mach 3. I never felt so helpless in my life, not being at the controls. That one missed, but John Alward was not so lucky.

John was my first student when I became an IP at Robins. John and his crew managed to be in the wrong place at the wrong time after the NVA took a SAM battery as far south as possible in their annual attempt to bring down a Buff. John had just started his PTT with the doors still open when apparently the bad guys weren't relying on radar, they just salvoed a bunch of SAMs at a Bullet Shot wave of cells and got lucky. The missile went off below John's aircraft and hit it with considerable damage. They had fires on several engines and had to shut them down. Many of the flight instruments were out or giving erratic readings. With known fuel leaks and fuel gauges reading zero, the copilot could not determine how much fuel they had left. John and his crew decided to attempt a landing at Da Nang even though weather conditions were marginal.

This was a D model and the gunner with damage in the rear was forced to come forward and join the rest of the crew. To make himself useful, he went up to the cockpit, sitting in the IP seat between the two pilots to assist running the emergency checklists as the pilots attempted to fly the crippled plane and deal with the numerous system failures. Coming down final with shitty weather, three engines out on one side and fires on several other engines, John remembered Sweed Brown and his crew from Columbus who had attempted a flaps-up landing at DaNang and went off the end of the runway only to be blown up in an old French mine field. Without fuel gauges, John's crew could only guess at the weight of the bomber to calculate their approach speed, so getting the rear truck to touch down first would be a matter of pure luck. On touchdown they caught the front truck first with the predictable bounce and porpoise oscillation that occurred when landing high on airspeed. After the third or fourth bounce and with the end of the runway coming up, even the gunner was calling for a go-around. John knew they were dead because a go-around was impossible with fuel likely near empty, the engines out on one side, engines on fire on the other, and major asymmetric control problems he would encounter when the power came up. But since he had no other choice, he gritted his teeth and . . .

A week later John was telling me this story at the bar on The Rock. After bringing up the power for a go-around with full rudder and lateral control into the good engines, the plane became airborne again just before passing the end of the runway. John said he kept it low but had little directional control; on its own the plane did a slow stable turn with 20 to 30 degrees of bank into the dead engines. I asked what happened next, as if I didn't know, and he said that after 360 degrees they rolled back out on short final and this time he was determined not to bounce the aircraft because there would not be another go-around. I gave John the best "You done good!" I could muster and he just smiled and said he had really done nothing compared to the guy who flew it back to U-Tapao because they didn't fix the plane. They counted 405 holes in it, some big enough to throw a cow out. When I asked who flew it to U-Tapao he said, "Adam Mizinski," pointing to him sitting at the other end of the bar. We moved on down to join Adam and I bought him a drink, because as you know war stories take on a life of their own, doubly so in a bar, and he told us a great story about how they got all but two of the engines running and the Marines refused to give him more than 53,000 pounds of fuel because with all the holes and leaks he was wasting it all over the ramp. Adam showed them and flew it to U-Tapao anyway.

IN EARLY MAY, WHILE we were still dealing with the buildup, operation Linebacker commenced with a step-up in mission intensity. This time we had a president with enough brains to let the military plan the targets and missions within guidelines, and our cells and waves began to hit enemy targets in the south to push them back from gains in the Easter Offensive. In the north near Hanoi we hit bridges, anti-aircraft defenses, Haiphong harbor where Soviet ships brought a steady stream of war materiel, switching yards and rail lines used to transport arms from Haiphong, rail lines that brought arms and supplies over the mountains from China, power plants, fuel depots and a host of other targets that had heretofore been off limits.

By the first week in June the Seabees were in a frenzy to finish new revetments to accommodate the G model Buffs coming into Andersen, squeezing them in so the wings hung over the sides. We reached a point where we had to keep 25 aircraft airborne at all times just to meet the parking needs. Inevitably, the day arrived for the obvious decision that came from General Johnson to close the north runway and use it for parking and stowage of ground equipment. With just one runway operational, a single mishap would not only shut down operations, it would make Andersen unable to recover returning mission aircraft that were going bingo on fuel. Redirecting Buffs to other Guam airports had the consequence that they were so big they would do things like destroy runway lights with their tip gear, never mind their inability to then take off again. We made every effort to discourage the General's orders just short of insubordination. He stuck to his decision and we turned the north runway into a huge parking lot.

In a few days Charlie's predicted disaster occurred. As we watched the routine recovery of Copper cell one afternoon, the second aircraft made the shortest landing roll of any B-52 in history. Copper two apparently landed with all main gear brakes locked for technical reasons never fully understood. To see a B-52 touch down and stop within less than six hundred feet was incredible to say the least. We could actually feel and hear it from the Charlie Tower several thousand feet away. When the smoke and dust cleared, it was obvious Copper two was going nowhere. The brakes were glowing red and soon the 300 psi tires started blowing from the proximity of extreme heat, each sounding like a cannon going off. Copper three was desperately looking for a place to land and within seconds we initiated our backup plan to recover aircraft we knew would be arriving at predictable times, all low on fuel and nowhere to go except The Rock.

Coordinating with the command post we immediately called for strip-alert tankers in Kadena to launch to provide refueling support for bombers en route to Guam that might divert elsewhere. We then began the big effort to move the aircraft parked on the north runway, many of them in different stages of maintenance. We were still in deep stuff for at least a couple hours with birds arriving every 10 minutes with about 30,000 pounds of fuel, stacking up, burning 18,000 pounds an hour even at loiter speed. Then turns become an issue since between 20,000 and 10,000 pounds they risk uncovering fuel pumps and flaming engines out, and a single approach takes 10,000 pounds.

Clearing the north runway was urgent. The locked-up Buff, Copper two, was unable to move even with full power and we were concerned too much engine blast might damage our one good runway. We were unable to launch our own tankers so the only other alternative would be landing at Navy Agana and Northwest Field. Both were extremely risky. The Navy shared their runway with the civilian international airport. With its narrow runway, the lights on either side would align almost exactly with the B-52s tip gear. Northwest Field, built during the Second World War, had been inactive at least since B-29s flew out of Guam.

By now, maintenance had hooked up a Euclid tug to the stuck bomber and was still unable to move it even with eight engines straining. The bombers coming in with under 25,000 pounds of fuel had less than an hour before engines began flaming out. We were talking with SAR personnel to coordinate their response in case crews had to eject and ditch their plane.

About the time the third cell arrived and were placed in a holding pattern, the first bomber waiting to land, Copper three recovered at Navy Agana, taking out some of the lights on the narrow runway and damaging one of the tip gear. We were at least 30 minutes away from clearing the north runway when maintenance troops with sledge hammers busted enough of the brake assembly away from the grounded bomber for the crew and tug to move Copper two off the runway. At this point we launched several tankers and started recovering the ten or so bombers stacked up over Guam. The Seabees must have been motivated by this since the new revetments were completed in a few weeks. By August, Andersen had 159 bombers on station. The wing had dodged a bullet with a lesson learned to take a serious look at Northwest Field.

ON THE ROCK A Buff would blow a tire now and then on takeoff and the crew wouldn't even know about it, which of course would be a big problem at landing. Pushing gross weight and stress to the limit, the tires paid a price on the coral-based runway frequently soaked in rain. Sometimes after a takeoff we'd do a runway check, or we'd see it from Charlie Tower, 12 feet of tire tread laying on the runway. These were tough tires with 38 ply in the tire and 5 ply in the tread but they had extreme wear and tear. The BF Goodrich tire supplier put info on the tire and tread that would identify the aircraft, so the Buff unknowingly in trouble aloft was frequently known from the remains left on the runway, and Charlie used that information to alert a crew they had a blown tire in the well. That meant the Nav, since he was junior, would have some fun just before recovery.

As they descended below 10,000 feet where oxygen is no longer required, the Nav, who hated this duty, would depressurize and crawl back on the 105 foot long eight-inch wide catwalk dripping with condensation and cold. Part of the pain in the ass was the Nav had to put on a chute to do this. He had to check the wheel wells on the front of and rear of the bomb bay, which was 12 feet across and 15 feet high. The rear well was a long crawl back. Nav would look over the tires and tell the AC which one was blown. With that information, the AC would pull the circuit breaker on that landing gear and use the others for a three point landing.

One day a Nav called the wrong gear to the AC because he forgot the gears cross each other and fold up into the opposite side of the aircraft. With a three point landing and one of them a blown tire, well, it didn't end well. The Charlies decided from that point on it would be better just to lower the blown tire with the others and land, bad tires and all.

EARLY SUMMER OF 1972 was turning out to be a period of innovation as we stretched everything to its limit. The effort to fly both D and G aircraft at the rate of over 60 sorties per day was a strain on every aspect of the biggest bomb wing in Air Force history by tonnage delivered. The normal accommodations at Andersen AFB for 3,000 personnel were now stressed to provide for over 12,000.

The equipment had never been worked this hard and for every aircraft launched at least four had to be generated. The crews were becoming accustomed to flying birds with problems they might have seen had they flown the plane a week before. Preflight crews assisted the generation process, rolling rejected equipment into a maintenance process as part of the generation cycle. The humid salt air didn't help with aggravated corrosion issues, though I have no memory of any significant material failure specifically due to corrosion. There were two incidents years later of very rough landings on the nose truck resulting in separation of the crew compartment from the aircraft just forward of the wing leading edge, both destroying the aircraft but leaving the crew with just minor injuries. Corrosion might have played a role.

Charlie Tower added specialists to facilitate the various functions. Uncle Ned was Charlie's maintenance counterpart. Between the two, an uncommon link was established between the flight crews and maintenance preparing the aircraft for combat sorties, uncommon in that each would be aware on a real-time basis of the weaknesses at each end. Within a short time these specialists developed the ability to look several moves ahead, like chess masters.

After a series of taxi screw ups, General Johnson declared that the next crew failing to follow directed taxi plans leading to interference with a cell launch would immediately be returned to the states, as if that would be a bad thing. Almost the next day, a brand new AC and crew, flying in from the states for their rotation, landed on 06L, were issued taxi instructions to Sugar row and proceeded to unwittingly block a D cell heading down the hill to the parallel taxiway. I watched with concern thinking this crew managed to fly a bomber halfway around the world and were about to be admonished for violating a chickenshit rule and possibly sent back the next day without the dignity of flying a single combat mission. The thought was a short one because Uncle Ned had already coordinated a tug to push the crew back into one of the new revetments; thank you, Seabees! Laughs went around Charlie Tower and the new crew was put to the test of providing the good cheer for the next few days. Funny how this stuff made it to the bar within minutes.

There was a need to fix the taxi problem and the main tower was completely out of their league. It is hard to describe an elephant walk of three cells of B-52s working their way to the runway hold point, loaded with weapons followed by twenty or more maintenance vehicles nursing the bombers every inch of the way. Uncle Tom was the new addition to the Charlie Tower to be in charge of taxi operations, with Cousin Fred assisting him. The main tower was now almost out of business, but who needs Air Traffic Control anyway when our intentions are to drop bombs on an enemy 3,000 miles away?

WHEN WE GOT ALL the revetments in place that summer and stopped using the north runway to park airplanes, we had close calls with three typhoons. Airplanes were getting waterlogged and we were starting to deal with airspeed problems. On a typical day, running 60 sorties, Charlie assisted Buff crews with three or four engine failures, a couple of fires, half a dozen hydraulic problems and a couple of tire problems, not to mention bomb/nav issues.

One rainy night a Cobalt cell of G models took off into the path of an oncoming typhoon which was about 200 miles from the island. Shortly after their level off, Cobalt two called Charlie to say he had lost airspeed. When I asked if he was having trouble keeping his cell position, he said, "We fell out of formation, lost position visually and on radar, what can you do for us?"

I told him to set 2.78 EPR for engine thrust and asked Cobalt lead and number three to assist in making radar contact. Cobalt two acknowledged setting the thrust just as we lost VHF contact. We would later learn that the pilots set the power which would normally have accelerated the bomber to around 280 knots indicated, but their speed was aft of the power curve. This meant that as the speed gets lower it takes more and more power to accelerate, much like a speed boat trying to get off the step onto a plane. After about six or seven minutes they gradually slowed until the plane started to buffet. Possibly confused by the high power setting I gave them they misinterpreted the buffet.

Part of a 52 pilot's training is to recognize high speed buffet when portions of the wing begin to go supersonic. As the plane exceeds the critical mach it will tend to tuck under causing a serious control problem. The procedure was to reduce power and apply back pressure. So when these guys felt the initial buffet as it was stalling, they mistakenly thought it was a high speed buffet and instead of advancing the throttles they pulled them back to idle. The aircraft went into a full stall, rolled off on one wing and then the other with the pilots struggling to level the wings. They finally ended up in a flat stall at 191 knots. With fire belching out both ends of the engines and the plane feeling like it was coming apart the AC ordered the crew to bail out as they passed 15,000 feet. They bailed out into a typhoon.

At this point other aircraft were trying to relay for us and someone asked if we got the distress call from Cobalt two. As soon as we heard that we initiated the SAR. We had a crew in the drink in a typhoon at night in tiny one man rafts with estimated 100 knot winds and 25 foot seas. It was to be a very long night and we had a new cell to launch every thirty minutes.

The next morning the SAR C-130 got to their area, spotted the crew and dropped a string of emergency 20-man rafts. By the end of the first day the winds were now 120 knots and it was impossible for an aircraft to reach them. We were trying to figure out a way to get to them, when the Navy came to our aid by sending two nuclear subs to the area. When we did some rough math those subs averaged 40 knots to reach our guys, but when they got there the seas were too rough to remain on the surface, making recovery seemingly impossible. The subs could see the survivors on their equipment and had a fix on their locator beacons but couldn't get to them. After a day of being unable to help, they were preparing to leave when a Chief Petty Officer sold the sub Captain on a plan where one of his crew, a Navy SEAL and strong swimmer, would go in the water after the survivors with a long cotton rope. The Chief, a big guy well over six feet tall, would lash himself to the conning tower and handle the business end of the rope. They put their plan to the test, with about 50 seamen below ready to pull on the rope at the Chief's signal. The SEAL swam to one crew dog after another, tied the rope to him, gave the signal and the crew dog was hauled in while the Chief let the rope slip through his hands as his men below pulled and flopped the crew dog up on the deck like a fish. Go Navy!

They got them all except the Radar Nav, who was floating face down, clearly not alive, and they didn't go after him because the risk was so high. After he was rescued the EW said the biggest mistake of his life was crawling into one of the 20-man rafts. With winds of over 100 knots and big seas, they drifted with the typhoon for two days while he rolled from one side of that raft to the other then back again endlessly while the raft beat the hell out of him and he swallowed lots of saltwater. The other guys actually fared better in the single man rafts

Some years later at the CFIC School, with every instructor candidate I would cover up his airspeed indicator for a good half hour while he hand flew the plane to include flying a holding pattern, penetration and approach with a touch-and-go landing. Invariably, with his airspeed covered up the actual speed and landing would be the best of the day. The airplane had ways to tell you what airspeed it needed in any given configuration if you knew the stabilizer trim settings and power requirements. Subtle changes in pitch forces would translate to power changes when holding an altitude or establishing a known pitch attitude with a predictable vertical speed. Even the sound told a lot.

HOUSING WAS A REAL squeeze with the buildup of Bullet Shot. First, all the available bunks were taken, then all the available hotels, then a tent city, then a tin city with lots of pressure for crew living space. Uncle Ned and I were officially billeted in a hotel in town, right on the water, two guys to a room. Despite the housing crunch, on this third rotation to Guam as Charlie, my wife, Margaret, joined me there. Uncle Ned's wife also came over, and we found an apartment in Dededo to share by rotation. We spent our off time with our wives, relieving some of the pressures of separation. Transportation was a problem because the buses were not reliable, and there was no room to be late for my Charlie shift of 12 hours for six days, one day off. So I bought a car. Maybe I should say so-called car, a little Datsun 1600 sports car that came with five gallons of rustoleum, the better part of the deal. With a car Margaret could get around while I played Charlie.

Guam, 1972: L-R: Margaret Stevens, Nick Stevens, Tom Bardsley, his girlfriend
Photo courtesy of Nick Stevens, all rights reserved

My car had a flight line pass and Margaret would drop me off at Charlie tower, then weave her way back between bombers and stacked bombs. One morning the Wing Commander was there, Jim McCartney. I didn't know him well yet and was unsure how he would react so I told Margaret to hunker down in the car until he left while I went to work. The roach coach showed up and McCartney strolled out to get a cup of coffee and donut, met Margaret and they chatted about 20 minutes. When he came back in Charlie Tower he said, "You actually married to that gal?" He thought she was too good looking for me. I didn't know much about him, but he had 1200 combat missions in F4s, 123s, 52s and God knows what else. We didn't know it then, but he was brought in to run the 43rd Bomb Wing for Linebacker.

With heavy losses in the south and bombed into near-submission in the north, our enemy returned to the negotiating table. Bombing the north was put on hold in hopes of a negotiated settlement. Linebacker had more results in three months than three-and-one-half years of Rolling Thunder in which McNamara turned up the heat just a little at a time, trying to find the enemy pain threshold his calculations told him was there. He never found it.

TWO THINGS OF NOTE happened in the fall of 1972. First, the North Vietnamese were talking but being very difficult, dragging out the process as calendar days, weeks and months slipped by. Second, my 179-day limit was approaching and I was rotated back to the states to attend Squadron Officer's School. I was disappointed and would have preferred an extension to continue the important work of Charlie, especially at this crucial point in the war where my experience could best be put to use. Instead, I would be sitting in a classroom learning things I had long known, an easy course with the luxury of volleyball in my spare time, hard to enjoy while fellow crew dogs were doing the real work of combat pilots at a time it really mattered. I swore a lot. And then it got worse, just for me.

The North Vietnamese kept playing head games. On December 14, while I was either in class or on a volleyball court, there was a stand-down on The Rock with crews gathered in a hangar so the Wing Commander, Gen. McCartney, could tell them President Nixon ordered Linebacker II, an intense bombing campaign of enemy assets in and near Hanoi and Haiphong harbor, and we were going to kick the living shit out of them this time. The wing had four days to get 159 airplanes in flying condition. U-Tapao was doing the same with their 50 airplanes. Finally, real missions to fly, and I wasn't there!

Launch of Linebacker II was on December 18, and I would have loved to be where I belonged, in Charlie tower choreographing the biggest elephant walk in history. The first wave took four hours to launch with 90 second MITO, not a single takeoff abort and only one bird aborted having, for the second time in history, to shut down four engines on one side en route to the target. This unprecedented launch performance was a testament to the motivation of crew dogs to shove the war down the enemy's throat. From December 18 to 29 we pasted the hell out of them in what the news media called the "Christmas Bombing."

The North Vietnamese played the propaganda card well, claiming we had killed large numbers of civilians, which was untrue. In fact, the Air Force recognized the potency of a B-52 strike approached that of a tactical nuclear weapon, and targets were selected with care. That said, we did bomb the dogshit out of enemy assets near Hanoi and shut down the flow of arms and supplies from the Soviet Union to Haiphong harbor and from China by rail at the same time we reduced bridges, airfields and other assets to piles of rubble. There were surely civilian casualties, but that's war and we didn't target them.

While I was playing ding-dong school, crew dogs in 52s flew in huge waves from The Rock and U-Tapao in Linebacker II, totaling 731 sorties, and with that many takeoffs there were a scant dozen aborts as crews did all in their power to make it fly. Air defenses around Hanoi were the toughest in the world and the enemy fired SAMs sometimes in salvos, unguided by radar, trying by chance to hit one of the B-52 bombers flying overhead in huge waves. The first six days our crew dogs followed normal target approach strategy, then they varied the line of attack and staggered altitudes to screw with the SAMs.

Out of an estimated 600 to 1,200 enemy SAMs fired during Linebacker, and who knows how many AAA rounds, they hit 30 of our 52s, SAMs accounting for 14 kills. Whether by enemy fire or accident, we lost 50 good men in those 11 days while we mined the harbor to shut down shipping, destroyed bridges, rail lines and switching yards, fuel depots, weapons stockpiles, power plants, manufacturing facilities and brought North Vietnam's industrial plant to its knees.

Somehow intelligence found the location of the enemy's SAM storage site, and we blew that place up, too.

AAA was a nuisance at 85mm because it usually didn't reach our altitude, but 100mm was a mean and different story. I know of only one B-52 shot down by AAA in Vietnam: John Mize qualified for a nickname like "magnet-ass" by taking hits three times in Linebacker II. The first time, on his 292nd mission, he took a SAM hit, held it together and recovered at U-Tapao. On his 295th mission, he was hit again by a SAM and recovered again at U-Tapao. On the last day of Linebacker he was the only one hit by AAA coming off the target in his PTT, lost all four engines on one side and a good fire going in engines and on the wing. He didn't have a lot of directional control but got it headed out of Vietnam toward Laos and U-Tapao. With four engines gone he couldn't maintain altitude so he improvised a *driftdown* maneuver in which he slowly traded altitude for airspeed to keep it flying and avoid the asymmetric problems at lower airspeed. With a wing on fire, he crossed Laos and nearly reached NKP, the *Nakhon Phanom* airbase in northeast Thailand just across the Mekong river from Laos, an air base nobody discussed because of covert missions staged there. By that time he had an entourage escorting him, two fighters and a Jolly Green SAR helicopter, and he had eyes in the rear with the Gunner who reported the wing burning through in spots to the ribs. At 15,000 feet, things were critical and with help close by they bailed out over Laos and were quickly picked up. John got the Silver Star for his escapades.

There were just two MIG kills that I know about during Linebacker II, one by our fighter pilots and one by an 18 year old B-52 Gunner. As Wendell Weaver was bringing his cell home post-target, the young Gunner in number two reported a MIG on their rear, hanging back steady outside of 2,000 yards, clearly out of ammo and knowing the Gunner's 1,500 yard cone of fire range with the quad-50. We knew MIGs that had expended their weapons would sometimes fly our wing to report our altitude and airspeed but one hanging on our rear was a new twist, apparently staying close between the number two and three bombers in relative safety after having been chased all over the sky by our fighters. If that was true, the Gunner surmised, when the MIG pilot was ready to peel off and dive for his base, he would probably hit the afterburner and jump within the quad-50 range. He set up to fire, and sure enough, the MIG went to burner, jumped in range and the Gunner shot him down with a three second burst. I don't think that young Gunner ever had to buy his own beer again. One of the best parts of this story was that the cap fighters observing all this and helpless to engage the MIG, had to confirm the kill.

During Linebacker II, the US was continually prodding the North Vietnamese to resume negotiations, with the President deciding at each juncture to continue bombing and turn up the heat. When the North Vietnamese capitulated after 11 days of intense bombing, B-52s had delivered over 15,000 tons of ordinance, fighter-bombers added another 5,000 tons, devastating 18 industrial and 14 military targets. The enemy's supply of SAM missiles was depleted and their losses were staggering.

I FLEW 118 BOMBING missions in and around Vietnam, but some guys ahead of me in the cycle flew as many as 400; a lot of crew dogs were weary by the end of doing the heavy lifting while brass and politicians failed continually to take the war to the enemy.

As for me, I didn't ponder the politics of the war much while I was flying missions. That reflection would come to me later in life. At the time I had a simple task to follow orders and carefully put the ordinance where it was supposed to be. The one mission that gave me real discomfort was when our CO ordered release when we didn't have proper control over where the weapons were going. That order was wrong on several levels.

In some ways the war and its controversies were background noise, the backdrop to what was most important to me, how our missions put men and a special machine to the ultimate test and in the process found the limits of their capabilities. The irony is that the B-52 was designed as a subsonic nuclear resource that would spend most of its time on the ground, tampered with and every week or so fly a little, and we only discovered what she can do by working her guts out on conventional bombing missions in Vietnam.

The longest I ever flew the 52 on a single sortie was 25 hours, though a friend, Zim Zimmerman, flew 43 hours in an H during surveillance out of Guam to the Indian Ocean. I had no idea the 6.8 gallons of oil in the engines would last that long. I have always appreciated the experience we had with the aircraft and how it continues to influence the way we fly large multi-engine aircraft today.

I'll never forget the landscape at the Ban Kari and Mu Ghia pass, where the Ho Chi Minh Trail crossed from Laos into South Vietnam. We bombed so much trying to stop the flow of enemy supplies, the network of interconnecting craters resembled the moon. Some craters were filled with water and others just big ugly blemishes we could clearly see from 38,000 feet. The Navigators told us they knew when they were near the passes because the heading system would respond to all the iron in the ground. Sometimes we could see the dirt roads winding around the craters that were not there a few days before. I sometimes thought how ineffective it was to spend our resources blowing up dirt roads in the jungle when we could easily take the war to their home where it would be over quickly with less pain for both sides.

After 22 years in the Air Force, most of it in SAC my father warned me about, I joined United Airlines and flew all their big airplanes including the 747-400. Some 25 years after my bombing missions I thought back to those days while sitting high in the right seat of a 747, drinking coffee as we coasted in over Vietnam. It reminded me of a movie about WWII B-17 operations, nostalgic, with distant singing and conversation muffled by time over a place where many battles had been fought and many lives spent, long ago forgotten.

The first time I spoke with Ho Chi Minh Center it was as though this fellow was in the center of a large concrete room speaking into a stand up microphone. The Asian voice had an empty echo as if there were no one else in the room except the controller, who was giving us clearance into his airspace over Da Nang, which we could see below with its long, familiar runway. It occurred to me that radio was probably one of the many things we left behind. The landscape was familiar after all those years, green and beautiful.

What a waste, I can't help but think so many years later. War is a nasty business and we prolonged the misery by pussyfooting for years when we could have applied overwhelming force to end it all. The hardest thing for me after all the effort, disruption, and lives lost was the idea of leaving without completion and knowing what we left behind – a broken promise and slaughter of millions in the communist cleansing process that would be ignored by the media and American public.

From Da Nang to crossing the Mekong River near Vientiane, the landscape below looked as pristine as it may have been 100 years before, the bomb craters long since grown over by the jungle, the scars of the war not visible from above. There were also no signs of any progress in the way of new roads or industry. This would be my usual run for twelve years and every time we made the crossing into Laos and Thailand the same emotion would hit me as if all the dead souls were there to remind us of what we all did on both sides. I wonder if – check that – I know we are doing the same today and what a corruption of political will it is.

After the fifth day of a six-day trip to Bangkok and Tokyo, 747 crews all tended to do the same thing, set the cruise control at .86 mach and run for home. We once made it from Tokyo to San Fran in seven hours and 34 minutes with a ground speed at one point close to 1,000 mph thanks to a strong jet stream. Like many other runs, that one gave me plenty of time to reflect on how much the technology and equipment had improved, and that my 747 passengers and crew may never realize how the B-52 had contributed many advances to large multi-engine aircraft design, and that a lot of it came from what we learned in the intensity of bombing missions in the war. Newer aircraft were better looking, more efficient, featuring rich appointments and lots of systems and protections to prevent the many emergencies we dealt with in the war. Once a 747 I flew was dragging an engine shut down for low oil pressure and that was a big emergency to the crew. I didn't have the heart to tell them crew dogs ate emergencies like that for lunch.

The B-52 lives on. In 1961 the last H models were built, a big step up at the time with TF33-P-3 turbofan engines that reliably deliver 17,000 pounds of thrust. The H is still flying combat missions today, acutely maintained and now in better shape than ever with a life planned to 2030 when it will be around 70 years old.

Something else lives on. No matter how many newer, sleeker, more elegant airplanes I encounter, the B-52 still owns my heart and I'll always be proud to have been part of the greatest show on earth.

Above, Nick Stevens (L) in 767-300 with crew. Below, Nick Stevens and wife, Margaret, with the biplane he helped a friend build, a Eugene Ely Curtiss D-4 pusher, the first aircraft to take off and land on a ship. This plane will be flown in the Navy's Centennial celebration in 2010 and 2011. Nick is now retired, living in Virginia, chasing fish on the Potomac and at Hatteras, NC. Photos courtesy of Nick Stevens, all rights reserved.

Postcard to America

ON THANKSGIVING DAY IN 1968, when I was eight years old, my Mom and my brother and sister received the heartbreaking news of my father's death in Vietnam. Since he is not here to write a postcard, I will write one about him.

Dad was a career soldier, a professional, an aerial rocket artillery helicopter pilot. He was known as Black Bart, respected and loved by his men. I'm told he led from the front and always looked for ways to protect the soldier on the ground and the pilots and air crews who trusted him with their lives. He developed flying techniques that are still in use today and he distinguished himself in the battle of the Ia Drang Valley in 1965 when he was Charlie Battery Commander for the 2nd of the 20th First Air Cav.

On his 2nd tour, Dad was Battalion Commander of the 4th of the 77th, 101st airborne. He could have been many safe places as Battalion Commander, but my Dad chose the dangerous cockpit of a helicopter gunship where he died serving his country. I not only love him as my father, I treasure his memory as a decent and honorable man who worked hard to make the world a better place.

If my Dad were here today I think he might say this to Vietnam veterans: "Be forever proud of your service in Vietnam, don't be discouraged by those who never learned the truth about that war, and don't let the lies pass without speaking the truth loud and clear."

Long after he died, in 1977 I sat in a Riverwood High School classroom listening to Mr. Gross, a history teacher, tell all my friends that my dad and his fellow veterans were baby killers. When I protested, he told me to go sit in the library until he finished teaching the Vietnam War unit. Mr. Gross gave me artificially low grades the rest of the semester.

I grew up amidst military people, learning by their example the meaning of selfless, sacrifice, duty and honor. They earned my respect and my love many times over and I will not tolerate idiots pushing a political agenda at the expense of our magnificent armed forces. Not then. Not now. Not ever.

Laura Armstrong, Roswell, GA
Army Brat, US Marine Corps wife who stands straight and quiet with misty eyes when the flag passes

Ghosts of Torment, Ghosts of Love
Donald Johnson
Atlanta, Georgia

AS WE PREPARED OUR rucks for an operation in the bush, sometimes the silence was so thick and heavy it could be cut with a knife. Each Marine would quietly select and pack ammo clips, grenades, C-ration boxes, socks, etc., lost in our own thoughts, unwilling to say out loud what we all knew, that some of us would likely not be coming back alive. I would steal a look at this friend, then that one, wondering if he would be the next to die. They still visit me in my dreams, the ones who died.

Donald Johnson, brand new Marine
Photo courtesy of Donald Johnson, all rights reserved

Sometimes we would say to a buddy, "Hey man, if I don't make it back, would you tell my family . . ." One Marine was a minister back in Texas, and he told me before an operation that he had a bad feeling about this one, that if he didn't make it back he wanted me to go through his stuff, make sure his wife didn't receive *everything*. I asked him what he was talking about and he said, "Just promise me!" So I promised. He was killed on that operation, and I did as he asked, went through his stuff and found the girlie magazines he didn't want his wife to know about. I guess he was human like the rest of us.

I tried to keep to myself, not to get too close to anyone because that makes it tough when they are killed. Since just two of us were black and the others in the unit white, keeping to myself wouldn't be hard, would it? But there were two things I didn't count on.

Growing up in Georgia in the 50s and 60s, I had little contact with white people and that probably worked both ways. But one of Momma's gifts to me was an open mind. She taught me not to judge anyone by the group others put them in, that each person is an individual deserving to be measured on their own actions. In the bush of Vietnam in 1966 and 1967, near the DMZ with heavy concentration of enemy forces, I was only 19 years old and it didn't take long for me to realize the white boys were a lot like me; they loved America, they didn't want to be in Vietnam, they were scared when the shooting started and their blood was red, too.

The second thing I didn't count on was how combat presses people together, forging a bond like a hot furnace making steel out of iron and carbon. Even when things were quiet, the tension and fear of waiting, sitting back-to-back in a foxhole at night staring frantically into the dark trying to see movement and trusting your buddy to stay awake to watch his half of the perimeter, those times pushed us together, too.

And so, slowly, without my realizing what was happening we became close, we became brothers. I realized only years later as I matured that I loved those men. I still think of them every day.

Like all Marines I am a Marine forever and fiercely proud of serving my country. But I didn't set out to be a Marine. This is how it happened.

WHEN I GRADUATED FROM high school, the football coach at Savannah State University encouraged me to try out for a football scholarship. During the summer, while I waited for football tryouts, I saw so many job opportunities in the paper that I said to myself, "College can wait; I'm going to make some money!"

I got a job at Ft. Benning, Georgia, helping with the supplies coming into base and I made some soldier friends there. After working there about eight months I received my "Greetings" draft letter. I almost didn't mind being drafted because I had watched so many young soldiers working their way through basic training and coming through the PX and Commissary, and I told some soldiers, Drill Instructors and officers I knew that I'd see them soon in basic training.

I returned to Atlanta for my induction physical and processing. At one point they lined us up and told the line to count off one, two, three, four . . . one, two, three, four and so on. Then they said everyone who counted off number four should step forward. I was one of those who stepped forward and heard them say our privileged little line would be drafted into the Marine Corps.

Uh-oh!

I told the man I wanted to be in the Army. He said, "Well, you can go across the hall and sign up for four years in the Army, or you can be drafted for two years in the Marines, your choice." I didn't like the sound of four years, so I became a Marine.

At one point in Marine boot camp I took an elective aptitude test, hoping and trying my best for a good score so I could be an engineer, an electrician, a typist, anything but a grunt in Vietnam where I had a great chance to be killed.

My Drill Instructor asked me why I took the test. He said, "Man, you're wasting your time, the Marine Corps isn't going to waste time training a draftee, you're going to carry a rifle!"

Uh-oh!

I would soon learn the meaning of 0311, the Marine Corps MOS for rifleman, and in my case there would be lessons which haunt me to this day.

AFTER INTENSE BASIC TRAINING and a 15-day leave, I departed for Vietnam through Oceanside, California, where we stopped for two weeks of jungle training. One evening we loaded on a plane and took off, refueled in Alaska and Okinawa, finally landing in Da Nang, Vietnam.

After two days at the giant Marine base at Da Nang I was flown on a CH-34 helicopter to Dong Ha, north of Quang Tri and very close to the DMZ. I expected to land at Dong Ha and be indoctrinated, learn something about the area and our mission and maybe phased in gradually to our combat mission. But the unit I was to join was right then engaged in a firefight up on the mountainside, and that's where they took me in the helicopter. I tried to tell them when I got there I was just a new guy, never mind that I was scared as hell, but the Lt. said, "We need bodies!"

I didn't know what to do, where to go, what the situation was and bad guys were already shooting at me. I thought surely there must be a better way to fight a war than just throw green guys into it. As it turned out I learned fast because we were up there on the mountain for over a month, and it was bad, really bad. And it was dirty. You don't know dirty until you soak your pants and shirt every day with a new layer of sweat, and those sweaty clothes are the only place to wipe when something wet gets on your dirt-encrusted hands, and there is no shower or change of clothes for a month.

We were fighting NVA regulars, the toughest enemy troops, not local VC guerillas. We set up ambushes for them and they ambushed us. They were aggressive, tough, well-armed and in uniforms, not like the sneaky VC shooting and hiding; these guys would come right at you.

ONE OF THE FIRST things I learned was not to get too close to anyone, not to care too much. We lost so many men in that operation that we were seriously under strength and the Marine Corps sent us back to Okinawa to re-build the unit. That's where we relaxed together, let down our guard, got to know each other, drinking beer and chasing women, laughing and joking and getting in trouble like young men everywhere.

One guy didn't want anything to do with blacks like me, but even he couldn't hold out forever. I had some records with popular songs and after a few beers we'd sing along and even cry together with a Marine who got a Dear John letter, so sad. We became friends. I wish I could remember his name.

We picked up the new guys as they came in from the states. They would look to guys like me as experienced combat veterans, asking questions about how to do things, wanting to know the tricks that might keep them alive. Hell, I didn't know much of anything except be careful and duck a lot.

We went through some training exercises and after about three weeks in Okinawa we prepared for our return to Vietnam on a ship, a troop carrier. That damn ship would rock this way, roll that way, and guys were sick and puking all over. Then, just to make sure we Marines are tougher than other troops, we did a beach assault landing. The ship was tossing all over the place while we climbed down those rope cargo nets with backpacks and rifles. The landing craft was bobbing up and down so violently that you had to time your release just so; some didn't and broke their legs.

We arrived on the beach like Marines, out the front ramp of a landing craft, but all that waited for us on the beach were sand fleas. No shots were fired.

WE RETURNED TO THE Da Nang area, but we were not on the base. We operated miles outside the perimeter around Da Nang, in the bush where we hunted the infiltrating enemy. It seemed that they decided when and where we would fight.

Mostly we operated here in three-to-four-man teams, away from the main unit, on an ambush or listening post. LPs were bad because we were just two or three men, dug in far outside the wire, back-to-back in a foxhole at night, doing our best to be invisible and quiet in the dark, because we were smart enough to know we were the early warning device, a few troops used as bait and a tripwire to alert the rest of our unit when the enemy was creeping toward them to attack in the middle of the night. If the enemy found us out there we would have little chance to survive. Invisible and quiet, our eyes franticly searched the night for what, sooner or later, would surely come and try to kill us.

A lot of our guys went home messed up or dead. Some were shot, some stepped on booby traps and triggered something like a Bouncing Betty, a grenade that popped up to waist level before it went off, or stepped in a punji pit where hardened bamboo stakes, dipped in feces, would penetrate jungle boot bottoms to infect the nasty wounds.

VIET CONG WERE PREVALENT in this area, and they recruited from local villages. The word "recruiting" had special meaning for the VC and for the village people, but I never witnessed a recruiting session, I just saw the aftermath. When we approached a village, we knew the VC had been recruiting boys when we found villagers crying and bloodsoaked spots where the village chief had been beheaded, maybe his wife had been disemboweled, maybe his toddler children had their hands cut off, to get the attention of the villagers who had been forced to watch, you see, to convince them the VC were serious people and when they said bring us your sons who are twelve and older they were not kidding around, you see.

I was amazed when later in the war back in the US people were talking about American troops committing atrocities all over the place in Vietnam. I never saw a Marine intentionally harm a civilian, but I did see Marines get killed because they were too cautious in trying to protect civilians or avoid hurting them. I did see plenty of Marines helping the Vietnamese people build schools, give medical care and teach their local troops how to use weapons and defend their own village against the VC.

When we had been fighting NVA regulars up in the mountains, we could see them in uniform. This was different. The VC were always in hiding, striking when they chose, disappearing after inflicting damage. The Vietnamese we passed on the road, the ones who smiled and waved at us, might be the same one shooting at us that night. We had all kinds of crazy rules about civilians, and we couldn't even stop them to search for weapons unless we had just cause. These were the *Rules of Engagement.*

When we received fire from a village, we couldn't return fire. First we had to prove there was a weapon in the village, meaning we had to see the weapon. If a buddy took a round in the chest and we watched him die as he drowned on his own blood or his heart stopped, maybe we couldn't do anything, maybe we couldn't even fire toward where the shot came from because we didn't have clearance to fire. We were told if we shoot somebody and they don't have a weapon, we might go to prison. So we were frustrated to say the least, but I never saw a Marine intentionally harm a civilian. We were the good guys.

Sometimes firefights with the enemy were very short, just a quick surprise attack on us then the enemy withdrew to hiding. Other times they were long, sustained battles, sometimes lasting hours. Even then, the enemy was good at dragging off their dead and wounded, while we were pressed by our higher-ups to find enemy bodies and weapons to justify the fight, especially when civilians were casualties.

Our enemy fought from tunnels or caves sometimes, and when the fight was over we would search the tunnels or caves looking for bodies, but they might be booby-trapped and kill more Marines so we sent in the German Shepherd dogs trained to sniff for bodies or weapons. One time, after a long firefight, we found six bodies, but we knew the force we were fighting had to be at least *ninety*-six!

It was demoralizing to be unable to do what we were trained to do, especially when one of our own was killed. This was politics, orders from above that the generals and the rest of us knew was bullshit in a war zone. We were sacrificing our own men to reassure politicians back home we were not murdering civilians, while our enemy murdered as a matter of strategy on a routine basis, ignored by the press because, apparently, it wasn't unexpected. The world had turned upside down. Nothing made sense.

We learned to not take any chance that we could avoid. When we approached a rice paddy with civilians working their field, I might hear the Lt. say, "Johnson, woman at ten o'clock" and that meant it was my job to watch her every moment as we crossed the rice paddy. Every civilian had a man assigned to watch them while we crossed rice paddies or passed nearby, to alert the unit to any unusual behavior that might mean they were VC in disguise, or giving a signal to the bad guys, because we couldn't tell which of the civilians were on our side and which were on our enemy's side.

DAY AFTER DAY, WE fought this insane war with our hands tied behind our back by all the rules. We didn't take territory to hold it as in other wars, we fought the enemy wherever we could find him to stop or at least discourage his infiltration into South Vietnam. That meant we might fight the same battle in the same place at different times, and some troops would wonder what the hell we were doing, whether it was all a waste.

LCpl. Donald Johnson explains his M-16 rifle to a Vietnamese Popular Force member
Photo courtesy of Donald Johnson, all rights reserved

After six months of patrols and combat I was tired, weary. I had an opportunity to join the CAC (Combined Action Company), a unit that would stay with a village and help them with civic improvements like drainage, crop enhancements, wells, etc. The CAC program was later renamed CAP (Combined Action Platoon) program in late 1967, because it turns out that, in Vietnamese, the acronym CAC, with a certain inflection, is a vulgar word for penis. What did I know? CAC sounded good to me, hard work, maybe, but without the daily fighting. I didn't know that other Marines called CAC "suicide squads." The goal was to befriend the peasants, help them improve their lives, teach them how to defend themselves against the VC, ultimately to prevent Viet Cong infiltration.

We had about 15 men in our Bravo 4 unit, set up in Duyen Son, seven miles southwest of Danang. There were other CAC units a half mile away and a couple of miles away was a Marine outpost, to reinforce us in case of attack, at least in our mind. But in the black of night it was just us in the bush near the village, wondering if some of the villagers had alerted the enemy to our position, wondering if they had casually walked off and counted paces from a landmark to our position to mark ranges for a mortar attack.

When a bunch of Vietnamese peasants watched your every move, how long could it be before the enemy knew, too?

We became tight, buddies; Richard Glaude and his cousin, Edgar Vallecillo are names I remember. Glaude was an 18-year-old from Maine who walked with a limp. He and Vallecillo were scheduled to go home together on June 15; they called it "D-Day."

We stayed inside our compound at night, just outside the village, behind the school, inside our perimeter of concertina wire with stones in c-ration cans attached to make noise, and claymore mines. When animals made the wire shake and the rocks rattle, we would fire a flare to take a look, nervous that we were vulnerable to the enemy. We normally didn't come out of our compound until morning light.

ON JUNE 6, 1967, after the night settled to its thick warm blackness, I sat my watch sweating on top of a bunker in our compound. Another Marine was in a bunker to my right, both of us straining to see to the wire perimeter. Was something out there? The night can play tricks on your eyes, your ears and even your heart.

Something seemed wrong that night. It was too quiet.

At 2AM he whispered, "Psssst, Johnson! You hear that?" I didn't hear anything now but my own heartbeat!

A few minutes later he whispered, "Hear that?" No, I didn't, but I was getting very tense. As I checked my M-16 to make sure it was ready to fire, he stepped out of the bunker and walked toward the wire to fire a flare to light up the perimeter.

"Pop" went the flare. I didn't know he was going to get out of the bunker to fire the flare.

Holy shit! There were about a dozen VC silently crawling toward us, inside the wire, just 10 feet in front of him. For some reason he yelled, "Don't shoot" and it seemed like a hundred rounds hit him and tore him up as he went down and as I dove into the bunker. I fired a 20-round clip, reloaded then fired another clip and scrambled to reload again while rounds threw up the dirt around my position. VC blasted holes in the wire and lots of them poured into our compound. This was doubly dangerous; first, we were vastly outnumbered and second, the VC were all over the place so we were at risk of shooting one another as we fired at the enemy.

A sapper threw a satchel charge into my bunker and it went off while I was scratching my way out - WHAM! - and threw me up in the air but I could still function.

I was down to my last clip. I started to run toward the sandbagged ammo dump to get more ammo but the VC were everywhere and I thought to myself through the concussions and gunpowder smell of the firefight, "I'm not gonna make it!" As I was running, a sapper threw a charge in the ammo dump and everything started detonating. That's when shrapnel tore up my knee about the same time another flare lit up the compound, illuminating a group of VC surrounding a bunker, pointing their AK-47s inside. Two Marines emerged with their hands in the air but the VC shot them in the head and chest, killing them.

I saw other VC overwhelm Marines and kill them when they should have taken them prisoner, though I'm not sure which would have been worse, and I desperately scrambled to some low brush, trying to hide, knowing if I fired again I might kill some of them but it would betray my position and I had little chance against the tide of VC. They already had us beat that night; there were 14 of us before they started shooting and maybe 100 VC. I was panicking as I lay there, even thought about shooting myself because I didn't want to die the way those other guys did.

I tried to be very still. I knew they were going to find me. I knew I was a dead man. A helicopter gunship circled overhead but we were so intermingled with the enemy they couldn't shoot to help us, so they just dropped flares to keep our compound lit up.

WHAM! WHAM! The earth shook as artillery started coming in, and that meant US support. I heard the clank of tank treads, the sound of US Marines coming to help. As I heard Marines entering the compound, and the VC were doing their magical disappearing act, I yelled, "I'm alive! I'm alive!"

But most of my brothers were not. All but three of us were dead or badly wounded. Glaude lay dying. It was his 19th birthday. His cousin was dead, their glorious D-Day just a few days away. While I looked at what had happened to them, something inside me died, too.

OUR RESCUERS TOOK ME to the hospital where doctors worked on my knee, but there was nothing they could do for wounds far worse deep inside. I guess each man has his limit and I had reached mine. How many people I care about can I watch dying a violent death and still go on living myself? My gut told me – no more, not one more.

If only I had shot my rifle when I had rounds left, maybe some of them would have lived. Why did that Marine have to step out of the bunker, why didn't he just fire the flare from the bunker? Why did I live while they had to die?

The morgue people came to me in the hospital for help. Some bodies were so torn up they needed someone who knew the men to help them identify with certainty which was which. I thought to myself, "I don't want to do this, please God, don't make me do this . . ."

But I did. I'm not sure which was the worst day of my life, that night when we were overrun by VC and my buddies were killed, or the day I identified their bodies, seeing them in terrible shape, their bodies torn up and their lives, their spark, their spirit gone, wrapped in plastic inside the rubber body bags. As they unzipped each bag and I saw the plastic covering them, I wanted to yell, "Get that plastic off my friends so they can breathe!" I was weeping and irrationally hoping that somehow they could get up and walk with me, talk with me, tell jokes like we did when we were waiting, and maybe tell me it was OK that I didn't fire my rifle when I had a few rounds left.

After that day I was ruined as a combat Marine. My body was alive but my spirit had died. I couldn't forget my friends, I couldn't get over how they died. I wouldn't be effective in a fight any more, I had no desire to kill and I couldn't watch any more dying. I didn't have much time left in-country anyway, and my officers gave me other duties in Da Nang because they recognized the signs of extreme stress. I wanted to go back to my unit just to see who was left, to hear their voices one more time, to shake their hands and to say goodbye before I went home, but I wasn't able to go and I never knew who made it besides me.

MY TRIP HOME WAS lonesome even though there were a few guys I knew on the plane. We were advised to change immediately into civilian clothes for our own safety in west coast airports, but that from the Dallas airport further east we shouldn't expect any trouble wearing our uniform. That sounded pretty strange to me but my thoughts were elsewhere.

I came home a broken man inside despite a somewhat normal outward appearance. I wasn't the same any more. I was cautious about affection, even to those I loved. When my Mom kissed me I would lean away. Hugs and kisses, intimate words were hard for me now. Maybe I was subconsciously keeping my distance, trying not to get too close.

I married my sweetheart, Lorene, when I got home. She noticed there was a new kind of distance between us. My smile was forced instead of radiating as she remembered.

The Marine Corps put me through a transition training program to become a postal worker and I started with the US Post Office as a letter carrier. But I always had more jobs to keep me running fast.

I chain-smoked and I worked a lot, usually held two or three jobs because the demons came in my sleep and I only wanted to sleep when I had to, when I was exhausted, maybe too tired for the dreams where my dead buddies talk to me, dreams in which I always wondered, "Why did I live while you died?"

I BECAME A WORKAHOLIC. When I got off work at the Post Office at 4PM or 5PM, I had a little time before starting my shift at the Georgia Highway Department at 7PM. When I got off work there at 1AM or 2AM, I threw papers on a paper route, then I would stumble dead tired home, drink for a while and then crash for a few hours of sleep. The next morning I'd get up to do it all over again.

I didn't spend the time I should have with my wife, my daughter and my two sons. When vacation time came I would make the plans and arrangements and even pack the car, then wave goodbye while my wife and kids drove off to enjoy a week while I did what I had to do, keep busy, keep working, keep moving and drink enough before sleep to pass out, hopefully too tired to dream. I was smoking four packs a day and drinking eight to twelve beers a day.

I didn't think or talk much about my Vietnam experience. In fact, since the country turned so hard against the war and our soldiers were treated like they were either criminals or damn fools, I didn't want anyone to even know I had been in the military; I kept it to myself.

After some years I noticed one manager who was always there at the Post Office when I arrived in the morning and still there when I left for my other job at night. I wondered if a guy like me, willing to work long hours, could get into management. That's just what I did after his coaching on what courses and exams to take to prepare for management. I became a supervisor and rose to Manager of the large and very busy Midtown Post Office in Atlanta. The demands of the job were a perfect match for my need to work long hours, and I tended to work 4:30AM to 9:30PM to keep the demons at bay. I frequently took home proposals and reports to review over a few beers.

Sometimes in the middle of the night I would hear myself say, "Did you hear that?" as I sat up in bed and startled Lorene, and she would assure me nobody was there. But I would get up and walk around the house, checking doors and windows, glad I didn't have a gun for fear of what I might do in my sleep. Eventually she slept in a different room because of my thrashing in the night.

One year I went to a fellow veteran's funeral. I didn't really want to go, and when I was there I ended up talking to the mortician and gave him just a tiny peek at what was bubbling inside me. I told him I didn't understand how he could do his work, that my dead friends visit me often in my dreams and I am tormented by how they died. He told me the difference is that when he sees the bodies they are already dead and he usually does not know them, whereas I knew and cared about these men, and I watched them die in a horrible way; that was the difference. He was right.

I tried to think of the good times, the way my friends used to joke and play around, how we would march into a town and get loaded and tear up a club, but the demons wouldn't go away.

A few years ago my life of evading my problem caught up with me. My cholesterol was through the roof. Constant drinking messed up my liver. My blood pressure skyrocketed and I had flashbacks of the night of June 6, 1967.

I even had visions. Sometimes when I was talking to a person I would involuntarily see them as if they were dead, in a rubber bag, with grey skin, eyes rolled up, but seeing and hearing their conversation . . . then suddenly I would snap back to reality.

I CRASHED.

My doctors treated my body and made me rest, stop smoking, stop drinking, and I got into counseling at the Vet Center for the things weighing heavy deep inside me. I gave up my resistance to seeing the mental health people at the VA about PTSD, and ultimately the medication they gave me helped me get some peaceful sleep. All these things helped.

I retired. I started spending the time I should with my family, and I participated in an experimental program at the Veterans Administration. They have built a virtual reality program that helps veterans suffering from PTSD to relive some bad experiences and in doing so to deal with them.

So I put on the virtual reality helmet under carefully controlled conditions, monitored by Dr. David Ready at the Veterans Hospital in Duluth, just north of Atlanta. I saw myself sloshing through rice paddies when all of a sudden tracers flew at me from a tree-line. Oh, shit! That feeling comes back in a heartbeat, trying to figure out the best direction to hustle while someone tries to kill you. Step by step we worked our way to the night of June 6.

We went through one layer of memories after another, reliving the things that have haunted me for years. When I took off the helmet after a session I was usually sweaty, quiet and exhausted, staring at nothing, weary and sad. Lorene was worried this experiment might make things worse. And she was right. I was very agitated for a while, but I kept going and through repetitions the demons subsided, though I don't think they will ever go away for good.

EVENTUALLY, RELIVING THOSE TERRIBLE times helped me find forgiveness from the ghosts and from myself. I had carried guilt for decades, believing I should have done something different that night; that maybe I had the power to keep some of them alive.

After one virtual reality session I asked myself and Dr. Ready, "Why did I survive? Am I really to blame for their deaths? I didn't fire my rifle after I got to a safe place." My agony was the guilt of thinking I might have saved another Marine, that I didn't die with the rest of them because I was hiding instead of firing at the enemy.

He told me something that I hang onto very tightly. He said, "You made the right choice. You lived. You defeated the enemy that night by living. It's not your fault."

It's not my fault. Finally, I believed that.

At last, I can face the ghosts. I can remember them without the burden of guilt. I can remember serving my country well, doing my duty with some men I had no intention of knowing and ended up loving them before I lost them. That they died so young and so horribly, that their country did not appreciate their sacrifice, those are terrible things.

But I now have the relief of knowing in my heart . . . it's not my fault.

Photo courtesy of Donald Johnson, all rights reserved

After being discharged from the Marine Corps, Donald Johnson worked for the U.S. Postal Service as a letter carrier and attended night school at Massey Business College where he received an Associate Degree in Business and Accounting. He was promoted several times over the years, finally to Manager of the very busy Midtown Atlanta Post Office, until his retirement.

Donald's two sons are married and have careers with the U.S. Postal Service and his daughter is teaching school. He spends as much time with them, and the grandchildren, as he can, making up for lost time and enjoying each day with those he loves.

Postcard to America

I RETURNED FROM VIETNAM in February, 1968 with quite a tan from the waist up and applied for a job where the most jobs were in my upstate New York town - General Electric.

As I entered the personnel area, I noticed many other young people there, obviously right out of college. I was looking for a manufacturing job as I had no real skills. I filled out the application form and handed it to the receptionist, who asked me if I wanted a job interview right away. I said yes and a few minutes later a man about 30 years old called my name and took me into his office to chat. He told me what job openings he wanted to fill and even offered on-the-job-training for some of them. We chatted about the different jobs and he offered me a Quality Inspector's job on one of the many manufacturing lines. I readily accepted the job and asked when I could start. He wanted me to take a physical the following day and report to work the next Monday.

As I was leaving, I offered my hand to thank him for his confidence in me. He could see my watch tan line and asked me where I got such a terrific tan, in February. I said I had just returned from Vietnam less than a week ago. The look on his face indicated ugly disappointment. He told me to forget the job because I was "not the person he thought I was" and tore my application in half right in front of me! He got up from his desk, opened the door to his office, gave me a look that said "git" and out the door I went.

I didn't let that get me down, even though I had never been that ashamed, pissed and humiliated in my life. The next place I tried for a job, I got hired. I retired from IBM in 2001 with a more generous pension and benefits than I ever would have received at GE. Maybe when God closes a window, he does indeed open a door.

Michael Leonardos, Endicott, New York
Co. C, 588th Combat Engr, Batt, 25th ID, Wolfhounds
Tay Ninh area III Corps, Vietnam, 1967-68

Just One Life
Donna Rowe
Marietta, Georgia

I HAVE ALWAYS BEEN troubled that the virtues of America in Vietnam, the good things, were almost never reported. While anti-war radicals spread the false story that our troops committed atrocities as a matter of course on a daily basis, the reality is Americans spent their time doing the things you would expect, building roads and schools, teaching improved farming and drainage methods, digging wells, providing security and medical attention and a hundred other things the people back home would acknowledge with pride, if only they knew the good works we performed in addition to fighting a war. But good deeds didn't seem to fit our news media's agenda, so the true record rests with those of us who were there.

I was a nurse in the life-saving business, a Captain in charge of the triage unit of the Third Field Hospital in Saigon. During my tenure, I am fiercely proud to say the wounded that came to my unit alive stayed alive while under our care, we didn't lose a single one. Some died before they reached us, some died later in surgery or from complications, but we were determined to move heaven and earth to keep them alive while under our temporary care, and my magnificent staff of corpsmen, doctors and chaplains performed miracles.

OUR NEWS MEDIA FOLLOWED any scent of a war crime like a voracious herd, so long as the crime was committed by Americans, but they seemed to ignore atrocities by our enemy. Maybe that was because there were so many, or maybe our enemy's misdeeds did not fit into preconceived notions about the war that drove news reports. One day in May, 1969, the VC committed an atrocity that would change my life.

Our radio room received a call from Dustoff, a medical evacuation helicopter, requesting permission to land with a "Severely wounded civilian baby onboard." The question was necessary because we were in the middle of the enemy's *May Offensive,* and our medical facilities had strict priorities for accepting casualties while we were in *Offensive Mode* due to the limitation of resources. The priorities were: (1) US armed forces, (2) US civilians, (3) Allied forces, (4) host country military forces and lastly, (5) Vietnamese civilians.

I didn't know it at the time, but in the mountains near Cambodia, American troops, whom we think were from the 1st Infantry Division, found a Montagnard village which had been completely decimated by the VC. Montagnards is a French term meaning "mountain people," a term they gave to the Degar people who are indigenous to Vietnam's Central Highlands, a small, dark, peaceful people who live in mountain villages in communal houses built on stilts, persecuted for centuries by the Vietnamese and Cambodians. The US considered Montagnards, or "Yards," to be an ally, their men frequently led by our Special Forces advisors in small groups to gather intelligence. Maybe that is why the VC slaughtered the civilian villagers.

Amidst a pile of bodies our troops heard the whimpering cries of a baby. They dug through the pile of dead to find a baby girl, still barely alive, clutched tight in the arms of her dead mother. The mother's body was in rigor, they couldn't move her arms, and so they called for a Dustoff for the baby and carried both bodies to the helicopter that landed under fire to make the pickup. The baby, in her rigid mother's arms, was loaded onto the helicopter.

Of course I didn't know these things when the radio call came in.

David Alderson, the Dustoff Aircraft Commander, said they thought the baby would die very soon. She was dehydrated, malnourished and had fragmentation wounds in her abdomen and lower chest, causing her abdomen to fill with blood and thus hampering her little body's ability to breathe. I had not previously heard David Alderson's callsign, so I knew he had traveled a long distance to reach our location in Saigon. I'm speculating, but he probably got passed on to 3rd Field as other hospitals along his route were being inundated by casualties from the various battles of the offensive. I discovered many years later what an absolute hero David Alderson was, not only for going in under fire to rescue a dying baby, but for pulling out of the hell of war thousands of our casualties during his three tours in South Vietnam.

For most of that day we had struggled to keep up with a heavy flow of casualties. When Dustoff radioed to ask "Will you receive a civilian baby?" we had several "full" Dustoffs on the way to us. I had to answer that question as my first sergeant, NCOIC SFC Grant, stood at our radio looking to me to respond. This is one of those moments when our American core values, those our parents ingrained in us, come to the forefront. "Do what is just and right," my parents always counseled. So I gave permission for David Alderson to land.

SFC Grant looked at me and we nodded to one another, a tiny admission that it broke the rules but was the right thing to do. He told me "Capt'n, you'll take hell for this!" I remember my reply was, "What can they do, send us to Vietnam?" Our ambulance met the Dustoff at the helipad across a busy avenue from the hospital. Specialist Geer was the driver who brought the baby to us and relayed to us the story of the destroyed village and its sole survivor . . . this tiny little baby. Her dead mother's arms were broken to release the baby on the helipad. I have no recollection of what happened to the mother's remains, but in all likelihood her body went to our morgue at Tan Son Nhut Air Force Base directly across the street from 3rd Field Hospital. Specialist Five Richard Hock, one of my best combat trained medics, took the baby from the ambulance drivers. He realized immediately that this little soul was in respiratory distress, due to her abdomen engorging with blood from her fragment injuries thus hampering her diaphragmatic breathing. We got a breathing tube into her with the smallest tube we had in the Triage area, put a manual breathing bag on it, and Richard took over breathing for this little one until we turned her over to the Operating Room staff several harrowing minutes later.

The Triage doctor ordered a "full body" screen on her so we rushed the baby to the X-ray room so we could locate pieces of shrapnel to be removed in surgery. On the way from X-ray to the operating room, I saw Father Luke Sullivan, our Catholic Chaplain, and pulled him into the crowd that was half-running down the hospital corridor. I told him "Father, come with us, you have to baptize this baby!"

He told me he didn't have any holy water with him. I was concerned that the baby might die any moment, and if she did survive, she could not stay for a long time in our military hospital. Under Father Sullivan's guidance, our hospital supported a Catholic orphanage on the outskirts of Saigon. I knew in my heart that if this baby made it through hours of surgery, the Sisters would take this baby if Father Sullivan asked. All of this flashed through my head in half a second as I pointed to an outside wash-off sink in the corridor adjoining the Emergency Room and the Operating suites, and said, "Father, any water you touch right now will be holy, God is watching, let's get this baby baptized!"

So Father Sullivan used water from the sink to sprinkle on her tiny little head and said "I baptize thee . . . " he stopped and looked at me for a name. A name, a name, a name I thought, then remembering the Irish song my father sang to me while dancing me across the floor as a child, *I'll Take You Home Again, Kathleen*, so I blurted out quickly in the rush, "Name her Kathleen Fields! *Kathleen* from the Irish ballad and *Fields* because we're in the 3rd Field Hospital." Father Sullivan's hand touched her little forehead while he stated the baptismal rights then he looked around this fast moving litter/gurney and said ". . . and your Godparents are Specialist Medic Darrell Warren (a Mormon), Specialist Richard Hock (a Catholic), and Captain Donna Rowe (a Methodist). We became Godparents on that day, joining with a Catholic Priest to do a tiny bit of God's work all while rushing the baby to life-saving surgery. Father Sullivan in his sermon that night referred to this act as an "ecumenical baptismal" as he told his flock "There are no denominations in war, but there is always faith in God."

KATHLEEN SURVIVED THE HOURS of surgery to remove shrapnel. She was so tiny - when I laid her on my arm she only went from the crook of my elbow to the palm of my hand - the surgeon had to use the smallest surgical instruments they had in the O.R.

The nurses on the post surgical ward and my Godfather medics scrounged up an orange crate and lined it with towels as a makeshift crib. Being a combat medical facility, baby diapers and baby formula were not in the supply room so they figured out how to make diapers from washcloths reinforced by sanitary napkins. With the help of the wonderful Red Cross gals assigned to our hospital they got soy-based baby formula shipped in from Hawaii, as baby Kathleen was not tolerating what we had on hand, which was cow's or goat's milk.

The next day the hospital commander, Colonel Chandler, walked into the Triage area as I was washing down after a wave of casualties and said

"Captain, I understand we have a civilian baby in this hospital." I remember thinking to myself, "Well! Here it comes!" but I answered "Yes, sir, we do."

His hands clasped behind his back and a gentle grin on his face, he said "Well done, Captain," and quietly turned and walked back to Headquarters. God bless him, he was a good commander and compassionate man who understood that war doesn't change those caring and compassionate core values we took into war . . . right was right.

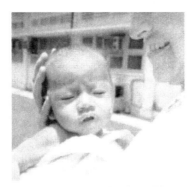

Cpt Donna Rowe with baby Kathleen
Photo by : Helen Musgrove, embedded reporter,
from the Jacksonville FL Journal, all rights reserved

Both of my medics, Specialists Warren and Hock, took their Godfather roles quite seriously, even though they were still in their teens. They painted the walls near her crib from army drab green to bright yellow and after her post-op recovery, they'd go get her off the ward and would walk her down to the Emergency Room and Chapel courtyard so we could see her progress and even hold her for a while. These were just a few of those peaceful home-like moments in the midst of all the mud and blood of war.

Kathleen became the darling of the hospital staff. This baby smiled a lot, didn't cry very much. She was like a magnet for us. Big, strong men who knew nothing about babies turned into mush around Kathleen and learned from the nurses how to hold and feed a baby, even how to change diapers. Darrel Warren and Richard Hock spent their spare time with her like all proud Dads would do.

This was a refreshing change. It was life beginning, a welcome relief from all the broken young bodies we dealt with every day.

A few days after Kathleen's arrival, three soldiers in full combat gear came walking in the back gate into the Triage parking lot and into the Triage area where I was doing supply check and washing down the floor. They asked about a wounded baby, and if the baby had survived. I told them she was on Ward 2 and directed them to her where they visited briefly, then headed out. As these men left to return to the war, one of the men said, "Thank you, Maam", and I whispered in my head "No! Thank you. Thank YOU!" I swear I remember these men having the "Big Red One" 1st Inf. Div. patch on their fatigues, as I rarely saw that patch in our area; if that is so they hitched a long ride into Saigon to check on this baby they helped save.

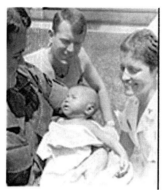

*Specialists Darrel Warren, Richard Hock, baby Kathleen and Cpt Donna Rowe
Photo by : Helen Musgrove, embedded reporter, from the Jacksonville FL Journal, all rights reserved*

AFTER ABOUT TWO WEEKS, Kathleen was healthy enough to be transferred to St. Elizabeth's, the Vietnamese Catholic orphanage outside of Saigon. We were sad to let her go, but we scrounged up extra food from the hospital mess to take with the baby to the orphanage since they were always short of everything.

Father Sullivan tried to celebrate Mass in the hospital's tiny chapel almost every evening for the new patients or our forces in or around Saigon. During Mass one night, shortly after Kathleen's arrival, his sermon was the "Ecumenical Baptismal", in which he tried to convey that denominations of our faith were put aside in war and only our faith in God and His Son prevailed. One of those present was US Navy Lt. Marvin Cords, who asked Father Sullivan to arrange for him to see the baby. When Cords first saw her, she had a wound dressing that just about covered her entire body. He and his wife, Sally, had already adopted three children, and they had talked about adopting a Vietnamese child. He hired a Vietnamese attorney to track down Kathleen's birth certificate. Weeks passed, but still no birth certificate came from the nuns at the orphanage. Cords then sought help from Father Sullivan, who told the nuns to get a birth certificate. Father Sullivan was their priest too. His support contributed to the care of the 300 orphans at St. Elizabeth's and the Sisters were anxious to please Father Sullivan. Days later, Kathleen had a birth certificate.

After more government red tape and delays, including a waiver that had to be signed by then South Vietnamese President Nguyen Van Thieu, Cords made arrangements with a USAID nurse who was headed home to the US to escort Kathleen out of Vietnam and deliver her into the arms of her waiting *American Mother*, Sally Cords, in America. Little baby Kathleen lived with her family on military bases here and there, growing up with five adopted siblings who were black, white, Vietnamese and Native American. I truly marvel at the Cords' compassion for all of these children of God that they brought up in a loving comfortable home in peace.

A couple of newspapers in the US picked up the story about baby Kathleen.

After a few years of Marvin Cords dropping us a card at Christmas, and two or three annual letters, the cards stopped coming and we lost touch over the years. But we never lost the memory of baby Kathleen.

Over the years I have often thought of baby Kathleen. She was kind of a symbol of the good things that can happen amidst so much misery. Even though the media of the time were portraying us as *babykillers*, I knew in my heart the truth of what my staff and the brave men in the field did to rescue our Baby Kathleen. I also knew that the *Kathleen Story* is just one in thousands of good and courageous stories that could be told about the men and women I was proud to serve with in this war in Southeast Asia, and that such goodness was not out of the ordinary, it was how young Americans conducted themselves.

IN OCTOBER OF 2002 my husband, Al, was President of the Georgia Vietnam Veterans Alliance, Chapter One. He got a phone call from a friend of ours in the Dustoff Pilots Association. Our friend told Al they needed Vietnam Veterans to be at Kennesaw State University on the 22nd to meet a refurbished Huey helicopter; a documentary was being filmed about this icon of the Vietnam War.

Well! October 22nd was a cold and rainy morning and I didn't feel like going to stand on some field to meet a helicopter, but my dear husband said, "The men would be disappointed," so I got dressed and there we all stood under umbrellas telling war stories and seeing many veterans we knew and meeting some we didn't yet know. There was a low ceiling of rain clouds. I had my doubts that any sane pilot would be coming into this tight LZ on the college's parade field surrounded by buildings. I should have known better, as Vietnam helicopter pilots were the best with tight LZ's.

Al was approached by the Director of the film, Patrick Fries of Arrowhead Films, who told him about the doumentary, *In the Shadow of the Blade (ITSOTB)*, and he asked Al to "signal" in the helicopter with some of his veterans standing on the parade field. Al agreed, and my first thought was "Oh! This is going to be great, fresh cut grass, in the rain, with the backwash winds from a helicopter!" I could see us all being covered with green wet grass!

Then we heard the Huey . . . whop-whop-whop . . . a sound all Vietnam veterans know down into their soul, a sound which cuts to our backbone, the sound of our war. That Huey took them into battle, brought them needed supplies and when needed, carried them out of war in one of three ways: alive, wounded or dead. To me that was the sound I heard every day for 369 days coming to our hospital with the dead or the dying or severely wounded casualties. A feeling of fear and anxiety settles on me when I hear that sound.

Sure enough, through the gray drizzling clouds dropped this Huey helicopter and the Director said, "Al, take your veterans onto the field" and the cameras started to roll. I, of course, was trotting right along with the men, when the Director said, "No wives, just veterans!" If looks could kill, poor Patrick would be dead, but my calm direct husband replied, "She is a vet!" Patrick grabbed my arm and said, "Do you have pictures? Are you a nurse?" To both of these questions I hesitantly replied, "Yes!"

After, the landing of the Huey, Patrick asked me to go get my pictures from home, about 15 minutes away from the college, and bring them back for him to see and film. Al and I had been married two years when we both volunteered for Vietnam, and I knew where we had the scrapbooks and the photo albums from our tour in Vietnam, but I had not looked inside them for over twenty years. When I arrived home, I went to our bookshelves in our finished basement and grabbed the black scrapbook. Why the black scrap book? I have no idea, but I believe it was a divine intervention; God wanted the *Kathleen Story* told and her story in letters and pictures are in that black scrapbook made 34 years earlier.

I met many people that day, including Terry Garlock, one of the speakers who told the crowd the story of the common virtue he saw in Vietnam, with the example of two fellow Cobra pilots who risked their life to rescue him when he was shot down. I talked to the ITSOB crew about my story, about the good works by Americans in Vietnam, and I told them about baby Kathleen as I showed them the pictures and letter from Father Sullivan in the black scrapbook depicting the full story of Kathleen, all captured on film. The end result of filming at over 42 LZs over eight states was to be a documentary composed of Vietnam veteran stories, and the story of baby Kathleen would be a moving part of this remarkable documentary.

THERE WERE REPORTERS AT the Kennesaw State LZ, and one of them, Bill Osinski, wrote a story in the Atlanta Journal Constitution (AJC) about me and baby Kathleen. I didn't know it, but Kathleen had been searching on the internet for over 10 years trying to find the persons on her Baptismal Certificate. She hit upon the AJC story by Googling "Vietnam Nurse," which found the Bill Osinski story in the AJC. Kathleen had been on several Vietnam veterans sites leaving her message and her story to see if anyone could help her.

Ed Russell, a Vietnam veteran, saw her postings on the message boards of some sites and made the connection. Ed had been Father Sullivan's assistant at 3rd Field Hospital but his tour ended before this whole incident and so he had never heard about baby Kathleen. He did, however, respond to this guestbook entry on a Vietnam veterans website in 2002:

> *"I'm looking for any staff and/or military personnel who may have been at Third Field Hospital in Saigon, 1969. Anyone who may have remembered a small Vietnamese girl brought in by helicopter. Her whole village was killed by Viet Cong? . . . I have very few names and no memories except*

the year and the place. Could you please contact me, if anyone knows anything? Thanks!"

Ed did a little research, found newspaper clippings written by Helen Musgrove, for the Jacksonville (FL) Journal of the original story about Kathleen published in 1969, made the connection to my name but didn't know where to find me. He told Kathleen what he had learned. Kathleen called Bill Osinski at the AJC and he directed her to Cheryl Fries, the Producer of *In the Shadow of the Blade*. Bill knew that before filming, Arrowhead Films had to get information and permission from all persons filmed. When Kathleen called Cheryl, I guess Cheryl almost dropped the phone, as they were in the final stage of editing the film. On March 8, 2003, I received a breathless call from Cheryl.

"Are you sitting down?" she asked.

In my casual way I said, "No."

She said, "Donna, I need your permission to give out your personal information."

I asked why and she replied, "I just got a call from THE BABY!" I went weak in the knees, then to total excitement.

Thirty-Four years had passed since that hot bloody day in Vietnam, now here she was back in my life again. Spinning is the only word that describes that moment. Within a few minutes from hanging up the phone with Cheryl, I got this soft-toned voice on the phone, saying, "Hi, I'm Kathleen."

Well! We talked for almost three hours. I will never be able to describe the feeling I had when talking to my God daughter, it was nothing short of a miracle; so much time had passed, but the heart still remembers. Then the wheels really started to roll! Bill Osinski, Ed Russell and Arrowhead Films all started looking for the remaining cast of characters in the Kathleen Story.

Richard Hock came home one day from his work as an Atlanta Fire Department Paramedic to a voicemail from a woman who wanted to talk to him about Vietnam. He didn't much like talking about his three tours in Vietnam or about his wartime experience, but he called back. The lady who answered the phone called out for someone else to come quickly to the phone: "Kathleen!" and he knew, in a flash he knew.

I also discovered that Richard Hock and I had lived within 20 minutes of one another for 20 years.

Cheryl Fries decided this was too good a story not to tell all the way to the end. She said "This is a story about humanity in the middle of war, about good people in a bad situation. Kathleen's future exists because good people in American uniforms cared."

Cheryl said they were stopping the editing right then to do a "reunion" shooting. The ITSOTB crew arranged for Hock, Warren and me to be at Ft. Sam Houston, Texas, where we all had trained as medical personnel on the way to Vietnam. They found Darrell Warren, a school principal in Las Vegas, Nevada, and David Alderson in Arizona. Things moved quickly and the next thing we knew Southwest Airlines offered us free tickets to Texas. Al and I met Richard Hock at the airport in Birmingham, Alabama, and Darrell was supposed to fly out of Las Vegas. All of us headed to Ft. Sam Houston in San Antonio, Texas. David Alderson, the Dustoff pilot who originally delivered this baby to us in Vietnam, was to fly the Huey to meet us at this last LZ, once again with Kathleen on board.

Sadly, a week before the reunion, David Alderson died of pneumonia at a Virginia hospital. His co-pilot in Vietnam stood in for him and flew the Huey from Austin to Ft. Sam Houston with Kathleen and her American parents on board. I had a feeling as I watched that helicopter approach over the old Brooke Army Medical Center that David was watching and guiding that helicopter. His presence was strong in all our hearts and minds. I like to think he witnessed from heaven our meeting and felt our thoughts of him.

At the LZ, Patrick Fries and his film crew positioned Richard and me on the parade field with active duty Army medics and nurses stationed at Ft. Sam. Unfortunately, Darrell Warren had some mix-up in Las Vegas with the airlines and couldn't get to San Antonio. We missed his presence. I gave the helicopter the landing signals as our troops did on many Vietnam LZs, and then there she was, all grown up. Kathleen's husband, Billy, and their three daughters, Mary-Ann, Jo-Jo and Sean were driven down to Ft. Sam and were also waiting on the side of the parade field. It was a wonderful moment and day; seeing Kathleen again was very emotional and inspirational for all of us. And Patrick caught it all on film.

Richard Hock, Kathleen, Donna Rowe meet after 34 years
Photo courtesy of Donna Rowe, all rights reserved

In August of 2004 the documentary *In the Shadow of the Blade*, had its premier in Atlanta, hosted by the Georgia Vietnam Veterans Alliance, Chapter One, and featuring the story of Kathleen. Kathleen and her family came and stayed with us for the premier but this time she handed me her brand new baby, Katie (Kathleen) Rose, so tiny, so fragile, that in a heartbeat I was back at the 3rd Field Hospital and holding Kathleen again. We had quite a celebration that night, and the circle of life seemed complete. The *Kathleen Story* along with the film's Huey is now part of the Vietnam War Exhibit in the Smithsonian's American History

Museum in Washington D.C., which makes Kathleen and me happy as it tells the true story to millions of visitors a year.

One life can touch so many, and as Kathleen's life unfolds the magic continues. That she is Montagnard is especially satisfying to many of us because the persecution of her people continues, virtual genocide sanctioned by the communist government of Vietnam.

THERE IS SOMETHING IN this story some readers might not recognize at first, something America needs to mull over, chew on for a while.

The Vietnam War was consumed by controversy and, in its later years and since the war ended, it has been shrouded in myths and half-truths, the real truth hiding in the hearts of those of us who were there.

The VC atrocity of slaughtering a village of civilians in the mountains made the US news only because of the cuddly story about saving a baby, otherwise it was not news; VC atrocities were so . . . ordinary.

The US troops who called in Dustoff for a wounded child were spending resources and taking risks for Vietnamese civilians. When the troops made a special trip to check up on the child they were revealing their soft side, which happened often because beneath the façade of combat toughness, they were good people with big hearts.

The Dustoff crew commanded by David Alderson pushed the envelope and risked the displeasure of their boss for a civilian child just like they bent the rules regularly, not for their personal benefit but because they were trying to do the right thing.

My staff and Father Sullivan went above and beyond for a baby because it was the right thing to do – and because she stole our heart.

The clerks and other authorities that pushed past red tape to get Kathleen's adoption done in record time were doing the very same thing.

The anti-war radicals sold the story that American troops committed widespread war crimes in Vietnam. I never heard of or saw any evidence of our troops committing any war crimes, though I know isolated incidents did occur. Rather, the vast majority of us who served in Vietnam were good, honest and decent men and women who went to war and took our strong core values of "Duty, God and Country" with us, to do the best we could in spite of the horrors we faced.

Maybe this is the real legacy of the Kathleen story. It reflects the fact that Americans in the Vietnam War were good people, sons and daughters, husbands and dads who kept their humanity under hard conditions and helped the Vietnamese people in many ways despite the war. They conducted themselves in a way that should have made you and their nation proud.

Our country turned on us, the ones they asked to do the fighting and dying in Vietnam. The irony is, no one hates war more than those that have to fight it, and those who served in Vietnam earned their nation's gratitude, not her disdain. I pray that my beloved nation and her people never make that mistake again. As Gen. Hal Moore (US Army, Ret), a true hero of the Vietnam War, said in the film *We Were Soldiers*, "You can hate war, but you must love the American warrior!"

Donna Rowe has a rich history of community volunteer service in addition to her successful career as a real estate broker. She lives in Marietta, GA, with her husband, Al.

Postcard to America

I DIDN'T EXPERIENCE ANY hostility - neither physical nor verbal - after returning from Vietnam the first time in early 1969 or the second time in late summer of 1971. What I did notice was . . . gross indifference. Even vets from other wars seemed to strain to avoid showing any interest whatever in what was happening in Vietnam.

Having been a police officer previously, I knew all about citizens who had not a clue of the world beyond their daily life, but the indifference from WWII vets puzzled me.

When I was on leave after my first tour, I was at a party with my former comrades on the police force, most older than me, many who had been in WWII and were quick to share their war experiences with me and others. However, other than a polite comment to my safe return home they had no interest whatever in any discussion of Vietnam. They would rudely change the subject whenever I tried to relate an experience.

On my trip home after my second tour through Seattle to Detroit, I sat next to a businessman who let me know that he was an Air Force pilot during WWII. He made no effort at conversation other than a quick question about the identity of one of my ribbons. He asked if it was the Distinguished Flying Cross. I said it was. His response was an uninterested, "Oh.hmm . . ." His attitude seemed to be, "My war was noble, yours was not!"

Years later I thought of that conversation again when Vietnam Vets led the public in praise of our troops. Unlike the vets of WWII and Korea, who seemed to join the public in being ashamed of Vietnam vets instead of proud of us, we actively supported vets of the wars in the middle east, determined that never again would our troops come home to insults, not while we are alive!

I expected bad behavior from hippies, draft dodgers, and other morons and didn't much care what they said or did. But, the indifference from the average citizen, and especially vets of other wars, was a huge disappointment.

Bill Stanley, Marietta, Georgia
1/9 Cav, Camp Evens, Vouc Vinh 1968-69
228/1st Cav, Bear Cat, Phouc Vinh, Chu Lai, Marble Mountain (Da Nang) 1970-71

Robin Hood and Brotherhood
Wayne King
Peachtree City, Georgia

THE ANNAMITE MOUNTAIN RANGE, paralleling the coastal flatlands and known to the Vietnamese as the *Trường Sơn* mountains, form a divide that sends water east to the South China Sea, and west to the Mekong River, which flows to the southern end of Vietnam to form a huge delta region. In the shadow of these mountains, on the dusty Bong Son plain in northern II Corps, LZ English was my first home in Vietnam. Not far away was the Vietnamese village of Bong Son, beside the flat and shallow Song Lai Giang river, surrounded by An Khe, Pleiku and Qui Nhon, a mountainous and dangerous area that had been a hotbed of Viet Cong activity for a long, long time.

LZ English was the home of the 173rd *Sky Soldiers* Airborne Brigade, and other units like mine, the 61st Assault Helicopter Company. Six plus one is seven, the magic number in the dice game of craps, and our 61st AHC symbol was a pair of dice, our takeoff pad called "The Crap Table." Our primary mission was to move troops and supplies from one place to the other, to give them the mobility they needed to gain an advantage over our stealthy, wily and tough enemy.

I was one of those helicopter pilots, but I had another job, too, and I sometimes wondered which was more important to me, and to my CO. I had no idea the construction experience I had in my youth would be useful in Vietnam, but in a place where we lived in tents and tolerated dust in everything, even gritted in our teeth, in a place where the Army was thin on resources, the ability to construct our own buildings was a vital matter of comfort and morale.

So, when I wasn't flying, I wielded a hammer and a saw because I knew how and I enjoyed building. I also enjoyed the camaraderie of Richard Benswick working beside me, and I think this distraction during our off-time was soothing to the soul. How many guys can indulge the pleasure of a hobby with a buddy during a war?

ANYONE WHO SAYS I stole things in Vietnam just doesn't understand the logistic system! I never stole a thing, though I did become a master of *value-added re-allocation.* You see, Uncle Sam had what seemed an endless supply of building materials, the raw goods of creature comforts in such short supply at LZ English. But Uncle Sam's stuff seemed always to be in the wrong place, and so Richard and I learned the tricks of moving certain items from where they were in excessive abundance to a place they were sorely needed. We didn't steal anything, we just relocated Uncle Sam's stuff within Vietnam to achieve the *highest and best use* for these goods paid for by the taxpayer, and it still belonged to Uncle Sam. Can you see the virtue here?

We needed more comfortable places to live, to relax and have a drink after a day of fighting the war. So when we were tasked to fly to a base to pick up 20 pallets of something, we'd con or persuade or weasel our way to bring back 25 pallets. Then we had something to trade.

Our daily wheeling and dealing kept a stream of construction materials coming for the billets we needed. Somehow the guys with steaks could always find the guys with beer and work out a win-win deal. But sometimes one side didn't win. Whenever I visited the Air Force and saw that they had way more good stuff than I thought they should have, well, I borrowed things like mattresses. I mean, nobody was watching their billets in the middle of the day and they had extra mattresses stacked against the wall while we just had miserable cots, and the Air Force guys had so much stuff they weren't much interested in our trade business unless we had LRRP rations, so why shouldn't I re-allocate those mattresses to where they were sorely needed? Uncle Sam just had not yet figured out yet where the need was, and I felt I was helping the war effort. Don't you think the folks back home would have wanted me to re-allocate those mattresses?

Wayne King, Greg Ressler and their hobby
Photo courtesy of Wayne King, all rights reserved

Richard and I got to know some other re-allocation specialists and we struck deals in dark corners. But sometimes more direct action was called for. Our company had two 50-gallon drums with an immersion heater for showers, which meant we almost never had a hot shower, and even cold water was in short supply. But we noticed the 173rd had a big re-supply operation going, with helicopters lined up in an airborne waiting line, coming in to a high hover, one after another, to slingload a 500 gallon bladder and fly it off to parts unknown. They sure had a lot of 500 gallon bladders! So Richard and I borrowed the one helicopter in our unit without 61st AHC external markings, or any other identifying marks – no need to get people stirred up - and we took off from The Crap Table and just, well, got in line with all the other 173rd helicopters. The aircraft in front of us couldn't see us, the one behind didn't know or care who we were, and when our turn came the ground crew wasn't going to ask questions, they just hooked us up to the lines of the net holding the bladder while we hovered, then we took off! That's how we got our 500 gallon water supply for showers.

Please! Hold the applause.

We constructed buildings for the enlisted men and pilots, far better than tents. I wrote my Dad and asked him to send a door kit with doorknobs, locks, keys, hinges, the works. When I finished, my room had a locked door, the walls were stained ammo-crate wood pima-primed with JP-4 (helicopter fuel, kerosene), and I even trimmed with the wood outlining the cardboard crates that were used to ship wall lockers. You'd have to spend some time in war zone tents to appreciate this; I had straw matting over the windows for shade and privacy, built a makeshift drop-ceiling, and had tile on the floor from our horse-trading with the Navy. It was the lap of luxury, relatively speaking, Richard's place and mine, twice as enjoyable because of how it was built.

Damn, we had fun in our off-time! It was like our own little band of Robin Hood and the merry men, re-allocating from the rich and serving the poor, laughing our asses off all the way. Besides, this is the only time Richard and I flew together, because I was a slick pilot and he was a gun pilot.

WHEN I ARRIVED IN Vietnam near Thanksgiving 1968 without final destination orders, I expected a plush assignment at the huge base beside the ocean at Cam Ranh Bay because I knew a few highly placed people and thought I would pull strings. After two weeks in my sleeping bag, I eventually had to face off with a personnel officer wearing the same Artillery insignia I wore. He listened to my plea for a Signals assignment that would eventually get me back to fixed wing flying, looked me in the eye and said, "Son, I am going to do you a favor. You won't thank me now for it but you will later. I am going to send you to LZ English with the 61st AHC. These guys are getting plenty of combat experience and you'll have the opportunity to put into practice many of the skills you have learned." He was right.

I was smart enough to know I didn't know everything, but I had no idea how much I had to learn.

My first flight was a checkride with a seasoned CW2 named Sam Kyle, a two-tour experienced pilot with a huge handlebar mustache who wore the aircraft like an extra-light comfortable suit of clothes; he was the finest helicopter pilot I ever knew. When Sam was flying he lowered his seat to make best use of the armor on the side of his seat, and he crouched low behind his *chicken plate*, down to his eyeballs, and held the cyclic with just the end of one finger. I thought, "How the hell does he do that?"

He showed me how, but it was impossible at first, like learning to fly all over again, since I was used to leaning my arm on my leg to steady my cyclic hand, and he was showing me how to control the cyclic with just a finger and no armrest so I could crouch behind the chicken plate. I would soon discover the virtue of crouching behind something that stopped small arms fire.

A few minutes into my checkride with Sam, we got a radio call directing us to pick up wounded in a nearby LZ. We diverted to the LZ, located and confirmed the color smoke they popped, and I set up my approach. Sam must have seen the strained look on my face as I descended into my first combat situation, and he said "I have the controls" as he took over and I let go. Sam flew it in cool as a cucumber while we were taking enemy fire, and overlapping traffic in my ears on the UHF (other aircraft), VHF (airfield tower control) and FM (ground units) radios confused the hell out of me. I tried to make myself as small as possible behind the chicken plate and nearly jumped out of the aircraft when the crew chief and gunner two feet behind us opened up with their M-60s to shoot back. I watched Sam with admiration as he calmly touched down nice and smooth while the wounded thrown in the back turned out to be KIA, the dead remains of what had been just a little while ago young Americans doing their duty, now smearing the helicopter floor with what had been their lifeblood. With no wasted motion, Sam nosed it over and took off, taking our grim cargo on the first leg of their journey home.

That was a memorable checkride, and a fitting introduction to flying slicks in Vietnam.

Most of our missions were combat assaults, meaning we loaded up a group of slicks with about seven American troops each, or ten ARVNs since they were smaller and lighter, and flew them in formation into an LZ that may or may not be hot with enemy fire. Richard and his gun buddies would fly with us and prep the LZ with suppressing fire or maybe just fly in an escort mode in case we received fire, ready to roll in with rockets and miniguns.

The sudden loud crack of rocket explosions out our window on either side of the LZ was comforting cover, but sometimes they just came too darn close. I asked them once in a while when I was edgy if they had a bet on who could shoot me down! I might have been more edgy if I knew then how bets and guns would coalesce in my future.

I've been asked if I was scared flying into an LZ. Well, that depends. Most times I was too busy to be scared.

When we flew formation with other helicopters, the idea was to keep it tight so we would all touch down together in an LZ of limited size, offload the troops, nose it over and take off together; that's better than going in one at a time and giving the enemy more opportunities to shoot us, and it puts more troops on the ground in one landing, a stronger unit instead of weaker pieces.

Flying formation meant staying within one rotor diameter of the next aircraft, usually on a 45 degree angle by visually lining up his near rear skid strut with his far front skid strut, and keeping his rotor plane visually on the horizon so we had three feet of vertical separation in case the rotors overlapped. If the lead aircraft was not sufficiently smooth in turns or deceleration or descent, it was like trying to stay steady in the middle of a slinky though the fun of flying the slinky was diminished by thoughts of certain death from a mid-air collision. I monitored several radios, listening intently for things like "taking fire!" or "go around" if the initial landing attempt was a no-go. As we went in I had to keep the aircraft steady with the others descending and bleeding off airspeed, watching for stumps or booby traps or trees too close to the rotors, always alert for VC popping up to punch our lights out, watching the other aircraft to stay in sync with touchdown, keeping an eye on RPM to make sure we have enough power to keep flying, knowing the troops are just as anxious to get away from the huge target they are riding in and will jump out before touchdown, and mentally preparing for the painfully loud report from my M-60s in the back when my crew chief and door gunner went hot to shoot back or to lay down suppressive fire.

Whether we were inserting troops into an LZ or picking them up in a PZ, who had time to think about being scared?

But if the LZ was hot, and the lead said "Go around" so that we stayed in formation and set up for a second try at a hot LZ where the tracers were already flying, well, the pucker factor escalated and my butt-cheeks sometimes took a big bite out of the seat as I crouched behind my chicken plate and used one finger on the cyclic like Sam showed me.

Public domain photo, no rights claimed

I've heard a lot of guys talk about hearing rounds punch through the thin magnesium skin of our aircraft or whack into the engine compartment, but I don't think I ever heard it through all that noise. We just found the holes when we got back.

On a 173rd combat assault, in a flat area out toward the ocean on a night mission, we had enough light and a big LZ so we flew formation of about seven slicks and set up to all land at the same time. Lead slowed down too much, like he lost his visual reference, and as our formation came apart and the aircraft scattered we started taking fire out of a hooch. I saw in the twilight a Vietnamese woman exiting the hooch. Our door gunners opened fire and I could see tracers from the aircraft in front of me walking toward her. She was knocked down and did not move, almost certainly dead. I suspect she was VC, but that scene is forever frozen in my mind and even now I wish she had lived. When the shooting starts things happen too fast, with consequences that last forever.

EXPERIENCED PILOTS COULD VOLUNTEER for LRRP missions, and I flew LRRPs a number of times. These covert missions deep into unsecure areas were always teetering on the edge of high risk, for the six-man LRRP team on the ground and for helicopter crews inserting them and trying to get them out.
The PZs for getting LRRPs out were often very tight for even one ship if the team was outrunning the VC and had no time to make it to a pre-planned PZ. I would slowly descend straight down while the crew hung out on monkey straps calling the clearance for rotors and especially the fragile tail rotor. There were times we maneuvered back and forth on the way down to fit between layers of trees and, looking up through the roof bubble we could see nothing but tree limbs. Liftoff was straight up and might bring the loud ch-ch-ch-ch-whack-whack noise of the rotor chopping tree branches, made worse if the VC was hot on the LRRP team's tail and shooting at us while we nursed that hog up, foot by foot, into the air. Even when we were a sitting duck and they were shooting at us, I had to stay steady to apply a very gentle control touch to keep the rotor plane steady so it didn't lose lift to excessive movement.

Wayne King bargaining with Montagnard kids selling carved wood helicopter
Photo courtesy of Wayne King, all rights reserved

Back at the base, the 173rd LRRPs were part of my circle of trading partners. It helped that they loved helicopter pilots, and they had high-value trading items like LRRP rations, freeze-dried vacuum-sealed tasty meals, prized everywhere, even by the Air Force guys. They also had enemy weapons they brought back from their covert missions. One day when I visited them to pick up an AK-47, one of their crazier guys showed me the jar of ears he was collecting from enemy dead. I grabbed the AK-47 and got the hell out of there, with the creeps crawling up my spine.

When LRRPs called us for extraction, sometimes they were calm and quiet, whispering as they did for days in the field, and sometimes they were in a shitstorm, crying and screaming, popping claymore mines, throwing grenades and firing weapons, a lightly armed team of six scrambling to stay alive until we got there, and the look on their faces when they jumped in the aircraft told a story of terror, relief, fear and loss. One day as we took off one of the LRRPs reached from the back and gave me a big hug. Some of them weren't yet 20 years old.

BETWEEN FLYING AND BUILDING and re-allocating Uncle Sam's stuff, our work schedule was intense and we were frequently tired, able to doze off anywhere for 15 minutes of winks. I think I could have done it standing up.

I flew what was commonly called ash-and-trash missions, hauling people, mail and cargo. I had the pleasure of flying our doughnut dollies from base to base and the rare opportunity to lead a formation of our Huey's armed with smoke grenades attached to the skids to a Bob Hope show. A rare opportunity for an Artillery officer, I called some fire missions from the air, which gave me a distinct advantage on the adjustments. I flew sniffer missions with a machine in the back and detection devices trailed below the aircraft, filling up a map grid in parallel tracks, with the specialist operating the machine trying to detect enemy movement.

A little variety helped.

Our living conditions weren't the best, but were so much better than the grunts humping a rifle in the field, or the guys stuck in a remote firebase secured only by a few mines and rolls of concertina wire when the sun went down.

By contrast, we had the luxury of getting away from the grind once in a while because of our mobility. We could stop at a nearby base officer's club for a beer or buy goodies at the PX. We could stop in to visit the Navy because they always had great chow. We could even go into a little town for a hair cut or a steam bath.

So when I would approach the landing pad at an artillery firebase in the middle of nowhere, I usually felt guilty because inevitably I ruined someone's day. My rotor downwash would send someone's air mattress to a new home in the top of a tree, or scatter the papers of some poor guy trying to write a letter home, or blow away a tent that was the only home a few guys had. There is nobody to blame. Where I would have liked to make their day a little better, I inescapably made it worse.

THE WATER WAS LOUSY in Vietnam and we doctored it with coolaid, which got old fast. But even if the water had tasted good, we frequently wanted something stronger. When we drank beer or booze after a day of flying, it made the war recede into a forgotten corner of our minds for a little while, put on hold while we joked around. And so we drank a lot, especially if the following day was a day off. That was our relief valve, which was probably important since we were young and macho and loathe to admit or talk about the turmoil bubbling deep inside.

One night I drank and laughed and joked until late into the night because I was not scheduled to fly the next day. In the morning, Operations rousted me from bed and told me they were short a pilot and I had to fly the C&C ship. I told them no way, I was still drunk, no way could I fly, it wasn't safe, and what about the "bottle-to-throttle rule?" The Operations officer told me, "We don't want you to fly, we don't want you to touch the controls, just get in the seat." He promised to make sure the other pilot was experienced and I could sleep in the other seat but he had to have two pilots so the bird was legal to take off.

Aw, shit! I was still woozy and sleepy, my mouth tasted like I ate a rat and my stomach was churning like a washing machine agitating something nasty floating in Tabasco sauce. While doing a preflight inspection I stooped to check something and fell over. Shit! And flying Command & Control meant I had to be on good behavior for a self-important Colonel who would probably fly around for hours doing and contributing . . . nothing. But this Colonel turned out to be a good guy. He picked up on my condition right away and didn't say anything. After takeoff when we were bumping around in choppy air and he saw my gills turn just the right color of green, he handed me his steel pot so I could puke in it.

And I did. I still owe that guy.

I WAS NEXT FOR bereavement duty when one of our crew chiefs was killed. It was my job to collect his personal belongings, to cull out anything that would be embarrassing to his family, pack his belongings for shipment, and write a letter about his death to the family. That was a hard day.

About halfway through my tour I met my wife in Hawaii for R&R. We enjoyed catching up, and that's all I'll say about that. Saying goodbye to return to Vietnam was not easy.

I took off from The Crap Table on the morning of March 24, 1969, shortly after my return from R&R. I was part of a flight of slicks on our way to a combat assault, and heard Richard say on the radio he had a caution light; he was returning to land and check it out. We kept going, knowing the guns would catch up while we made slow turns in formation to give them time.

Richard called a few minutes later to say it looked OK and he was taking off from The Crap Table again. Shortly after takeoff his transmission froze in mid-air, he plowed in and died a violent death along with his co-pilot, crew chief and gunner.

I don't remember too much else from the rest of that day, I was flying and walking around feeling cut in half, as if my guts were hanging out. We lost guys now and then and the grunts we flew were sometimes wounded and killed, the normal ugly stuff that happens in a war, but Richard's death was like a stab right through my heart. Intellectually I always knew it could happen to any of us, but I guess emotionally I just wasn't prepared when it happened to my best friend. I suppose nobody is ever prepared.

I went through the motions of doing my job, and I tried to put it behind me, I really tried, but I couldn't. Food didn't taste good. Booze didn't help. I didn't care about the fancy hooch we had built; everything in it reminded me of Richard. I was pissed off at the utter stupidity of the war.

My world had changed, and I took it hard. I couldn't do this any more.

I told the CO I wanted to fly guns like Richard flew. I wanted to shoot back. I wanted to kill something. My CO told me the 238th Air Weapons Company at An Khe, just down the road, was new in country and needed experienced pilots. So I not only transferred to the 238th, I also was sent to the two-week Cobra transition school in Vung Tau, the beautiful coastal city near Saigon that was used for in-country R&R.

Anyone from a place like LZ English couldn't help but enjoy Vung Tau, with tall palm trees, cool ocean breezes, decent restaurants, relative safety, daily scenes like guys skiing near the beach behind a civilian or US Navy boat, sunbathers slathered in oil and dozing under sunglasses. Oh, and I did learn to fly the Cobra, but it would turn out our unit never got the Cobras while I was there and we continued to fly the underpowered B-model Huey gunships for the rest of my tour.

One thing I remember from Vung Tau is that guys who were stationed there seemed to complain a lot. Maybe people complain everywhere. I noticed it in Vung Tau because, compared to LZ English, these guys really had it good, living in hotels, a pretty town at their fingertips, etc. They complained that it was too hot outside, or their air conditioner wasn't working right, or the Army did this or that wrong, or the food in a restaurant was no good and so on. Weenies!

When I returned to the 238th I nearly killed myself and a newbie. I had over 500 hours in the cockpit so far and had just learned to fly Cobras. I was Mr. shit-hot pilot! When I took a newbie out to the free-fire range and made a rocket run on a target, I forgot how underpowered the B-models were compared to a Cobra and I waited too long to pull out of the shallow dive. The aircraft mushed in without sufficient lift to stop our descent, and I barely got it flying again after dragging the skids through the trees, straining the guts out of it in the bargain.

I not only gave the new pilot a very bad impression, I had to confess to the maintenance officer, a friend of mine, that I probably ruined his aircraft and that he should pull an inspection on it to be safe. It was down for a week.

FLYING GUNS WITH THE 238th was like flying in a different war. We didn't fly combat assaults covering slicks; we were hunters. Our AO was much the same as the area I already knew, but our primary mission was patrolling highway 19 from Qui Nhon through An Khe and 50 more miles to Pleiku through the Mang Yang Pass, aka "Ambush Alley." The Mang Yang was the site of many ambushes by the Viet Cong because the terrain made convoys sitting ducks with steep hills all around. Parts of the pass were dotted with graves of French soldiers from their battles in the 1950s, buried standing up and facing west, toward France.

Sometimes we wished the enemy would shoot at us so we could shoot back, and my wingman and I would cover each other while one of us flew low trying to draw fire. Since our clear mission was to kill them, the VC kept their heads down when we were in the area.

When we weren't patrolling, making ourselves visible and threatening to our hidden enemy, we were on standby, waiting to be called when a unit in the field needed close air support. Standby was relaxing until it got to be boring.

My new lifestyle was completely different than the 61st. At the 238th we didn't fly nearly as much, and we had our own aircraft and our own crew instead of the daily assignment rotations we had at the 61st. That change meant we got to know our crew chief and gunner very well, and it meant our crew chief could tune the aircraft to get the most out of it. We cooked and ate together, slept in the same Alert Tent when we were on stand-by, and we even got to know the families of our crew.

But the waiting was boring, and being itchy for action as young men are, we had to do something with our time and so we made bets, usually winning or losing a can or six-pack of beer.

We bet on whose crew chief could drop a smoke closer to a rock. We bet on who could put a rocket closer to a tree, and I became a pretty good shot with rockets. We bet on whether a chicken can really fly, and we had the wherewithal to prove the answer by taking the chicken up and giving it a fair toss. Could the chicken fly? Well, the chicken could certainly flap its wings frantically, but in a word – splat – no, the chicken couldn't fly. We bet on how a fly lands on a ceiling; does it loop or does it roll? I don't remember the answer, or how we proved it.

OUR B-MODEL HUEY gunships had seven-rocket pods and mounted miniguns on each side. With the weak engine and narrow rotors, everything we did was about weight. We were cautious about how much fuel to take on, and our loaded takeoffs were usually getting the bird light on the skids and sliding down the PSP until we hit translational lift.

When we engaged the enemy in a firefight, shooting back never satisfied the emptiness of losing Richard Benswick, but it did feel like I was doing something worthwhile that day.

The VC struck by ambushing a convoy on the highway one day when we were already airborne, patrolling in the area, and we caught them in the act, in the open, and hosed them pretty good. When we were low on fuel and ammo our other gun team relieved us, and while we were re-armed and re-fueled some guys handed us a tray of food for lunch and we wolfed it down as we sat in the cockpit with the rotors still turning, took off and returned to the battle to relieve the other gunship team. After that battle, our ground troops lined up about 35 torn up enemy bodies on a bridge so the locals could see what happened to VC. I didn't much want to look at the dead men, not even from the air. They all looked so young and I figured some of the locals were probably seeing their own family members lying there dead and could not react naturally for fear of what might happen to them. War sucks. Or maybe I'm not sufficiently macho.

MY TOUR ENDED WITH my body intact. The trip back to the US was not eventful, except for my barely contained excitement that I would soon be with my wife and my sons, Wayne Jr., who turned five while I was gone, and Scott, who was just two months old when I left. My family gave me a warm welcome at the airport in Wichita, Texas.

I would go on to an Army career of aviation, both rotary and fixed wing, which pleased my Dad, Buren, because he always planned for me to fly. Dad was a veteran of WWII, Korea and Vietnam. He was a P-51 crew chief during WWII, stationed in England. Dad painted my mother's name "Allie Mae" on the nose of his P-51. That was fitting because the P-51 was a remarkable fighter and my mother was a remarkable woman, president of the American Business Women's Association in Burk Burnett, Texas, had her own frame shop, was a 50 year member of Eastern Star and Cub Scout mother when I was that age. She sent me to dance lessons, piano lessons, taught me how to eat with the right fork at the dinner table, and when I was thinking of West Point or aviation she told me not to waste my time but to set my sights on President of the United States. She thought I could do it, too.

Wayne King's Dad, Buren King
Photo courtesy of Wayne King, all rights reserved

My love and admiration of my father led me at one point to search for his P-51, trying to locate the remains if it was still intact. It was a labor of love that led me to meet some great people from Dad's generation. I never found his P-51, but one of those people I met custom made a high-quality P-51 model for him with the markings on his bird as shown in the photos he brought home from his war. I was so pleased to get this done for him before he passed away.

I disappointed both my wife and Mom, not with my career, but with my foul mouth when I came home from Vietnam. Obscenities flowed naturally from me whenever I spoke a word, partly the habit I picked up in Vietnam, and partly because I was not the same person when I came home. But I worked on it and re-learned how to speak in a civilized manner in polite company and around my own kids.

Wayne King's Mother, Allie Mae King, with Wayne and brother Bob
Photo courtesy of Wayne King, all rights reserved

Sometimes we don't realize what we have until we take the time to look back and reflect. So, as I tell this story about my Vietnam experience, I have to say I drew my strength from my wife, Cookie. When I committed to the challenges of US Army Rotary Wing flight school as a young man, she committed, too. With four year old Wayne Jr. to take care of, pregnant with Scott and sick in the mornings, she still took to heart the flight school instructions and got up early to make me breakfast before a long day of training. That duty and morning sickness often required her dash to the toilet and to this day she has no interest in breakfast.

Cookie quietly endured the crazy schedule and constant moving of flight school, and when I got orders for Vietnam she had the worry of my being hurt or killed in a war we wished would go away. Like all the other guys in Vietnam I had my macho warrior exterior, and also like them I quietly yearned to be home with my family. A bad day in a war zone can make the strongest man heartsick, but what sustained me was that, however busy she was with the kids, Cookie wrote to me every day. Some guys received a letter from home just a few times in their year in Vietnam. Cookie's daily letters were not only my connection to the family I loved, they were my lifeline, my refreshing reminder that there was a peaceful world somewhere, with the soft breath of a baby and the trying antics of a toddler instead of shooting and killing and dying. She sustained me.

So many Vietnam vets I know went through a tough adjustment period and a high number of them lost their wife to divorce. I guess I had my own adjustment, too, when I came home and Cookie had to teach me to speak in a civilized manner again. I'm so glad I didn't lose her; my wife, my sons and their families are the center of my life, my purpose for living.

I TAUGHT STUDENTS HOW to fly TH-55 helicopters in Ft. Wolters for a couple of years. Some days with new students were as dangerous as a combat assault under fire. I was helping keep the supply pipeline for the war full, each man with their own story to tell if they were lucky enough to come home. With a yearning to fly fixed wing again, I transferred from artillery to military intelligence and trained at Ft. Stewart, Georgia, in T-41s, then at Ft. Rucker for instrument and twin engine training in T-42s. I had a variety of assignments and flew helicopters and airplanes all over Europe in a plush job I managed to hang onto for five years.

Later I commanded a task force on a number of hush-hush missions with special ops guys in Honduras, including Ollie North, and I got used to wearing civilian clothes with pockets full of cash for emergencies and bribes and no ID of any kind. My job was to command and deploy a hand-picked group of military intelligence specialists highly trained in the operation and maintenance of airborne electronic warfare equipment. The aircraft I flew were unique, equipped with some of the most advanced electronic equipment at that time including inertial navigation devices developed for space flight. I loved the part where they told me if I had engine trouble halfway between the US and Honduras, to protect the secrecy of the equipment I was to ditch in the Caribbean and not land in any other country.

It was a good career, doing what I loved, flying.

My part in the Vietnam War was my aviation baptism of fire, an intense learning experience in all the things a pilot has to do to stay alive, even with bullets flying. Aviation is all about safety, and that introduction served me well.

In my memories of Vietnam I often think of Richard Benswick. Though we would never say such a thing as young men in a war, losing him left a hole in my heart that never really healed. Years later I visited the Vietnam Memorial wall in Washington, DC. Like for innumerable other veterans who lost a close friend, seeing Richard's name on that wall brought back the pain and tears as if it were yesterday.

You never forget.

Wayne King and wife Cookie
Photo courtesy of Wayne King, all rights reserved

Wayne King retired after 20 years of military service. Upon retirement he flew for Eastern Airlines and then with his two sons Wayne and Scott he built a business in manufacturing and supplying compost and specialty soils to commercial and industrial customers. Wayne lives in Peachtree City, Georgia with his wife, Cookie.

Postcard to America

I FLEW 187 F-100 MISSIONS in Vietnam between June '67 and '68 out of Bien Hoa Air Force Base. We saw the battles of the Tet Offensive from our cockpit, we were part of those battles, and we watched US media reports about it on Armed Services TV. We wondered what war they were depicting because TV news reports did not resemble our experience.

After I returned from Vietnam I tried to watch all three evening TV news broadcasts, ABC, CBS and NBC. One evening the subject was an interview with the mother of a US soldier who had just been released by the Viet Cong in a rare prisoner swap. Apparently all three networks interviewed her simultaneously, but only ABC broadcast the final exchange, when the reporter asked the black mother how her son felt about his Vietnam service, clearly fishing for negative comments about the war. She said, "He was proud to serve his country."

The other two networks cut that part out of the interview. I suppose those remarks didn't advance their agenda.

The TV news reports I saw accusing US troops of widespread atrocities did not reflect what I saw. As just one example, I rode in a convoy with the Top Tigers, an Army helicopter company volunteering their down time on a Sunday to deliver muscle, tools and lumber to a Vietnamese village that had held off a VC attack the night before. We went to help them rebuild, and I distinctly remember the Top Tigers CO denying my request to ride in the lead jeep because, he said, "The lead jeep finds the mines!"

I later flew that same CO in the back seat of an F-100F on a routine fighter mission. About half way through, as we pulled up from a bomb run he started moaning on the intercom. I asked him what was wrong? He said, "I knew I should have gotten into the Air Force!" I'm glad he had fun.

Our news media twisted the truth out of all recognition and the American people ended up believing myths and half-truths about the Vietnam War.

William E. Haynes, Rancho Palos Verdes, California
Lt Col USAF (Ret), Commander, 90th Tactical Fighter Squadron
Bien Hoa, Vietnam 1967-68

To Return With Honor
James Warner
Rohrersville, MD

MY TOUR IN VIETNAM was longer than most. I was involuntarily extended when I was shot down on a mission over North Vietnam, captured, then held for five years and five months as a prisoner of war.

In February of 1964, after four and one half years of college, I had made little progress toward a degree. Instead, I had spent my time reading on subjects that caught my interest and ignoring those which did not.

My reading passion changed my life in a way. Like most college students of the time my thinking was on the liberal side, and then I began to read different books focusing on communism. Arthur Koestler's *Darkness at Noon* captivated me and I read others like *Up From Liberalism* by William F. Buckley, Jr. and I began to think. As I read about the Roman republic I wondered about the apparent lack of patriotism in the American left as compared to Roman patriotism, which seemed to be the strength of the Roman state. With all these readings I drifted to the right. When President Kennedy was assassinated in 1963, notably by a communist shooter, the entire nation was jolted to the right. I came to believe that international communism was the greatest evil ever to threaten humankind. As I learned to think for myself, I became a hawk and left what I considered to be the foolishness of liberalism behind.

I had a wonderful education, but no degree and no serious life prospects. The time had arrived for structure and discipline. I joined the Navy. Before I shipped out, friends in college who were veterans, some still finishing degrees on the GI Bill who had fought in Korea, took me aside and warned me not to volunteer for anything. But I didn't listen closely enough.

The Navy sent me to a one-year electronics school at Great Lakes Naval Training Center in Illinois. When I checked in at the Training Command at Great Lakes, there were no slots for us for a few weeks. As a result, rather than leave young enlisted men at loose ends, we were put to work doing things that required muscle instead of brains. On the second day, they asked if anyone knew how to type. I figured that the Navy already had the rating of yeoman, who was an enlisted man trained, among other things, to type. I figured that a temporary office job meant fetching coffee for officers and answering phones. I was right. While my pals sweated at hard labor in the yard, I drank coffee, read books and answered phones. I had learned the wrong lesson about volunteering.

While in electronics school I applied for flight training. I was accepted into the program when I finished the electronics school and reported to Pensacola Naval Air Station for training as a Naval Flight Officer. As I was enrolling in flight training, the US participation in the Vietnam War began to escalate. Shortly thereafter, the Marine Corps offered to guarantee training in the F-4B Phantom fighter/bomber for Navy trainees who would cross over to the Marine Corps. I jumped at the chance. This would assure me of a tour in Vietnam and a chance to do my part fighting the communists, unless the war ended before I could get there.

I received my wings in April of 1966. In March of 1967 I arrived at the Marine Corps Air Station at Chu Lai, on the coast of South Vietnam about 55 miles south of Danang, where I was assigned to Marine Fighter Attack Squadron 323. I was required to sit through a day of lectures and to sign a statement that I had read and understood the rules of engagement and acknowledging that I would be responsible if I failed to abide by them. I would learn those rules were a major part of keeping us from winning the war.

Training in the F-4B Phantom at MCAS Beaufort, in South Carolina
Photo courtesy of James Warner, all rights reserved

Because of my electronic training in the Navy, I was assigned collateral duty as squadron avionics officer. With responsibility for more than one hundred enlisted men in the electrical, radar and communications divisions, I had the largest section in the squadron. As the junior officer in the squadron it was inevitable that I would also get an additional collateral duty as coffee mess officer.

As the back-seat Navigator or RIO (Radar Intercept Officer), my job was to get us where we were going, and then to share duties in the attack with the Pilot, who had his hands full. We both kept track of our location and the target, and the bombing run setup. I closely watched the dive angle and airspeed, calling them out and making corrections as required. The Pilot did most of the flying.

We flew bombing and ground support missions in a wide area, but our authority to bomb targets was strictly controlled. Hot pursuit of enemy forces across the borders of Laos or North Vietnam was not allowed, so our enemy knew where to seek sanctuary. The lunacy of this policy is that we were assigned bombing missions in Laos and in North Vietnam, but we could only cross the border if specifically assigned to do so.

Surface to Air Missile (SAM) sites, the radar-controlled anti-aircraft enemy missile batteries that shot down our jets, could not be hit while under construction for fear of killing Russian advisors. Once they were operational and firing missiles at us, our Wild Weasels were equipped to detect their location and fire radar homing missiles back but our F-4Bs didn't have the equipment to find them and didn't have anything to fire at them.

Our bombing target selection sometimes seemed like nonsense. For example, for a period of time we were assigned to fly the A Shau Valley, an enemy stronghold, and bomb any trucks we could find. If we didn't find any trucks we were to drop our 750-pound bombs on either ford of the Rao Lao river running through the valley. These heavy bombs buried deep in the river mud and blew big craters when their delayed fuse went off, but every day when we returned we could see the truck tracks just going a little further around the craters to cross the river. I proposed bombing the main road in the valley at its entry point through a pass because, if we hit it with 750s, the enemy would need weeks to reopen the road. But I was told to keep on looking for trucks and bombing the river fords; we never saw trucks because they must have heard us coming and darted under cover. The pass, where I thought we could do real damage to the enemy, was strictly off limits.

I've come to accept such stupidity in war; it seems inevitable if you read enough military history. Maybe there are too many variables in war to expect common sense to prevail. Consider the recent example in Iraq, the mistreatment of prisoners at the Abu Graib prison. In the first place, the prisoners were stupid to follow orders to remove their clothes. The stupidity in our chain of command in not specifying expected prisoner treatment was compounded by a failure to ensure they were treated properly. Finally, the stupidity of the news media in equating these sophomoric hazing games with real torture boggles the mind. I always expect to find stupidity in war.

SOMETIMES WE HAD A success. One of those was a close air-support mission for an ARVN platoon that was pinned down by a company of NVA regulars near the DMZ. Some of the NVA had gone up into a church tower, shooting at the ARVNs and at my jet with heavy fire. We could not return fire because it was a church. We dropped Snake Eye bombs on other enemy positions at extremely low levels. They had fins that would spring out and spin like helicopter blades retarding the fall of the bomb. It was important at low level that these bombs not have their fins banded else they would drop quickly, detonate right underneath the aircraft and blow us out of the sky. We dropped them at 400 feet, 400 knots and at a 10 degree dive angle. They were very accurate, which is vital in close air support. We dropped one bomb at a time, taking great care to not hit our friendlies on the ground, 12 passes under heavy fire. We prevailed, and our allied platoon survived.

F-4B Phantom, radar-guided bombing run over North Vietnam
Photo courtesy of James Warner, all rights reserved

AS WITH OTHERS WHO flew in the F-4B, I loved the airplane. However, as with all love affairs, the object of our affection was not perfect. The first imperfection of the F-4B we had noticed in training. When seeking to join up with another F-4B, we found that the airplanes were easy to spot because the J-79 engines, the most powerful engine in any fighter plane in the world, did not fully burn its fuel when not in afterburner. This resulted in twin trails of faint black smoke behind an F-4B. When looking for another F-4B, this made it much easier to find him in the open sky. Only after going into combat did we realize that these trails of faint black smoke also made it easy for an enemy gunner to track an F-4B.

The second design defect in the F-4 was brought to my attention in my role as avionics officer. In early August of 1967, the sergeant in charge of our avionics section came to me saying that he had bad news. He told me a wire bundle had been cut in a plane which had taken a hit. When I assured him that this should not be a problem to fix he took me out to the airplane to see for myself. Unlike wiring bundles I was familiar with, the wires were neither color-coded nor numbered. We had several men working around the clock for several days to trace down each wire. The airplane was basically a sound design, but the engineers who designed the F-4B had not thought of everything.

Finally, on October 10, 1967, I learned of another and more serious design defect in the F-4B. The squadron was briefed on the flight control hydraulic systems, designated as PC1 (primary flight control hydraulic system) and PC2 (secondary flight control hydraulic system). The F-4B was advertised as having the capability to fly at Mach 2.3, that is, 2.3 times the speed of sound. At that speed, flight control surfaces would require more power than a human arm can generate to move them. Thus, there is no mechanical link-up. Instead, there was a hydraulic system to operate the flight controls. Since the F-4B, as indicated by the letter "F," was a fighter plane, there is the chance that it would be in combat where people would shoot at it. Therefore, the contract called for a redundant flight control hydraulic system. What the contract did not call for was for the back-up system to be situated separately from the primary system, and it was not. The back-up system was placed side by side with the main hydraulic lines. Furthermore, I have since been told, they both used the same hydraulic fluid reservoir. I cannot confirm this last bit of information, but it may help explain subsequent events.

We were told to be alert for a PC1 warning light when pulling off target below 3,000 feet since small arms fire can reach that altitude and still be effective against an aircraft. The proper response to a warning light under those conditions was to immediately hit the "Jettison All" button. This button would release everything external on the airplane, including armaments, auxiliary fuel tanks, and the racks on which such things were hung, leaving the aircraft aerodynamically clean. That would reduce the hydraulic power required to control the airplane and give us a little more time to determine if PC2 was still functioning. If not, then the controls would have completely failed requiring a bailout.

THERE WAS ONE PILOT in our squadron whose judgment was suspect. I will call him "Dilbert," a name used in the Navy for one who is slow-witted. When Dilbert had been a maintenance officer he would go out to the engine test stand without hearing protection. Since the mechanics were running engines at full power, this helped to destroy his hearing. Since he frequently made poor choices while flying, his hearing deficit made his backseaters uncomfortable since we couldn't tell him what to do with any confidence that he would hear us.

One of my roommates had collateral duty as an assistant operations officer. It was his job to write the squadron schedule. On the night before my last flight, he explained that Dilbert's regular backseater had taken leave and that he was having trouble finding backseaters to schedule to fly with him. He asked if he could put me on the schedule with this pilot. Without a thought to the advice I had received about volunteering, I said, "Hell, Dick, you schedule me with him. I'll fly with anybody." That would be a fateful decision.

The next morning my roommate woke me up to tell me that I was scheduled to fly with Dilbert that day, that it was Friday the 13th of October, so I would probably get shot down that day. He asked me to promise that if I were shot down I would say hello to his old friend John Frederick when I got to Hanoi.

I took his comment about Friday the thirteenth as a joke. However, when I stepped outside of my hooch to get on my bicycle to go to breakfast, my bike was missing. Well, maybe that would be my bad luck today, I thought to myself. I walked to breakfast.

Every day, when we briefed for our first flight, we would brief on a selected emergency procedure, chosen by the operations officer. Today it was the hydraulic system problem he had lectured to us about on Tuesday. Special emphasis was placed on the correct response to a PC1 emergency light appearing when pulling off target. That response was to immediately push the "Jettison All" button, making the aircraft aerodynamically clean, then, if in a turning climb – the preferred way to come off target, remain in the turn until the nose came to 050°, then level off and use power to control altitude. To the extent possible, do not touch the control stick again - since any control movement would drain some of the remaining hydraulic fluid - remain at that heading until approximately 10 -15 miles off the coast, then eject.

This emergency procedure was simple, but in a fighter plane, in combat, you cannot repeat emergency procedures too often. I was certain that by this time, everyone had the drill down pat. I was wrong. As we walked out to the plane, Dilbert turned to me and asked me to take out a map and show him again where the area was to jettison everything if we had the emergency of the day. Friday the thirteenth was appearing to be more of a threat than I had imagined. Still, I was sure that nothing could happen.

Our target was a tracked vehicle just north of the DMZ. On our bombing run we bottomed out at 1300 feet and Dilbert announced a PC1 light as we pulled off target. I was glad we had spent so much time on this emergency. I replied, "Roger, hit Jettison All now!"

My heart fell at his response. "Negative, I still have six bombs and I'm going to make another pass." By God's grace, as soon as he got the nose down the hydraulic system went dry and he ejected. If we had stayed in the bomb run and went dry as we tried to pull out, we would not have made it.

Like his failure to understand the emergency procedure, and like his refusal to punch the button when I told him, his bailout was a failure. It was his duty as Pilot to reduce airspeed, direct the Navigator to punch out first, then to follow. He didn't even say goodbye when he left.

With a wind of a little more than 600 knots in my face, I followed him out. To my surprise, when I hit the outside air, I began to spin, and remained spinning until my parachute opened. Dilbert had left the airplane at full throttle, and may have put it in afterburner since I recall seeing flames behind the plane as it plummeted into the ground. By the time I got out the plane was going so fast that I tore two panels out of my parachute when it opened. I was lucky. At 600 knots, aviation safety experts thought it was probable that I would sustain severe injuries. However, the only injury I received was in my neck, and that was from the position in which I held my head while I ejected, not the airspeed.

My chute was opened and I was stabilized, I looked past my feet several thousand feet to the ground. My first thought was that people pay money to do this! My second thought, which still sticks with me, is that I had all of the coffee mess money in my flight suit.

As I descended, people on the ground were shooting at me. You know, it makes no sense under the circumstances, because I had been dropping bombs on them, but the fact they were trying to shoot me really pissed me off. I decided to return the favor. I had stopped carrying the standard issue .38 caliber revolver in favor of a .45 caliber auto pistol. I put a round in the chamber and began searching for shooters. I shot at one, but doubt that I could connect at that range.

I hit the ground and rolled as we had been taught. As I got up on my hands and knees to stand up, something hit me in the back of my helmet and drove my face into the dirt. I assumed that this was debris from the ejection seat but have been assured by representatives of the manufacturer that this was not possible. When I finally took my helmet off to inspect it there was a groove in the back. Apparently, a round grazed me and the flight helmet saved me.

I then tried to disentangle myself from the parachute shroud lines. We were issued a pocketknife similar to a Swiss Army knife except the drop point blade is spring loaded, and one of the blades is a razor sharp hook which was designed to cut shroud lines. It was carried, with the shroud line cutter blade open, in a pocket in the center of the chest in the parachute harness. To show what stress and adrenaline can do to one's performance, I couldn't find the shroud line cutter. I had to pull out my K-Bar, the large fighting knife that all Marines carry in combat, to cut myself free.

The airborne controller was overhead. I spoke with him on my emergency radio, but it was not clear what he could do. I was captured about ten minutes later. Only after my capture did helicopters show up. They were too late, but I remain forever grateful to their crews for risking themselves to try to save me.

WHEN MY GEAR HAD been stripped off, my arms were tied and I was pushed into a trench. There a gunner, with a Degtyarev Pekhotny (DP-28) Soviet made machine gun, would fire a burst at the aircraft attempting to rescue me, then come over to me, put the muzzle against my temple, jabber at me in Vietnamese, then I would hear what sounded like a hammer falling on an empty chamber followed by the gunner guffawing with laughter. I never saw him clear the chamber before putting the gun against my temple, nor charge it after what sounded like a hammer on an empty chamber. I am familiar with this model firearm and have fired it. I have never been able to reproduce the sound of a hammer on an empty chamber except by letting the hammer fall on an empty chamber. I have, therefore, no idea how he was conducting the mock execution. He did this five times and each time would laugh like hell, then go back to shooting at the aircraft trying to rescue us. My problem with the joke lay in the fact that this weapon is noted for heat buildup which could allow a round to cook off without being ignited by the hammer.

Dilbert broke his ankle when he landed. When the communists brought us together I had to carry him. We spent most of the first night trudging around the area just North of the DMZ. This area looked like the Western Front during World War I. There was no vegetation, the soil had been churned up by constant shelling from the Marine bases South of the DMZ, and it was criss-crossed with trenches. As we walked, we quickly learned how to distinguish incoming which seemed headed for us. When we heard this I would scramble over to a trench and put Dilbert in it, then, before I could get in, the round would hit. Many were very close, and one round knocked me down. However, not even a sliver of shrapnel hit me and I did not suffer from the effects of concussion so I cannot say how close the rounds were when they hit. I can only say that they seemed very close.

I was put in a bunker, with my feet in stocks, my wrists tied in front of me and my elbows tied behind me. At night, in order to keep me from attempting an escape, I was put on my stomach with my feet in stocks, tied up, with a wire from the logs of the bunker around my neck. This way I had to keep my back arched to keep from choking myself.

We stayed in the area near the DMZ for five days. Four days after we were captured, on October 17, as my guards were letting me out of the bunker to relieve myself, we were hit by an *Arc Light* mission of four B-52's, each dropping its load of bombs in line with the plane in front of it. There is nothing in my experience that remotely compares with a B-52 strike. Awe-inspiring does not come close and I have no other words to use to describe it. During Operation Desert Storm, I told my colleagues in our office that the Iraqi soldiers would be thoroughly demoralized after being bombed by B-52 strikes for 42 days. I said that I knew this because I was only hit once and I am still demoralized.

We moved at night once we started North. Each night we rode for two or three hours in the back of a truck. We were tied as I have described, wrists in front, elbows behind, and ankles hobbled. Without the use of our hands, we could not steady ourselves as the truck pitched from side to side traveling on the bombed-over roads. After our truck ride, we spent four to five hours each night walking. Since Dilbert had a broken ankle, I had to carry him. We traveled this way for four weeks.

About 75 miles south of Hanoi, the soldiers turned us over to an officer from the political department of the Army of North Vietnam, and we soon learned to wish for the soldiers again. We were kept in bamboo cages, constantly harassed and threatened with torture. We arrived in Hanoi two weeks later, on about November 21, 1967. I was black and blue from head to toe.

IN THE INFAMOUS "HANOI Hilton," the name American prisoners had given to Hoa Lo prison, I was put in a cell with Air Force Captain Ken Fisher, Air Force First Lieutenant Lee Ellis, and Dilbert. Dilbert had decided that the communists were right and we were wrong. This made the time we spent together very unpleasant since the other two and I did not agree with him. I expressed my utter contempt for communism and this led to constant squabbles in our cell.

The interrogations centered mostly on Dilbert because they really liked him. Apparently he was giving them lots of information and they were trying to corroborate through the rest of us. More important, he agreed with them and they had him make tapes to be used to influence others. John McCain told me that while he was in the hospital they played, and replayed, a tape which Dilbert had made.

Finally, after eight months, Ellis, Fisher, and I were transferred to the prison at Son Tay. We arrived at Son Tay on July 18, 1968.

OUR FOOD WAS CONSISTENTLY bad and meager. We were ever so slowly starving, susceptible to poor health conditions that accompany malnutrition. We received a bowl of soup twice a day, foul-smelling liquid turned black by what we called *sewer greens*, an unrecognized plant that apparently was aquatic in nature because it was hollow, with a smell as if grown in sewage. It was terribly bitter.

Sometimes we got a small bowl of rice with sand and tiny bits of rock mixed in, making it hard to eat. Sometimes there was bread, old and moldy bread with rat droppings in it. We often found bugs of various types in our food, and we ate them, too, because we were hungry and we knew we needed the protein.

However terrible the food was, there was never enough, especially in the winter when our bodies were trying to get warm.

DESPITE THE MISTREATMENT, WE regularly resisted the efforts of our captors to obtain propaganda from us. If they told us to do something, we might refuse, or we might even laugh at them, which made them furious. With all our might, we tried to deny them everything they were trying to force out of us.

On May 5, 1969, I was taken out to interrogation. The interrogator accused of me of planning an escape and demanded details from me. I was actually not planning an escape, but I would have if a plausible idea had come to me. However, the interrogation lasted 10 days until 15 May, 1969 – these dates are important. On the 15th, after ten days of interrogation, I was not happy with what was happening to me. I decided to get even with the interrogator, our political officer, whom we had named "Louie the Rat," because he looked, and smelled, like a rat. Rat was the dumbest officer I have ever known in anybody's army. I would not trust him to clean latrines in a place where there was no running water. Because he was stupid, I knew that I could make up an escape story, no matter how preposterous, and he would believe it. He would tell his superiors, whom I presumed to be smart enough to see that the story was so improbable that they would do to him what he was doing to me. Things do not always turn out as we plan them.

I told Rat that I would confess to escaping, but only to the "true means of escape." I told him that no one could escape by the means which he accused me of planning to use, which, in fact, was correct. Remember, I said he was stupid. My plan was to make up the most implausible means of escape and confess to that. Accordingly, I told Rat that the true means of escape consisted of stealing an entrenching tool – a small military shovel – and use it to chop a hole in the ceiling. With a hole in the ceiling, we would move aside the heavy, terra cotta roofing tiles, at night, and go out through the roof. You must understand that in a country too poor for much machinery, there was little background noise and it was dead quiet at night. Moving those heavy tiles, I thought, would make such a racket that it would alert the night guard to investigate.

Rat was pleased. As I expected, he believed this unlikely story. I was put back in my cell and was there for three days. On the morning of May 19, at six in the morning, I heard the voices of many people, as well as the rattle of keys, outside the door of our cell. Air Force General Charles Boyd, at the time a Captain, once told me that he had believed that the emotion he felt when people shot at him was fear, but he now knew that it was not fear. Fear is the emotion you feel when you hear keys outside the door and it is not time to eat.

The door opened and Rat stood there, surrounded by a posse of several guards. He pointed at me and they charged into the room, grabbed me, and dragged me over to the interrogation room.

For the first twenty four hours Rat did not bother to ask me questions. He came into the room on the morning of the twentieth. He shook his finger at me and told me that I must tell him, "fully and accurately" how I communicated "with the criminals in another camp." They always insisted that we were criminals, not prisoners of war, and were not covered by the Geneva Convention.

I asked Rat what he meant. He laughed and said, "Don't be surprised." Now the protocol on torture is that the torturer has to indicate what he will believe, then the victim has to take torture long enough to make it appear that when he repeats what the torturer wants to hear, it will appear to have been beaten out of him and, therefore, will seem credible. The trouble here was that there was no lie that I could make up which would seem credible. Further, even if I were creative enough to devise such a lie, they would then go to another camp and ask those men how they communicated with Warner. There was no chance we would ever come up with the same lie, so it did not appear to me that they would ever stop the interrogation. Since there was no way that I could answer his questions, I just said "The Hell with it."

I must interject here that I did give it a try. I considered telling him that cockroaches have been around humans since the birth of the human race.

Since they were so used to us, I wanted to tell him, and we had so many cockroaches, that we would put a cockroach under our arm for a while, to get him used to the smell of Americans, then write messages on his wings and throw him out the window. I realized that even Rat was not stupid enough to believe that; there was no lie I could make up to make him stop the interrogation. At the time, I was kneeling on the concrete, holding a stool over my head. When I realized that there was nothing I could do to make them stop, I put the stool down, stood up, and sat on the stool. Rat screamed at me to kneel. I told him that I would not torture myself. If he wanted to torture me, he would have to use his imagination and do it himself. He screamed, even angrier, that he would punish me severely if I used the word torture. I told him he was already doing that and that, since I could not answer the question, and he would not explain it, I would not talk to him any further. For the rest of the summer of 1969, Rat continued his interrogation. The only thing I would say to him was that I did not know what he meant by the question about communicating with Americans in other camps. He insisted that I answer the question, as well as other questions which were equally inexplicable and unanswerable.

On June 1 he put me in a small concrete box. The box stood in the sun, although there was a tree beside it which provided some shade. There was a steel door on the north side of the box. Since Vietnam is in the tropics, during the summer the sun is north of Son Tay. Therefore, the door was in the sun and it would get so hot that I could not touch it without burning my fingers.

In July, after two months of interrogation during which I had not given him what he was after, he put me in leg irons and wired the leg irons to a small stool. Leg irons consisted of two horse shoe shaped devices which fitted around each ankle. Each device has two loops in the back through which an iron bar can be passed. The iron bar has a small loop at the end that passes through the horse shoes, and which allows a padlock to be attached to assure that the leg irons remain fastened.

I could not sit squarely on the stool, since this would require bending my legs so much that the leg irons would dig into my ankles. On the other hand, I could not stand fully erect, for the same reason, and could not otherwise sit on the stool because I would hang over the back so far that I would fall over backwards. All I could do was crouch and balance on the balls of my feet. I was not allowed to sleep. After one day my feet began to swell, since the circulation was restricted. Then I got dysentery.

Every morning Rat would come to see if I was ready to tell him how I communicated with Americans in other camps. Since I had not washed in two months, and had dysentery, and was confined in a small, hot, space, I found a way to make his morning visits unpleasant for him. About three in the morning, every night, I would take the lid off of the honey bucket. When he came to see me, usually at seven in the morning, he would stand as far away as possible and ask me if I was ready to answer his questions. I would smile and say that I could not hear him.

From time to time I would provoke him, on the theory that he was already torturing me and there was nothing more that he could do. On one such occasion Rat got angry and gave me an uppercut to the chin. Given the weight distribution with the leg irons and stool, his blow lifted me from the floor and flipped me over, so I came down on the top of my head. This broke some back teeth.

Finally, with my feet swollen now to the size of footballs, Rat released me from the irons. They had to pry the leg irons out of my swollen flesh to get them off.

I knew that walking would help to reduce the edema in my feet, so when he closed the door I tried to take a step – in the box I had room to take two steps forward and two steps back. I fell on my face. The weight of the iron bar on my Achilles tendon had stretched it until I was hamstrung. It took three days before I could stand or walk.

Coercion almost always was accompanied by sleep deprivation. I had observed others deprived of sleep and had seen that for most men, after four days without sleep their thinking became disordered. After five days, most had hallucinations. By seven days, they began to suffer from paranoia in addition to hallucinations, and they suffered terribly. We feared sleep deprivation, not because of the discomfort, but because we feared what we might do when our thinking had become disordered and we were gripped by a psychotic paranoia.

By God's grace, I learned that when actually deprived of sleep there were some of us who did not experience the psychotic disorder of thought. I have no explanation, nor do I know of any research on sleep deprivation which might explain why some of us did not suffer severe hallucinations and paranoia. I can only say that, in my case, it was not heroism on my part. There is just something in my physiology that makes sleep deprivation less effective on me than on others. Nor can I say whether this immunity was a blessing or a curse. After all, if I had broken down they may have quit there. As it was, during the summer of 1969 I went as long as twenty one days without sleep. When I say that we did not suffer as others suffered, it does not mean that it was pleasant to be deprived of sleep. On the contrary, after ten days it was extremely unpleasant, and after twenty days it seemed unendurable. I say that it *seemed* unendurable because, as I had learned being interrogated for information which did not exist, there is no bodily discomfort which is unendurable, merely that which one would prefer not to endure.

The interrogation continued until September 3, 1969. On that day the communists announced the death of Ho Chi Minh. As we were to discover, there was no systematic torture after his death. A few men were brutally beaten, but there was no torture as it had been used before that time. Over the summer, as a result of the interrogations, I had lost about 30 pounds, and a few teeth.

Two years later I learned what prompted Rat's interrogation. It seems that on May 10, in the middle of my interrogation, two men escaped from another camp. They escaped exactly as I had told in the story that I thought was not credible and that was why Rat tried, all summer, to find out how we communicated.

I also learned how close of a call I had. The two men were recaptured and in the ensuing interrogation, one of them was killed.

In July of 1970 we were moved from Son Tay to another camp a few miles away. On November 21 of that year, commandos under the leadership of Army Col. Arthur D. "Bull" Simons, raided the camp at Son Tay attempting to rescue us. From our new camp, we could hear the small arms fire. We knew what was happening because we could hear helicopters and knew that the few helicopters in the North Vietnamese Air Force did not fly at night.

Shortly after that raid, the communists closed all outlying camps and moved us back to the Hanoi Hilton, this time to large cells. The first cell I was in held 57 men. We were so crowded that we had to sleep on our sides because there was not enough room for every man to sleep on his back. In January I was removed from that cell and put into another.

At this time there were a series of confrontations with the communists which we have since called the "Church Riots." In early January of 1971, the communists forbade all religious observances in our cells. With each step taken to stop religious observances, we increased our resistance. Finally, on March 19, St. Joseph's day, the communists took 36 of us and put us in two refrigerator trucks, 18 men per truck, and drove us to another camp.

THE REFRIGERATION WAS NOT working, but the trucks were airtight. As a result, when we reached our destination, after about a half hour in the back of the truck, the air was going bad and the temperature was rapidly rising. By the time they opened the doors, we were considerably more docile and easy to handle than we had been when we were put into the trucks.

We were told, repeatedly, that this camp was for men with "bad attitudes" and was intended for punishment. This caused us some concern because the term "punishment" had always been a code word for torture. However, except for the cramped quarters – the cells measured about six and a half feet by three and a half, and many of these cells had two men per cell for the first six weeks that we were there – while we were in this camp no one was hurt by our captors.

Near the end of May, on a Saturday, I was called out to an interrogation. This was unusual for two reasons. First, they had never interrogated us on weekends. Second, up until this time, only the senior American had been interrogated in this camp. I was only a First Lieutenant. I wondered what they wanted with me.

The interrogation was the longest I ever had without being tortured. I sat and listened as the interrogator droned on for more than an hour. Then I was given a clipping from a magazine or newspaper which had been pasted onto cardboard. I read that my mother had given testimony at something called "the Winter Soldier Hearings." As I recall, she said that she hoped that I was being treated well and that the war would end soon. I read on. There was testimony from people claiming to be Vietnam veterans asserting that it was official US policy to commit war crimes.

I was then given another story board. On this one, I recall there were some typewritten pages with the commentary of a Navy officer named John Kerry. I cannot recall if it said that this was his testimony in the Senate Hearing, but what he said was astonishing. He claimed that he witnessed a long string of war crimes, that he himself had committed some war crimes, and that it was US policy to encourage these crimes. I recall that he wanted unilateral and unconditional withdrawal of US forces from Southeast Asia. Unilateral withdrawal got my attention.

Even through the distorted viewpoint of communist propaganda, we had concluded that the 1968 Tet Offensive had been a disastrous defeat for the communists. That meant the only thing they had to bargain with was us, the prisoners. If the US were to pull out unconditionally, what would happen to us? If the US were gone, we would have no value. They always told us that we would be put on trial and executed. I had no doubt what would have happened to us if the US were to follow Kerry's advice.

The interrogator pointed to the story board. He kept repeating that this American officer admits that I deserved punishment for my *crimes*. As he spoke his voice became more agitated and his face got red. These were usually signs that something bad was about to happen. In 1969 I had been tortured for four months, and spent five months in a small box out in the sun. We thought that most of the torture ended when Ho Chi Minh died, but were not certain. I was devastated by these words, spoken by an American officer, which looked as though they were to be used to justify whatever was going to happen to me.

Nothing happened. I was returned to my cell, but those words haunted me. It was known in the US, thanks to Jeremiah Denton, that we had been tortured, and no one could have known that the communists had, for the most part, stopped it. Kerry must have known that his words would be used against us, and he had every reason to believe that we might be tortured because of his words. Further, he should have known that such words would offer hope of winning, politically, the battle the communists could not win on the battlefield, a fact which the communists openly admitted. Finally, his words were patently false. Knowing this, why did he say them?

I can't answer that. I can only say that he was either indifferent to the consequences of his words, or that he used poor judgment. Out of charity, I choose to believe the latter.

Years later I heard that in 1971 John Kerry surreptitiously met with the communist negotiators at the Paris Peace Talks, incredibly, while he was still a US Naval Reserve officer. If he did meet with the enemy, perhaps they told him they had stopped using torture. If so, he should have told our government. If not, he should not have said what he said because his words put us at risk.

IN AUGUST OF 1971 we were taken back to the Hilton and put in the area we had named "Heartbreak Hotel" because this was where airmen were kept during their initial interrogation. Late in September we were put back in the cells from which we had been taken in March.

One of the three men in my cell this time was Mike Christian. He had scrounged a bamboo needle, and colored thread from some packages we had been allowed to receive, and he sewed into the inside of his shirt an American flag. We started every day facing Mike's flag and reciting the Pledge of Allegiance.

In my misspent college years, among other jobs I had worked as an orderly in a medical ward of a mental hospital. There was a medical library in the hospital and I would read about every medical condition we treated. I became fascinated and began reading medical texts generally. By the time I went into the service, I had sufficient knowledge to provide elementary medical care if called upon. In prison I was called on to do just that.

Of course, there was little that I could do with what little we had. However, there was one man with us who was gifted with his hands. In the spring of 1972, one of my cellmates, John "Spike" Nasmyth, developed impacted wisdom teeth. I decided that it was necessary to intervene, somehow, because he would not survive otherwise.

The man gifted with his hands stole a razor blade and an old toothbrush from the communists and made a scalpel for me by embedding the razor blade in the handle of the toothbrush. He then made a brace to hold Spike's mouth open. Finally, when time for the surgery, he made a fire to sterilize the scalpel.

In the meantime, I had arranged for several men to complain that they had "ringworm" which prompted what passed for a medic to give them iodine soaked cotton balls. I used these to sterilize Spike's gums. Then I made vertical slits on his gums to free the teeth and expressed as much pus from each side as I could. Spike had a sore mouth for several days, but the problem did not recur.

IN MAY OF 1972, 208 of us were taken North to a camp on the North slope of a karst formation, a steep volcanic mesa, near the Chinese border. We speculated that the purpose was to hide us from reconnaissance aircraft. The US had established a thirty mile buffer zone to keep American airmen from straying into Chinese airspace. Although conditions were primitive, we thought this was the best camp we had been in. There was no electricity so there were no loudspeakers, the single most ubiquitous item in totalitarian countries. Further, without a naked light bulb glaring all night, we slept much better. Finally, we were locked in blockhouses, in individual cells. The guards did not unlock the buildings to come in and inspect individual cells, so we were much less subject to the scrutiny of our captors.

On the day I was shot down, when my roommate woke me up joking that it was Friday the 13th so I would probably be shot down, he told me that I should say hello to his old friend, John Frederick, when I got to Hanoi. I met CW4 John Frederick in this mountain camp. Gunner Frederick joined the Marine Corps in April of 1942 and was one of the finest men I have ever known. In the summer of 1972 men in his building reported that he had a high fever. Some time later, the guards took Gunner Frederick out of the building. We never saw him again. Vietnam was his fourth armed conflict, and he had been held as a prisoner of war for seven years when he died.

Our life in the mountain camp remained good. In the Autumn, there was a long Indian Summer during which the days were sunny, the skies were clear, and the cool, dry nights made sleeping wonderful. However, the idyllic times were not to last.

Near the end of October, the weather turned cold and overcast. The humidity was near 100% and the temperature hovered around 40° F. We were always cold and we began to suffer from an ailment in our feet. At first it seemed like athletes foot, but it got continually worse. Eventually, our feet were discolored, and intensely painful in a manner which cannot be described. After we returned to the States, we learned that we had beri-beri, a nutritional deficiency which destroys nerve tissue and can be fatal if not treated. In fact, I have severe neuropathy in both legs, resulting in a condition known as Charcot's joint, which destroyed my right hip joint. Many of the others continue to suffer from residuals of beri-beri.

During this cold period our captors seemed to anticipate our release, and they tried to fatten us up so we wouldn't show too much evidence of poor treatment. They gave us all the rice we could eat, and I remember being so hungry while my stomach was painfully full of rice. It seems rice just doesn't have enough nutrition to satisfy the body's needs, and certainly not enough for the body to keep warm.

While we were in this cold place, Mike Christian was in a large cell with a number of other men in Hanoi. One of those men was John McCain, who tells the story of the guards discovering the flag sewn into Mike's shirt, and how they beat him for a couple of hours and then dumped him back in the cell. Mike's cellmates cleaned him up, and while every movement must have brought him great pain, Mike wasted no time in digging out his bamboo needle and colored thread from a hiding place and began immediately to sew another American flag, not for himself, but because he knew how much it meant to his cellmates to face that flag and recite the Pledge of Allegiance every day.

In January, we were taken back to Hanoi, to the camp known as the Plantation. Here, after a few days, we learned of the signing of the Paris Peace Agreement. The communists essentially left us alone after that. Six weeks after the Agreement was signed, we were given fresh clothes and shoes and taken to Gia Lam airport, North of Hanoi, and turned over to American forces for repatriation. At long last, we were going home.

SOME NAÏVE TALKING HEADS on TV believe John McCain speaks for all Vietnam veterans, I suppose because he a well known Senator. Others have said, particularly during McCain's 2000 presidential bid, that he conducted himself poorly as a POW, that he accepted special favors.

Neither of these things are true.

Vietnam veterans, even former POWs, are not monolithic. We each think for ourselves and our opinions, viewpoints and politics vary.

John and I were cellmates for a while, and we shared five years as POWs and came in occasional contact. Furthermore, there was no place to hide in our small community and any breach of our code of honor was known to all. John McCain conducted himself honorably. For all of us, the communists offered freedom and our return home if we would only cooperate with them by *apologizing for our crimes*. Our senior officer and therefore our CO, US Navy Cmdr. Stockdale, made it clear to all that we would meet our Code of Conduct obligation to return with honor, and that meant withholding any cooperation of any type from the enemy, to attempt whenever possible to escape, to hold out as long as possible during torture sessions and to accept no special favors. Nearly 600 of us did just that, including John McCain, who received repeated pressure from our captors to accept special treatment because they knew his father was a very important Admiral and they sought to use that for propaganda purposes. John refused the enemy's pressure to accept special treatment and he suffered with the rest of us.

I teased John in 2000 about one politically-motivated rumor that while a POW he was allowed to have Vietnamese women in his cell. I asked him, since I was in his cell with him for a time, why the hell he didn't share his women with me! But of course it wasn't true.

Some of our POWs did accept special treatment. Dilbert was the worst I know of, and one friend reports that Dilbert was his turnkey, the trusted one who locked up fellow prisoners. Dilbert apparently tried to get early release. A comrade who was in the same camp with him reported to me that Dilbert was told that he deserved early release but that his opinions in support of the communist cause were so extreme that people would think he had been brainwashed, so in order to help the cause, he would have to remain in prison.

It should be said, however, that of the 12 who took early release one deserves an honorable exception. Doug Hegdahl, a US Navy Seaman, was captured when he washed ashore after falling overboard from the cruiser USS Canberra (CA-70). Because he was not an officer, and he pretended to be not very smart, our captors treated him as unimportant and let him roam the camp. Doug helped us communicate and he memorized 356 POW names using a cadence, a rhythm and his own organization for his memory. Doug was ordered to take an early release by Cmdr. Stockdale. Doug asked in the strongest terms to be allowed to stay because he wanted to do things right, as most of us were struggling to do, but his orders remained and he completed his mission in 1969 by returning home and reporting to the US military, and to our families, who was known to be still alive in the POW camps.

Cmdr. Stockdale became an Admiral upon our return, and he led an effort to prosecute the worst collaborators and turncoats, Dilbert and one Naval officer. Those of us who were witnesses gave testimony, but there was a precedent: President Eisenhower decided about collaborators during the Korean War that punishment cannot be more severe than the most lenient branch of service decided, to prevent disparity. Ultimately President Nixon decided these two officers would not be prosecuted, that like the rest of us they would receive the promotions we all received in absentia while we were

POWs, and they would be allowed to retire and collect their pensions. The base commander in Dilbert's area, however, limited his benefits in this way: Dilbert would be able to come on the base to visit the JAG (legal) officer and the Chaplain, but that was all. He could not take advantage of the PX, the Commissary, sick bay or any other facility or service normally available to retirees.

James Warner, March 1973, at Clark Air Base in the Philippines, as released POWs boarded a C-141 medevac airplane to return to the US Photo courtesy of James Warner, all rights reserved

All 11 who accepted special treatment by the enemy and came home early, ahead of the sick and wounded, are not now, nor will they ever be welcome when we ex-POWs gather to enjoy one another's company and remember our service together under difficult circumstances.

I HAVE BEEN ASKED, on many occasions and in many ways how I feel, knowing that it was all in vain and that my years in a communist prison were wasted. The answer is that it was not wasted.

Perhaps Vietnam can best be understood by comparing it to a conflict which occurred more than two thousand years ago. The Persian Wars lasted from 490 BC to 479 BC during which time the Persian Empire repeatedly invaded mainland Greece. The Persians were ultimately defeated.

The key to the ultimate victory of the Greeks was the naval engagement in the Bay of Eleusis, off the island of Salamis. This victory, in turn, was made possible by the delay of the Persian armies at the pass known as the Gates of Fire, or *Thermopylae*. Here a small band of Greeks, led by Spartan King Leonidas, held the pass for several days which impeded the progress of the Persian armies. Finally a local resident named Ephialtes showed the Persians a way over the mountains to get behind the Greeks. Forthwith, the blocking force at Thermopylae was killed, to the last man, and the Persians advanced. However, the delay forced by Leonidas and his men permitted time for Athens to be evacuated, saving the population, and for the Greek fleet to be assembled at Salamis, where the Persians were defeated.

The defeat of the Persians was important because the conquest of Greece would have cut off the legacy which the ancient Greeks left to western civilization. That is to say, our civilization was preserved, ultimately, in a small battle which our forces lost, for it was the delaying action at Thermopylae which made possible the salvation of Greece. Of all the lessons which can be learned from military history, this is one of the most important. As former Prime Minister of Great Britain, Winston Spencer Churchill said: "Success is not final; failure is not fatal, it is the courage to continue that counts."

We lost the battle for Vietnam, just as the West lost the battle at Thermopylae. And just as Thermopylae made possible the ultimate defeat of Persia, the Vietnam War made possible the ultimate demise of a world wide communist threat.

The communist movement exhausted itself in Vietnam. It was exhausted materially, intellectually and spiritually. While the American economy could easily handle the load of the war, the communist countries could not. Socialism cannot produce the way capitalism can produce. They were materially exhausted.

Sophisticated arguments about socialism which had been subscribed to by intellectuals in the West during the twentieth century showed themselves increasingly barren. By the end of the Vietnam conflict, although vast numbers of young people in America had been seduced, the basis for socialism was clearly and decisively rebutted.

Finally, they were exhausted spiritually. After a twenty year harangue in which they appealed to noble sentiments concerning the *liberation* of third world countries, they capped off their victory with the Cambodian holocaust, brutal reeducation camps and the long spectacle of the escape of masses of wretched boat people from Vietnam.

American resistance to the conquest of Vietnam by the communists was essential to the prevention of the spread of communism to Malaysia, Thailand, Singapore, Indonesia and the Philippine Islands. The conflict also gained time for the world to strengthen itself. After the end of the conflict, a handful of countries fell into the communist orbit, but these were not prizes and their commitment was fleeting and transitory. The exhausted hulk of communism awaited the arrival of President Ronald Reagan, British Prime Minister Margaret Thatcher and Pope John Paul the Great to finish it off.

Knowing all this, I say that my time was not wasted. It was an experience like no other. I repeat the words which Admiral Jeremiah Denton, the first POW off of the first plane out of Hanoi spoke on February 12, 1973: "We are honored to have had the opportunity to serve our country in difficult circumstances."

Photo courtesy of James Warner, all rights reserved

James Warner is a retired corporate attorney living in Western Maryland. He served in the Reagan White House, and authored notable articles including one of stiff opposition to the campaign finance legislation sponsored by his friend and former cellmate, Senator John McCain.

Postcard to America

HERE IS ONE REASON I have no forgiveness in my heart for those who did not support us while we were in Vietnam and now want to reconcile.

I was a fire team leader in E Co, 2nd. Battalion, 9th Marine Regiment, 3rd Marine Division in Vietnam. Around Christmas 1966 I was walking by the company office and a Marine said he had some mail for my squad, so I delivered the letters and a few packages to our hooch.

Sorting through the mail, I found a package addressed "TO ANY MARINE." Since our new guys wouldn't get any mail for some time, I handed the package to one of the new guys and went on sorting the mail. I turned around to ask what was in the package and saw him sitting on his rack looking very sad. I asked what was wrong and he showed me a pack of dog food with a note that said "EAT WELL TONIGHT, ANIMAL."

This young Marine was away from home probably for the first time in his life. That package hurt more than the wounds I received in September 1966, and I will never forget it.

To the people who regret the insults they hurled at our own troops in their anti-war zeal, I would say this: Your injury and guilt is self-inflicted and you will not receive from me the redemption you seek. Live with it, learn from it, and I hope you see today the shame of our modern day protestors encouraging the enemy trying to kill American troops in the Middle East. If you truly want to make amends, tell those people the damage they are doing to America's young men and women now in harm's way.

Bob McFadden, Salem, New Hampshire
E Co, 2nd. Battalion, 9th Marine Regiment, 3rd Marine Division, Vietnam, 1966-67

A Story Never Told
John O'Neill
Houston, Texas

MY CLOSEST FRIEND was Elmo Zumwalt III. He was involved in one of the most heroic Naval actions of the Vietnam War but there is not now, nor will there ever be, an official record of that action. This is the story.

Elmo III was the son of Admiral Elmo Zumwalt, Jr., the sort of naval hero who comes along every half-century or so and becomes the stuff of Naval legend. As a young naval officer during WWII in the Philippines, Elmo Jr. refused to abandon a burning destroyer and led a party which saved it. Near the end of the war, he took a captured Tug through minefields, with Japanese pilots on the bow guiding him into Shanghai, China to receive the surrender of millions of Japanese troops in North China before advancing communist Chinese armies could receive their surrender and their arms.

While in Shanghai, he met and fell in love with a beautiful White Russian refugee named Mouza,, who didn't speak English. He married her and their love endured until their deaths nearly 60 years later. He would later become the crusading Chief of Naval Operations (CNO, 1970-74), a proponent of equal opportunity in the Navy and the then–revolutionary use of what later became ship-borne cruise missiles.

During my tour in 1969, Admiral Zumwalt was the Commander of the Naval Forces in Vietnam, legendary for riding brown water Navy boats into danger.

Since their introduction to the Vietnam theatre in 1966, Swift Boat duty had been relatively safe as they patrolled the offshore waters of the South China Sea, intercepting and inspecting small craft for contraband weapons. Then in 1968 Admiral Zumwalt ordered Swift Boats into the rivers and canals of South Vietnam on a mission to take the war to the enemy hiding in the difficult terrain of the IV Corps delta region, to search them out and destroy them. Swift Boat duty suddenly became very dangerous with the enemy able to fire from cover at us like sitting ducks and our ability to maneuver in small rivers and canals severely limited.

OUR SIX MAN SWIFT Boats were 50 feet long, aluminum and without armor but we did have astonishing firepower; an automatic grenade launcher up front, twin .50 caliber machine guns on top, and a .50 caliber with 81mm mortar on the stern, usually loaded with anti-personnel canister-type *beehive* rounds.

Admiral Zumwalt's son, Elmo III, was a young North Carolina graduate who entered the Navy not to make it a career but because he believed with a war in progress he had an obligation to fight for his country. Chance brought Elmo III and me together as commanders of Swift Boats at Ha Tien, on the border of Vietnam and Cambodia at the junction of the Gulf of Thailand and a long canal marking the boundary.

The duty was gruesome. We often survived on ancient C-Rations, some from WWII. Some of us contracted terrible diseases, of which malaria was the most benign, and we were often shot at and mortared by an unseen enemy in circumstances where we could not effectively respond.

Elmo III had an extraordinary crew. They were loyal, proficient and brave, as if they saw in their commander a leader worthy of his post. I can see in my mind's eye today Elmo III, Narimore and all the others as young men of 18 to 23.

We knew large numbers of enemy NVA regulars were coming through our area on their way to their base camps in the Uminh and Nam Can forests. We seemed powerless to stop the infiltration from Cambodia, just serving as occasional targets.

Each night we would go to pre-assigned ambush locations along the banks of the canal, seeing little or nothing in a hot humid hell of mosquitoes. These missions were tense because if and when we found the enemy, our exchange of fire could be very close and intense.

Elmo III became convinced that the NVA had spies advising them exactly where we were going each night, allowing them to avoid us and infiltrate into South Vietnam. He talked about going somewhere different – not planned – to a location the NVA would not know because none of us would know about it in advance. It was a very dangerous plan because of numerous American bombing raids, the possibility of collision with other friendly forces, and the absence of reinforcement or rescue support by American forces who would not know his position. So that is exactly what Elmo III did.

He took his crew to a promising ambush location miles from any other Swift Boat and several miles from the position assigned to him. The force of serendipity, which has been with the Zumwalts from that burning deck in the WWII Philippines to Elmo III's daughter as an embedded reporter in 2003 Iraq, ran strong that day. Elmo III's small boat and six man crew collided with a powerful heavily-armed force of at least 40 NVA regulars crossing the canal. In the dark canal, they fought a fierce battle in which the Swift Boat crew was grossly outnumbered and unable to maneuver, finally ending with shooting from the decks at point blank range.

No one will ever know how many NVA were killed that day because the enemy was adept at quickly removing their weapons and wounded and dead from the battlefield, but we counted at least 25 rifles and many bodies the morning after the firefight. With classic understatement Elmo III said it was quite a fight. He then filed a combat action report making clear he had not gone to the assigned position and describing accurately what had occurred.

HAVING HIS OWN SON directly defy his orders created an obvious quandary for Admiral Zumwalt, despite the fact his son turned out to be right. Because of the conflict of interest in any disciplinary action, he assigned evaluation of the matter to his Chief of Staff, instructing him to ignore the father-son relationship. The Chief of Staff returned an interesting decision. He concluded that Elmo III had knowingly defied orders in a war zone and to some degree endangered himself, his crew, and his boat. Normally, this would result in a letter of reprimand or even a Courts Martial. On the other hand, the courageous and brilliant action of Elmo III and his crew was one of the notable naval exploits of the Vietnam War. Under normal circumstance Elmo III would be awarded a Navy Cross or a Congressional Medal of Honor and his crew similarly decorated.

The Chief of Staff recommended that neither occur. Instead he recommended that the Navy withhold both the Courts Martial and the medals, discard the combat action report and wholly ignore the incident, treating it as if it had never occurred at all. That is what Admiral Zumwalt did. There was no medal, there was no Courts Martial. There was no laudatory message nor even a mention of Zumwalt III's victory on the Cambodian Border.

Instead, Admiral Zumwalt came to see his son. After making clear his love and pride, as well as his relief that he was unharmed, he told Elmo III:

> "Elmo, every good Naval officer must stand up once or twice in their career and say 'No, that is wrong and, even if it costs me my career, I will not do it.' The reason you will never be a naval officer is you do it three times a week."

The guns of 1969 are long silent now. The young sailors of Ha Tien are becoming old men. Elmo III, who saved my life and others more than once in faraway now unknown places like the Dam Doi and the Rach Nam, is long gone, dying at a tragically young age in 1988, possibly from dioxin-induced lymphoma caused long ago. Admiral Zumwalt, terribly saddened by his son's death, passed into legend in 2000. In 2003, I saw Maya Zumwalt on TV reports as an embedded reporter with tank units in the Iraq invasion and was stunned to learn it was Elmo III's daughter, an adult I still somehow thought of as a child.

The Navy did acknowledge Elmo III's great fight of 1969 quietly. If you visit the U.S. Navy Memorial in Washington D.C. you will see there a magnificent bronze relief memorializing the accomplishments of the brown water Navy in Vietnam. The PCF-35 shown is Elmo III's boat, but in keeping with that long ago Navy directive there is no mention at all of the 1969 fight. I think that is exactly what both Elmos, united now in death with each other and many shipmates, would have wanted. It seems they each believed that deeds themselves and the self-respect they engender speak far louder than medals, words, or empty praise.

LIKE ELMO III, I accepted I would never be a great Naval officer, for the same reasons his father pointed out to Elmo in 1969, and I long ago turned to other things. Now in the autumn of my life, my mind often turns back to those early days and friends of my youth, and another important lesson I learned from those times. I met many very young men in Vietnam I consider to be heroes, real heroes, not cheap imitations like the vain self-promoters who shape a career around their own glorious and exaggerated exploits. The heroes I know are modest men, ordinary men who did extraordinary things in their youth trying desperately to keep one another alive, men who would tell you the real heroes never came back at all but gave their lives for others.

The heroes I met in that war have families and friends that even today hardly know what they did, likely unaware of the heroic feats they once performed. Like my best friend Elmo III's great 1969 battle, that part of their life is a story never told.

Bronze relief entitled "Inland Engagements – U.S. Navy River Operations - Vietnam" by sculptor Serena Goldstein Litofsky, on display at the U.S. Navy Memorial in Washington DC. © All Rights Reserved. Used with the permission of the United States Navy Memorial Foundation. www.navymemorial.org

Photo courtesy of John O'Neill, all rights reserved

John O'Neill is a litigation attorney in Houston, Texas. He became nationally well-known in 2004 as spokesman for the Swift Boat Veterans for Truth in active opposition to John Kerry in the presidential campaign. Earlier that year O'Neill donated one of his kidneys to his wife, Anne, who bore her illness with courage and grace and died in early 2006.

Postcard to America

I WAS STATIONED ABOARD the USS Saratoga in the South China Sea off Vietnam during 1972. I returned to the US before the ship started back due to transfer orders. I flew from the ship to DaNang, then to Clark AFB in the Phillipines to await my flight to the US. That flight made stops in four places, making it a very looooong flight. When we arrived at Travis AFB in California, most of us had been on that plane for over 18 hours.

The bus took us to the airport in San Francisco, where I got a haircut, my dress uniform cleaned, a much needed shower, and a place to stretch out for a few minutes. After that long ride on the airplane, I was not ready to be in a crowded seat again, so I purchased a First Class ticket to my destination, Boise, Idaho where my family was waiting.

When I was seated in my first class seat, I was interrogated by a little old lady, I swear she was Granny from the old Sylvester and Tweety cartoon, hair in a bun, umbrella and all. We were the only people in first class, and she asked a million questions on that short flight. Where was I coming from? Why was I going to Boise? Did I have any children? My youngest son had been born while I was gone, and he was only 6 weeks old when I got back.

When we landed in Boise and everyone was getting ready to exit the plane, she stood up and shook her umbrella at the crowd, telling all of them that "This nice young man just got home from Vietnam, and nobody is getting off this airplane until he gets off and sees his family!" With that said, she blocked the isle with her umbrella and didn't let anyone off until I was all the way into the terminal.

I must have spent two weeks in Idaho, before flying on my way to my next duty station, and I couldn't spend money for food or drinks the whole time I was there. I had a hero's welcome, just because I wore my uniform, not because of anything I had done.

Many other vets say they had an unpleasant homecoming, but I certainly did not.

Mike Minor, Jacksonville, Florida
USS Saratoga, South China Sea off Vietnam 1972
AQ-2 (E-5) member of the weapons team

Sideways
Steve Crimm
Roswell, Georgia

LONG BEFORE I DISCOVERED the virtues of flying sideways in a Loach, I learned the comfort of close gunship support in a slick, even though at first it could rattle our cage. The first few times rockets scream past our open window and detonate with a powerful WHAM! in the tree-line near where we would touch down, our butt cheeks just might take a bite out of the seat. But slick pilots, flying in formation and carrying loads of troops for insertion to a hot LZ, learned to love great gun pilots.

One morning just after daylight we were inserting 30 infantry troops. The LZ was out of artillery range from the nearest firebase so there was no artillery prepping the LZ. We had three pair of Cobra gunships from our sister company, D Co 158th Avn Bn, the Redskins. One pair had been prepping the LZ and immediate area with rockets and miniguns just before our ETA.

I was *peter pilot* in Chalk One of a six ship trail formation, one behind the other. The LZ was a grassy clearing on a low ridge that was only large enough for one ship at a time. We were able to fly straight in but had to depart to the left front to avoid many tall trees. The other five aircraft in the flight were maintaining about a 20-second separation to give each other time to offload in the LZ. When we were about 15 seconds out of the LZ the first set of Redskins stopped prepping the LZ. At the same time another pair of guns were flying formation with our slicks, maintaining a position a little to the left and right side of our flight and a little below us. The third set of guns was providing overhead cover for the troops being inserted into the LZ in case it became hot or the other aircraft started taking fire.

As I was concentrating on our approach to the LZ I realized that the set of guns in formation with our aircraft was firing rockets on each side of us, even some rockets below our aircraft into the LZ. From my vantage point it looked as if the rockets inbound to the LZ had just missed us. They continued maintaining their position and firing rockets into the LZ until we were on a very short final and any further rockets would have caused shrapnel damage to our aircraft.

For Chalk One that day, the first aircraft into an unknown LZ, the alarmingly-close rockets were a great comfort, and they did the job of keeping the enemy's head down instead of aiming and firing at me.

When I arrived in Vietnam in January 1971 I was still 19 years old. I soon turned 20 as I learned the tricks of staying alive flying with the Lancers, B Co. 158th Avn. Bn. which was part of the 101st Airborne Division located at Camp Evans in the northern part of I Corps. We were a slick company supporting US and South Vietnamese ground forces. We flew combat assaults and resupply missions and were heavily involved in Lam Son 719 supporting the South Vietnamese Army's excursion into Laos.

TOWARD THE END OF May of 1971 I was re-assigned to Headquarters Company, 2nd Brigade of the 101st Airborne, call sign Brandy 43, located in Phu Bai. For the remainder of my tour most of my cockpit time was as a scout, a Loach pilot. Loaches were called White, Cobras were called Red, so a Cobra and a Loach was a Pink Team, while two Cobras with a Loach was a heavy Pink Team, very effective in rooting out the enemy. Our normal mission from 2nd Brigade Intelligence for our heavy pink team was an assigned map grid for us to recon based on reports of enemy activity.

We usually flew Loaches low, sometimes slow, following trails, looking for bunkers or weapons or food stashes, snooping and darting here and there, blowing tall grass apart, sometimes flying under very tall trees, looking for sign of enemy activity, trolling for their fire to make them reveal their position.

When we found them we marked the target with colored smoke to identify the target location for the Cobra to attack. The Loach was a tough little helicopter, able to absorb extensive small-arms fire and still bring the crew home safely which is a good thing. Our scouts flew with a door gunner and an M-60 machine gun hanging from a bungee cord, sitting behind the pilot. Sometimes we used an eight-round-pump 40MM grenade launcher, like a pump shotgun. Before we discovered that weapon we used frag grenades in a glass jar with the pin pulled, so the jar held the trigger in place. When the glass jar hit the ground it broke and the grenade detonated after its built-in delay. It was a little risky but let us get the grenade farther down than a normal fuse allowed. Instead of detonating in the tall trees it went off on the ground where it did a lot more good.

We flew from the west side of QL1 in Phu Bai called *Heaven Pad*. Our AO was south from Phu Bai to just north of DaNang and west to the Laotian border, a very large area covered in thick jungle, hiding lots of the enemy.

One day I was teamed with two Cobra gunships from C Battery, 4th/77 ARA, call sign Griffin. We had just finished our recon for the morning and were returning to refuel when we received a call from the Brigade TOC that a LRRP Team was in enemy contact and needed gun cover and extraction ASAP. That meant they were in trouble, and LRRP teams were just six men, lightly armed. After hot refueling, we took off south to Fire Support Base Rifle, 10 nautical miles south of Phu Bai, an eight-minute flight at 125 knots. While en route to FSB Rifle I was in contact with the brigade TOC on a secure radio frequency to get the latest reported map coordinates, radio frequency and sit-rep for the LRRP team. The team was about six miles further south of FSB Rifle in the Roung Roung Valley. Uh-oh. The Roung Roung Valley was just east of the Ho Chi Minh trail and a frequent spot of concentrated enemy forces, a very bad place.

After crossing over FSB Rifle both Cobras and I switched to the FM frequency monitored by the LRRP team. On my initial radio call to the team, the RTO returned my call in an out-of-breath soft whispering voice, and I knew we had a team in big trouble, whispering because the enemy was so close, out of breath because they were running, their position compromised and evading a superior force trying to kill them. We would later learn the LRRP team had been discovered by an estimated NVA platoon in reinforced bunkers.

Steve Crimm with eight-round pump action 40MM grenade launcher
Photo courtesy of Steve Crimm, all rights reserved

Because the team was in close contact with the NVA they were not able to pop smoke so that I could identify their exact location. Instead I had to have them vector me in by radio and describe the terrain nearby. When I finally got their location identified, they were hiding behind a large log. Visibility into the foliage in the floor of the valley was very good at their location, vegetation only 10 to 15 feet tall and not very thick compared to the 100-foot thick triple-canopy jungle we normally encountered.

Now that I had the team positively located and they told me the direction they heard the NVA coming from, I flew in that direction. When I was about 75 meters away from the team I had my door gunner throw smoke to mark the NVA's location, then hauled ass because as soon as the Cobras saw the smoke they would be rolling in hot. The Griffin Cobras carried 2.75 inch rockets with 17 lb high explosive warheads, 40 mm grenade launcher and a 7.62 minigun. Anyone who has ever seen a Cobra work on a target knows the death and destruction they can deliver. It is impressive to watch a pair of Cobras flying their race track pattern with one of them constantly able to lay down their destructive power.

After the first rockets landed, LRRP Team Gavin gave directions to the cobras, adjusting fire based on what they were seeing and hearing on the ground. The team wanted those rockets to come as close as possible to them in an effort to kill or drive the NVA away from their position so they could be extracted. They, too, knew the comfort of air support so close it scared the shit out of you.

After the Cobras made several runs on the suspected enemy location they held fire while I flew over the target area. There are times when doing things the way we are trained is not enough, you have to adapt, improvise and overcome. And so, after passing over the team heading in the direction where the last rockets landed, I flew sideways so that the door gunner and I both had the best view through open doors toward the enemy. As we flew sideways the door gunner fired a constant volley on the enemy with his M-60, moving it back and forth from the left to the right in front of us trying to hit the enemy, make them keep their head down or to drive them away from the team.

Of course we didn't want to be hit, either, and I stayed just above the trees by a couple of feet moving at 30 to 40 knots so the enemy didn't have an easy shot at us. Since I was moving sideways that oddity might have screwed up their aim, because they missed. The LRRP team radioed to let us know when we were receiving fire because we couldn't hear ground fire over the roar of the M-60. The only time we could really see incoming fire was muzzle flashes if the enemy weapon barrel was pointed directly at us or if we saw tracer rounds.

Each time the LRRP team told us we were taking fire, we immediately broke back toward the team's location to clear the area because the Cobras were listening, too, and rolling in hot to cover us, ready to lay down several pair of 2.75-inch rockets as soon as we were clear. After the cobras had saturated the area, we turned and started back, once again flying sideways over the team, laying down M-60 fire toward the enemy to keep them away from the LRRP team until the Ready Reaction Force (RRF) arrived.

After the initial call for help went out from Team Gavin, our heavy Pink Team wasn't the only resource put into motion, we just happened to be the closest. The brigade's command and control aircraft, a Huey, was redirected to the team's location and the RRF was mobilized, consisting of several Hueys, additional cobras and an additional Loach, all at Camp Eagle, about 25 miles north of Team Gavin.

My door gunner and I were very lucky, we didn't take a single hit in the aircraft that day. When the RRF arrived to pick up the LRRP team, we cleared the area so they could do their job without worrying about mid-air collisions with us.

Because there wasn't an LZ near the team they were extracted by a ladder, a 25-foot-long, four-foot-wide, roll-up aluminum ladder attached to the Huey's cargo floor. The LRRPs climbed up onto the ladder, attached themselves to the ladder with carabiners and were flown to a suitable LZ where the Huey could safety land and the team could then climb into the aircraft for the continued flight to FSB Rifle.

The Griffin Cobras and I followed the RRF team as they left the area. I was requested to return to FSB Rifle for a debriefing with Team Gavin and the Cobras were released to return to Phu Bai.

When I stepped out of the Loach after shutting down at FSB Rifle, I walked over to the LRRP team. I was shaking hands and introducing myself to the team when I realized that I knew one of these guys! As I was shaking his hand with a big smile on my face, I said:

"If I had known it was you, I would have left your sorry ass out there in the bush!"

That, of course, is a guy's version of "Good to see you!" in a war zone. It had been just 19 months though it seemed like 100 years ago, since Sgt. Wyatt had been my Drill Instructor at Ft. Polk, LA when I was a raw recruit. He had taught me a few things, and I had just had the rare opportunity to return the favor.

The LRRP Team was inserted in the same AO again the next day, an operation that led to further NVA contact.

Just seven weeks earlier on August 1, 1971, my hooch mate WO-1 Bill Bradford, pilot, along with SP-5 Andy Bordes, door gunner, were killed when their Loach was shot down by hostile fire just west of FSB Rifle while working as part of a heavy Pink Team. We suspect they were killed by part of the same NVA battalion. Flying, and sneaking around in the bush as LRRPs do, in the FSB Rifle AO called for extreme caution.

Nearly six weeks later at dinner in the brigade commander's mess, I was ordered by the Colonel to come forward. The mess was called to attention and the Colonel's aide read the orders awarding this 20 year old Warrant Officer from Tennessee *The Distinguished Flying Cross* for my part in helping protect and extract Team Gavin. For that I was honored and awestruck. I thought what must have gone through the minds of countless helicopter pilots from that war when a medal was pinned on their chest . . . I was just doing my job, just another day as a helicopter pilot in Vietnam . . . even though I was doing it that time . . . sideways.

Steve Crimm with wife, Bobbi
Photo courtesy of Steve Crimm, all rights reserved

Steve Crimm was born in Savannah, GA, grew up in Tullahoma, TN. He started flying lessons in high school before training as a US Army helicopter pilot. After five years in the Army Steve flew helicopters in Canada as a "bush pilot," set up a TV program as a pilot-reporter in Salt Lake City, sold helicopter services in the petroleum industry, established a helicopter services business in St. Thomas and developed leading edge IT skills which he now uses in a non-flying job. Steve lives with his wife, Bobbi, in an airstrip community in South Carolina where he can take off in his fixed wing airplane after morning coffee.

Postcard to America

DEPARTING VIETNAM WAS BITTERSWEET, *good to leave and hard to leave my friends still there. The last leg of my flight home, after a layover in Denver, took me to Wichita, KS. As I made my way out of the airplane the attendant was bidding everyone good-bye. My uniform, dark tan and bleached hair were a giveaway where I had come from. She asked, "Are you home now?" I replied, "Yes." And I was. Finally.*

My family and friends didn't ask questions or talk about Vietnam, as if doing me a favor by ignoring an unpleasant topic. While I watched a TV news report on Vietnam a family member asked, "Why are you watching that? Now that you're home I don't care what happens in Vietnam."

I ran into a high school friend in a local beer joint, and in our conversation he asked, "Remember Bill? He went to Vietnam and came home a while ago. He is all screwed up. All the guys going to Vietnam come home screwed up." I didn't reply. He then said, "Where have you been? I haven't seen you in quite a while." I looked at him but didn't reply. The expression on his face told me that he figured it out for himself.

I got married and years later, when my daughter was four, I took her to the zoo in Wichita. They had just opened a new rain forest-jungle section. As we exited the area one of the teenage boys behind us said to his buddy, "Can't you see a Vietnam vet going in there and then flipping out?" I turned, looked at the boy and said, "Yep, you never know where those Vietnam vets might be." He got my point.

When a Vietnamese man came into a store as we were leaving one day, a friend I was with said, "You must hate seeing those bastards." No, I don't hate anybody.

America never did get it about Vietnam, about doing our duty for our country.

Gary L. Noller, Kerrville, TX
Infantryman RTO, DaNang area
B-1/46th Infantry, 196th Light Infantry Brigade, Americal Division, 1970-71

Going Home
Alan Walsh
Peachtree City, Georgia

MARCH 18, 1971, WAS a hot and steamy day in Vietnam, like all the others, but this day was special. Like every other guy serving in that war I crossed off time on my calendar, fantasizing about the final day that would come, the last wake-up, my 365 days in-country completed and going home alive. Since we flew out in the field for a week at a time we had the tiny thrill when we came back to our hooch of crossing off a block of days at a time, making no difference in the time served but deeply satisfying nonetheless.

Like many others I worried about making it through, but the last day finally came. There was no sendoff at the company area since the crew was in the field most of the time. We just packed up alone, turned in our weapons and flight gear and found a ride to the 90th replacement company a short distance away in Bein Hoa.

Out processing ate up a few hours then it was time to board the *freedom bird*, the chartered Pam Am DC-8 parked on the hot concrete of Bein Hoa Air Force Base, our ride back to *the world*. I had been anticipating this moment every day for the past year. I climbed up the external stairs to the aircraft, paused at the cabin door to turn around and look back through the rippling heat, humidity, and red dust at Long Binh in the distance, my home for the last year.

Suddenly I had the strange urge to walk back down the stairs. I didn't want to go.

Where did that come from? Why did I feel so connected to a place I had longed to leave every day, a place where a guy in my job would surely, sooner or later, get himself killed?

AFTER DINNER WITH MY family in Albany, New York, on Christmas Day, 1968, I departed for my adventure in the US Army. After Basic training I survived Warrant Officer helicopter flight school at Ft. Wolters and Ft. Rucker, with a high enough class standing to have a choice of special schools. I chose to fly medical evacuation missions known as Dustoff or Medevac, not because I had any interest in medicine but because I didn't like the formation flying slick pilots would have to endure. That was a flyspeck reason in the greater scheme of things, and I would discover only late in my Vietnam tour that I had by chance chosen the role just right for me.

As a shiny new pilot and Warrant Officer, I was not the same guy I was the day I joined the Army. Now I was a little tougher, more serious, more capable, more focused and determined. Still, I knew there was much to learn.

The medical crew training school at Ft. Sam Houston in San Antonio, Texas, was one of the finest military programs I ever attended. We covered

CPR, surgical air ways, broken bones, gunshot wounds, burns and much more and put into practice the procedures we learned in the classroom by working on goats with actual gunshot wounds, though I don't know who shot the goats. We also learned how to administer injections, start IV drips, draw blood, and we practiced on one another, learning the true meaning of *payback*! These EMT lifesaving skills would be valuable to anyone, except maybe Dustoff pilots like me since I couldn't possibly fly and tend to patients at the same time. But the training did help to understand what the medic and crew chief were doing behind me in the aircraft, especially under extreme pressure from enemy fire, lousy flying conditions or in-flight triage to decide which hospital destination was most urgent since hospitals differed in their specialties.

I ARRIVED IN VIETNAM on March 17, 1970, and had a choice of assignments between a *Medevac* unit or a *Dustoff* unit. The difference was the call sign *Medevac* flew for the 1st Cavalry Division in Hueys with M-60 machine guns on each side just like the normal Huey slicks, with an extra crew member Medic. *Dustoff* units were unarmed except for personal weapons and had a crew of four.

I chose Dustoff and was assigned to the 45th Medical Company (Air Ambulance) based in Long Binh, just east of Saigon in III Corps, right there in the same huge complex where we were in-processing. The company location was no measure of safety, though, because the unit had field standby sites in Tan An to the south, Lai Khe to the North, Loc Ninh further north near the Cambodian border, Xuan Loc to the east and Nui Dat with the Australians down near Vung Tau and the south China Sea. Ben Luc, with the Navy river boats and SEALs just north of the Mekong River, eventually replaced Tan An as a standby site. Most of our missions were flown from these standby sites, and most of our missions were short by design, since we tried to spread ourselves in a way that minimized the time from troops being wounded in the field to pickup and delivery to a hospital. Our typical schedule was four and one half days in the field – with no showers - followed by one day flying out of Long Binh and then one day off.

Initially flying as a new copilot I struggled to keep up with what the AC was doing. Even so, the ACs were very tolerant and got us all up to speed quickly. Our goal was to learn fast and earn our AC orders in as little as 90 days and 300 combat hours. Not all upgraded that fast but most did.

There was plenty of contact with the enemy in our AO, plenty of wounded and calls for help, so flying the 300 hours and learning how to survive came fast.

I DIDN'T REALLY EXPECT to make it through that year. I came to believe I had probably seen home and family for the last time, though I could never tell them that in letters. That wasn't any dramatic sense of morbidity; I was just being practical. My first two weeks in Vietnam 11 of our aircraft, nearly half our compliment, were either shot down or had to make some kind of emergency landing. Even though the pilots survived all of those incidents, how could a new guy expect to live through a year of those odds?

What does a young man do when he looks around him and sees that he is likely to die doing his duty on this assignment? He just keeps doing his job day by day, and probably keeps his morose thoughts to himself. There was no other option.

Unlike the other guys, I didn't take lots of photos for the future or buy stereo equipment or china at cheap prices to take home. I didn't expect to go home alive, but I did learn to love to fly and I did like my job. Unlike many others in this miserable war, our job was trying very hard to save lives and quite often we did just that.

We learned to do our work under trying conditions, landing in LZs that were far too small, hauling loads on the very margin of our power limits, hovering a couple hundred feet above the ground to hoist wounded up through the jungle canopy and meeting the challenge of night and lousy weather stealing our visibility and orientation, never knowing when enemy rounds would interrupt everything by hammering through the thin-skinned aircraft. Day by day, I learned more and morphed into a confident Dustoff pilot. I'll tell you about a few days that stand out in my memory.

WHILE ON FIELD STANDBY at Nui Dat, an Australian Army base located between Long Binh and Vung Tau on the coast, we received a mission to pick up ARVN wounded soldiers in the mountains just northwest of Vung Tau. We launched and found the pick up site on the side of a fairly steep hill in a dead-end valley not far from the mountain tops. Further down the mountainside a friendly unit was in contact with the enemy, supported by the door gunners of Australian H model Hueys, and we could see the tracers bouncing around from the exchange of fire.

Our mission was called in as "secure." Our ARVN patients had wounds from small arms fire but the unit said they were not in contact with the enemy at the time. Sometimes they lied about enemy contact to get their wounded extracted despite the risk to the Dustoff aircraft and crew.

Flying up the valley, we found the ARVN unit. The pickup zone was clear of any trees but it was on the side of a steep hill with scattered boulders. The AC was flying and I was working the radios as he came to a hover over the troops and set the left skid against the hill with the main rotor very close to the ground on the uphill side. At one point someone came down the hill and nearly walked into the main rotor but the AC spotted him and quickly yanked the aircraft back up into the air to miss him.

Setting back down, we had to load the patients on the downhill side of the aircraft which meant lifting them about five feet in the air. The crew chief and medic then had to haul them up onto the skids and in the cargo door. This was tough and slow going. After loading one or two the ARVN soldiers stopped handing up patients and all lay down to hide behind rocks. On my radio I asked the ground unit why they weren't loading any more since we could see more injured men lying on the ground, and then I discovered that the guy I was talking to was not even with this unit but a couple klicks away with another unit and working through an interpreter. He finally came back and told me to leave since we were taking fire.

We heard nothing and weren't taking any hits, but the ARVNs wouldn't get up to load more patients so we lifted up, turned around and headed back down the valley opposite the way we came in. The heat, altitude and power limits were working against us and we gained airspeed ever so slowly. We flew down to the aid station at Vung Tau, dropped off the patients we had and headed back to the mountains since we knew we still had more to pick up. We were again told the area was secure, that we should come back in to pick up the remaining patients left behind.

We made the same approach as last time since there was only one way in and one way out of the valley. I was flying this time since we normally took turns at the controls alternating missions. I set the aircraft down in the same place as before, one skid in the air and the main rotor dangerously close to the uphill mountainside, and the loading process began again. As they were loading the last patient I noticed dirt kicking up in front of my chin bubble and the ARVN troops again took cover. This time I could feel the hits the aircraft was taking, a very distinctive feeling in the airframe, though I couldn't hear the gunfire. Things seemed to move in slow motion; on the intercom the crew chief said he was hit, the dirt continued to kick up around the helicopter, and the radio came alive with calls to get out. The Aussie Hueys said they were on the way to help as we started the slow acceleration down the valley for the second time. With the power maxed out at 50 lbs. of torque the airspeed rose at a very slow 10 knots at a time. As we paralleled valley walls I saw the enemy soldier doing the shooting, dug into the side of the hill with a light machine gun, his tracers crossing in front of the windscreen from left to right as he led us out of the valley but I guess he expected us to accelerate much faster since he led us too far and we never took another hit.

The aircraft was damaged and shaking but still flying so we made it out of the mountains and down to the beach near the South China Sea where we set down in the sand. We were leaking fluids of some sort and the ship was vibrating even at flight idle so we were going to shut it down and leave it. One of the Aussie Hueys had followed us out of the valley, landed behind us and came up on the radio and in typical Aussie fashion told us we had landed in a friendly mine field and he didn't recommend stepping out at this point. We decided to take the chance of flying the machine very carefully to Vung Tau, not far away. It was a very long ten minute flight.

As it turned out we had a hydraulic leak and one of the four transmission mounts had been broken by a small arms round, a break causing all the shaking and rocking as we flew. The crew chief had taken a small arms round in the center of his chicken plate and the bullet jacket ricocheted up into his neck; it was removed with no problems at the aid station in Vung Tau.

The aircraft was not flyable so we hitched a ride to Long Binh for a replacement and returned to Nui Dat to complete our standby tour. The broken aircraft was recovered and eventually repaired and back on the line and flying again. This was my introduction to the adrenaline of being shot at and missed, and then hit, while flying. I was amazed at how fast things happened but how everything seemed to go into slow motion when rounds punched through the aircraft.

FROM A SELFISH DUSTOFF standpoint, the best units we worked for were Korean or Australian. When they secured an LZ, it was secure! We could count on it. They were very organized and when they told us they were ready for us, we could expect to spend only seconds on the ground. The Aussies seemed to love their job and the ROKs were highly disciplined.

The worst were ARVN or Thai units with no US advisors with them. They would be working through a remote advisor and interpreter by radio. No one at the LZ spoke English so things could get screwed up, especially at night.

I landed once to a strobe light at night and found one ARVN soldier holding the strobe light in the middle of the road by himself with no one around, and the patients nowhere to be found. At that moment I hoped he was on our side! I guess he wanted a ride out; some ARVN soldiers seemed more concerned with avoiding combat than anything else. We left him there to go back to his unit since we couldn't stay and risk the mortars that would surely start to fall soon trying to catch the prize of a helicopter.

Sometimes US ground units played loose with the truth about whether the LZ was secure. The RTO would whisper on the radio to tell me the LZ is secure, the bad guys were a kilometer away. I would ask him to speak up and he wouldn't. That didn't even pass the laugh test. He was afraid if he told me how close the enemy was I wouldn't land, so I had to convince him I would pick up his wounded but needed to know the tactical situation so I could plan my approach. Sometimes the exchange would get heated but it usually got worked out.

I RECEIVED MY AC orders after three months in country and 300 hours. That's what we all wanted, that's what we worked toward. AC orders were the demarcation point of saying goodbye to new guy, beginner, second-in-command. As AC we were now in charge of the crew, the decision maker, responsible for the aircraft, crew and patients. I was pleased to receive my AC orders, and yet I didn't feel ready to take on the responsibility. There seemed to be so much more to learn, and there was, but it was OJT by design! I remember a nurse using the term "AC syndrome" to refer to new ACs because they were quiet and serious, seemed to lose their sense of humor for a month or so, until their confidence increased and they became comfortable with the new job. Then they became cocky and their head started to swell. I can't say I was much different.

One day, with my new callsign *Dustoff 18*, I was headed to the Tan An field standby site in the Delta just north of the Mekong River with a co-pilot on his first assignment in the field. He was a good guy, decent pilot, and only a couple weeks in country.

We received a secure, "not in contact with the enemy" mission to pick up ten ARVNs with shrapnel wounds from a command detonated claymore mine. The ARVN unit had no English-speaking person with them and was working through a US advisor and interpreter in the town named Ben Tre about three or four klicks away from their position. These missions were frequently screwed up by communication problems and third parties.

Since my co-pilot was brand new I was teaching him how to do a tactical approach into this area as if it was hot, not secure. I figured a cold pickup zone was a good place for a little training so I showed him how to identify the LZ, analyze the smoke and wind direction and simulated tactical situation and how to plan the approach. I flew a couple miles away from the LZ at 1,500 feet then slowed to about 20 knots, bottomed the collective and nosed over toward the ground descending about 3,000 feet per minute - six times the standard descent rate - to tree top level and 120 knots. We would stay in the tree tops and race toward the smoke, which in this case was purple. We must have flown directly over the bad guys who got lucky with their fire since all of a sudden we started to take small arms fire directly underneath the aircraft, taking hits up through the floor as rounds punched through the aircraft. We had just refueled and had about 1,400 lbs. of JP4 onboard which was now flowing all though the aircraft from punctured fuel cells. Fuel flowed between my feet and down into the chin bubble. The fumes were so strong I thought surely the ship would catch fire before we went down. My legs suddenly felt real hot and someone yelled that they were hit about the same time the master caution light flashed on. In my peripheral vision I could see the individual caution lights coming on but wasn't sure which ones they were. In my mind I knew that we were going down, I just wanted to crash near the LZ since no one seemed to be aware of the bad guys in the area.

The thought of being captured was foremost on my mind since there were many horror stories of what the VC would do to prisoners. The Viet Cong, who in many cases were civilians by day and VC at night would find it very hard to explain why they had US soldiers in a cage under their hooch, so unless they had a convenient way to securely hide a POW or transfer them to the NVA, they didn't have much incentive to keep POWs alive.

I concentrated on the purple smoke and continued toward it hoping the helicopter would fly long enough to make it to the LZ and the friendly forces. I made it into the clearing, flared as we came through the small palm tree tops nicking small branches along the way and the engine failed on short final, turning my maneuver into a hovering autorotation followed by a decent touch down into the grass very close to the ARVN unit waiting for us.

We rapidly turned off all electrical switches, shut the engine down and scrambled out of the aircraft since I was certain it would catch fire and blow at any moment. It never did burn, just sat there dripping fuel while the blades swished around slower and slower.

Then the comedy began!

The ARVNs didn't know what had happened, did not know the aircraft was damaged, didn't know why the crew had scrambled out of the aircraft. They started loading their wounded on the Huey. When we saw this happening we ran back to pull the wounded out of this potential fire hazard. It must have been quite a sight! They would put the patients through one cargo door and we would pull them out the other door and drag them away from the aircraft to safety. None of the ARVNs spoke English so they must have thought we were nuts.

We finally got the point across by pointing to the bullet holes and leaking fuel. I saw a small Vietnamese soldier with multiple frag wounds lying under the tail near the skids with JP4 running all over his legs. I was not sure if I had landed on him but in any case he was still alive and in pain with fuel running into his open wounds. I leaned over to pick him up and felt a tearing sensation in my lower stomach. I didn't think much about it at the time since I was just doing what had to be done. Later on I learned what a hernia was!

I was still concerned that this ground unit had no idea the enemy was so close; they were conducting themselves casually, clearly unaware of the enemy nearby. Fortunately, prior to landing I put out a mayday call on 243.0 UHF guard with our approximate position and the fact that we were going down. In a short time an OH 58 scout appeared overhead and started circling. He was too small to pick us up but he relayed our position to a pair of Charlie model gunships which orbited the LZ for our protection until two slicks were able to land and pick us up along with our ten ARVN patients. We took the radios out of the downed Huey and took them with us to keep them out of enemy hands.

The slicks flew us all back to Tan An where we left our patients and then we hitched a ride with another Dustoff Aircraft back to Long Binh. After all the reports were filled out I was given two days off to wind down since I was approaching 140 hours in the last 30 days anyway. I lived with the hernia for a few more months until I had it operated on prior to leaving Vietnam in the spring of 1971. At least the operation got me a week of R&R in my own hooch!

WHEN YOUR DAILY JOB is carrying the wounded and dying, the aircraft floor accumulates blood, gore and fluids we would rather not think about. Some Dustoff pilots tell of landing in a shallow river to wash out the aircraft, but I never did that. Sometimes, if it got too bad while in the field the crew chief and medic would borrow a water truck to hose out the cargo floor. Usually it just accumulated and we brought the filthy aircraft back after 4 1/2 days then cleaned it in Long Binh.

Generally, we were a pretty rough-looking group coming in from the field; no shower or shaving and all that dust, etc. For the pilots, the first priority was to shower and clean up, then get something to eat. The crew chief and Medic would usually clean and wash the aircraft before tending to themselves.

We normally did not pick up the dead because our mission was focused on keeping the living alive. But rules are made to be broken. One day we picked up five ARVN KIAs which had been in body bags a few days in the sun out in the field. This was a special favor for a US advisor whose ARVN unit had been lugging these bodies around and needed to get them to normal KIA processing. We picked them up, and the nose-down attitude on take off brought a lot of unspeakable black fluid out of the body bags forward and down into the chin bubble, probably the worst and strongest smell I have ever experienced. When we got back to our field site, Tan An, water would not wash it away. The smell had gotten into the cargo floor. The crew poured peroxide over everything and finally did kill the odor. I wondered what the long range effects the body fluids and peroxide had on the aircraft structure. I bet things under the floor were starting to corrode!

Now let's go get some chow.

ON A RAINY NIGHT we were on standby at Fire Support Base Mace just east of Xuan Loc. I was the AC, lying on my cot inside a conex container covered with sand bags listening to the monsoon rains when I heard the radio operator receive a request for an urgent hoist mission for two US soldiers with gunshot wounds. *Urgent* meant loss of life or limb within two hours. A hoist mission meant it would be more risky than normal, doubly so at night in lousy weather in the mountains. That didn't even count the enemy.

It was a dark night with moderate rain and occasional lightning as I took off and dead reckoned northeast toward the pickup site in the mountains. Company SOP required gunship cover on all hoist missions but the guns were unable to make it out due to the poor weather. The pick up area was mountainous so finding the friendly position was difficult. I took advantage of lightning flashes lighting up the valleys, with the mountain tops obscured in clouds and reminding any sane pilot that weather kills as many helicopter crews as enemy fire.

As we approached the pickup site the ground unit fired flares to mark their position. The jungle was fairly dense with trees from 100 to 150 feet high. As we came to a hover above them, I used a tree top in front of the nose as a reference to maintain position over the pickup point. With the rotor wash causing the trees to sway back and forth, rain on the windscreen, and the wind I could feel the effect of vertigo come and go. I had to concentrate on the directions I was receiving from the crew chief and medic to stay in one place. We learned to trust our life to those guys in back giving directions and I did exactly what they told me to keep the main and tail rotors from hitting anything that would quickly destroy the helicopter.

Usually hostile fire is a real concern on a hoist mission since we were stationary, out in the open and unable to leave quickly but this night the weather was a more serious threat.

The two wounded men were brought up separately, occasionally snagging tree limbs on the way up and taking about 15 minutes to complete the pickup. Hovering is inherently intense work, and under these conditions 15 minutes was a very long time while struggling not to lose my orientation. Finally, the wounded were in the aircraft and we departed the area flying under the clouds, dodging mountains well hidden in the dark, to a small airfield on the coast called Ham Tam for fuel. We found a fuel bladder in the dark with no troops around anywhere for security, which was not uncommon, and did a hot refuel.

After refueling we took off and flew along a road between the trees since the ceiling had dropped to no more than 100 feet, using both landing and search lights at speeds from 20 to 40 knots. With the driving rain it took intense concentration to keep from becoming spatially disoriented. The weather must have been an issue for the bad guys, too, since we were a low and slow target but received no fire in more than 40 miles. Reaching Xuan Loc, the sky became broken to overcast so we climbed to VFR-on-top conditions and continued to Long Binh to drop off our patients. Arriving over Long Binh, above the clouds, I had the copilot fly since the right side of the aircraft has the better flight instruments and I gave him headings, decent rate, etc. through the clouds to the glowing lights of Long Binh. We broke out below the clouds with over 800 feet to spare and landed at the 93rd Evac Hospital. Piece of cake!

After refueling we shut down, totally exhausted, and slept a couple hours until first light and improved weather. We then headed back to Xuan Loc to finish our four-day field standby.

MY CREW MEMBERS AND I depended on and trusted one another with our lives. They trusted my flying skills and I trusted them in the back to do an in-flight job on our patients that would make a doctor proud. That trust transcended rank, age and race, which seemed to disappear as a factor in a combat zone; we were white, black, Asian and Hispanic and it didn't matter. What mattered among us was competence, reliability, what we could expect from one another because lapses by one could get us all killed. But you can't assume a man is good before he proves himself, as I was reminded one day near Xuan Loc.

I received a hoist mission near Xuan Loc for a US ground unit for a man with a *self-inflicted* gunshot wound to the hand. Since this was self-inflicted, probably just to get out of the field, I would not jeopardize my crew by treating this mission as *urgent*. Actually, hoisting him up through the jungle was pressing the limits of the risk I was willing to take for him, but since they said the area was secure I decided to do the unit a favor and hoist the dumb-ass out.

My assigned aircraft was the *dog of the fleet*, old and underpowered and never carried a hoist due to its minimal hovering power. The SP4 crew chief was a young black man I did not know. When the mission was called I asked the crew chief if he had flown hoist missions and his response was, "Yes Sir, many times."

We connected the hoist, launched, found the friendly position and I found my hover spot. The jungle was triple canopy with the ground far below the treetops. As I hovered for the hoist we all went up "hot mike" so we could talk to each other instantly without pressing a button, and the crew chief lowered the jungle penetrator. This would turn out to be the highest hoist I ever performed, nearly 240 feet, the limit of the hoist cable, and we were well into the red cable marking the last 15 feet as a precaution. I had to settle the helicopter down as far into the trees as possible, and the space available meant I had to hover with a tail wind, which took more power. The underpowered aircraft was a handful to keep in one place.

The ground unit radioed that the patient was hooked up and to haul him up. We hoisted him about 10 feet off the ground when the hoist motor died. It would not go up or down. I kept asking the crew chief what was going on and there was no response. The aircraft was close to critical, almost at max power and the transmission temperature was rising into the yellow band. Still no response from the crew chief.

The ground unit kept yelling in the radio to "haul him up" because they knew nothing about jammed hoist motors. I looked back quickly over my shoulder and saw the crew chief out on the skids with the hoist control in one hand and holding on to the aircraft with the other, no monkey strap on and his helmet bobbing up and down. No matter how much I yelled into the hot mike there was no response from him. I looked down to his audio panel and saw the ADF switch in the ON position . . . he was bopping to AFVN radio music with the volume all the way up. I went from concerned to pissed off in an instant!

I switched off the ADF and his head stopped bobbing; I finally had his attention. At that point I was thinking about shooting him. The hoist was stuck, the engine RPM was starting to fluctuate between 6400 and 6600 rpm and the transmission temperature was now at the red line. The ground RTO kept yelling in the radio to "pull him up" because he couldn't hear me respond that the hoist would not work with all the noise from above. Two hundred forty feet is a long way to fall in a helicopter. I had no choice but to blow the hoist, in other words, flip a guarded switch and electrically cut the cable. The guy with the self inflicted gunshot wound fell about ten feet and now had a broken leg! I admit that didn't break my heart.

Now the ground unit was yelling at me on the radio that the hoist cable broke! This turned out to be one fouled-up mission. I finally got through to them that I had cut the cable on purpose due to a hoist malfunction and would return with a new hoist. Flying back to Long Binh to get a new hoist I asked the crew chief again how many hoist missions he had done and now the answer was that this was his first. It was a good thing my two hands were occupied in flying because the temptation to shoot him with my pistol was growing more powerful.

With a replacement hoist we returned and got the guy out and to the hospital. As it turned out our hoist motor had just popped a circuit breaker which could have been reset. I should have thought of it but didn't at the time since it had never happened before. Another mistake not to be repeated, like assuming an unknown crew chief could handle a hoist mission.

The vast majority of the guys in Vietnam were good at their job, worthy of the trust placed in them. But that trust had to be earned. Each of us had to prove we would do the right things, even under fire, to minimize the jeopardy to our brothers.

VIETNAM VETERANS HAVE CARRIED the stigma of baby-killers for decades, an undeserved insult sold to the public by the anti-war left. The vast majority of our troops were like you and me, honorable people just doing their job and frequently trying to help the Vietnamese people. But there were a few bad apples in so many barrels, and there were occasional moral lapses, sometimes guys taking advantage of the chaos of war to commit dark deeds. Those who committed crimes smeared the rest of us, and I would be dishonest not to admit what I know.

I was called on a mission early one morning, just before sunrise. Near Lai Khe I was to pick up a Vietnamese civilian, a teenage girl maybe 13 to 15 years old. As I coordinated with the ground unit radio operator before arrival he reluctantly told me the girl had been raped by US GIs. I could hear in his voice he was not proud to say it out loud. The young girl was put on board with, I believe, her mother. I looked back into both their hopeless staring eyes and I felt sad that we had abused these people so badly. I was not proud of my uniform that day. It is much easier to handle wounded combatants, but the innocent victims, especially the children, were hard to take.

In bed that night I wondered if I had been on the ground with those guys, would I have put a stop to it or would I have looked the other way? I can only hope I would have done what was right but will never know.

I met a Cobra pilot over a few beers with the guys one day. He told us of a time near the Cambodian border at Muc Hoa, returning from a mission with unexpended ordinance he came across a "care package" tractor parked near a paddy dike. He rolled in on it and "smoked it" with a pair of 2.75 in. rockets, said it was hilarious. I guess I was way too sober to see the humor in destroying some poor farmer's tractor, but I did think of what some of us had become. I did not want to be one of those few who lost his humanity.

Were the red crosses on our Dustoff aircraft used as targets by the enemy? Well, I don't know. Yes, we were shot at now and then, but with the hundreds of missions flown we took fire just sporadically. I wonder if there were times the enemy held their fire because of our humanitarian mission? Even though our enemy seemed to use atrocities as a terror strategy, maybe some of them refused to let go of their humanity, too. I like to think so.

NEAR THE END OF my tour when I finally realized I may actually survive this experience, I reflected on what my role in this war really accomplished. I could see, through the eyes of a 21-year old, that we would not win this war the way it was being fought but at least my job had a meaning. The risks I took had a real purpose, but for most guys, this war meant surviving the tour and going home.

Many of those soldiers are alive today because of Dustoff crews like mine. I would learn many years later that so many of the wounded we hauled to hospitals are still searching for the Dustoff crew who took them to safety. There are no detailed records to answer those questions, and the urge to find their particular Dustoff crew might not be entirely rational, but for some there is a deep desire to shake the hand that picked them up, sometimes under fire, started medical care in the back of a bouncing helicopter and kept them alive on the way to a hospital. I take great satisfaction from knowing we made such a difference.

I matured many years in my one year tour and learned things about myself I never knew. I slowly came to understand how I would not hesitate to risk my life for a total stranger and feel confident that someone else would do that same thing for me.

I became aware of some of the changes in me when I became accustomed to the gore, the traumatic injuries beyond description, the sweet smell of massive amounts of blood on a hot humid day running across the cargo floor. I will never forget the smell of smoke grenades and bug repellent, agent orange and burning wood or JP4 burning latrine waste, the horrible smell of death in a body bag that had been in the sun a couple of days, odors etched forever in my memory.

I STOOD AT THE doorway to that airplane waiting to take me home, looking back for just a few seconds with a line of other guys behind me, and I wondered why I didn't want to go.

I entered the airplane and took a seat, lost in my thoughts.

Maybe it was a feeling the job was not done, that I was finally very good at flying Dustoff and could teach so much to the new guys.

Maybe it was the connection I felt with my crew, with other pilots, with the troops on the ground who depended on me to get them to medical care quickly when they were wounded.

Maybe it was because I left home as a boy and would return a man, a different person than my family knew.

Maybe it was because the people who understood me now were not the ones at home but here, in this miserable place of death.

Maybe deep down I knew something I would discover later, that the work I had done in Vietnam was not only a difficult job well done, it was important, perhaps the most important accomplishment of my life. And yet my friends and even my family would never know, never fully understand what I had seen, what I had done. Most never wanted to know. But even for those who asked, how could I find the words to convey life in a far different world compressed into a year?

I think for many of us, the lessons we learned in Vietnam would not truly be recognized until years later, through the lens of some years of maturity and reflection. One thing bubbling inside us was the unspoken affection we learned for one another, a bond among strangers forged in shared adversity that seems to last forever.

The airplane full of 198 other guys like me was strangely quiet, as if the others were puzzled by their own thoughts, like me. We were quiet as we taxied out for take-off. I expected that on take-off, the aircraft would erupt with cheering but the quiet continued. On liftoff the only sound was the whining of the engines, nothing else. The plane ride all the way to Travis Air

Force Base was very quiet, much the same as the trip to Vietnam a year earlier.

I didn't expect any fanfare on our arrival in the US and there was none. We all out-processed and since the war was starting to wind down I was released then and there, 22 months before my commitment to the Army expired. I was instantly a civilian, instantly unemployed!

One day I was a Dustoff aircraft commander, with skills honed to do things with the aircraft that new pilots might deem impossible, the next day I was just another jobless veteran wondering how I would pay the rent.

WHEN I ARRIVED HOME in my small upstate New York town it was great to see all my old friends who did not serve in the military. They were happy to see me, they were polite, but they weren't interested in what I had been doing the past year.

I went through flight school at age 19, became a combat pilot at 20 and Aircraft Commander at 21, but nobody cared.

I flew over 1,000 combat hours, 1,300 Dustoff missions taking about 3,000 patients from the field to hospitals, saved many lives, but my friends would rather talk about last week's football game.

I earned the Bronze Star, 29 Air Medals for combat missions and the Distinguished Flying Cross for heroism while my friends at home were still focused on cars and girls and the usual for kids.

It was good to walk down a country road and not worry about trip wires, mines, VC or anything like that but something was missing. Maybe when I complete my next war I will take some time to unwind with the people who fought beside me, get my head clear along with those who were similarly committed to the struggle, before I return to life with the American public, where we are consumed by our own comforts with little thought of the sacrifices made by others to keep us safe and free.

Photo courtesy of Alan Walsh, all rights reserved

Alan H. Walsh was born and raised in upstate New York. When discharged from the US Army upon returning from Vietnam, he continued his love for aviation with a B.S. Degree in Aeronautical Science and various flying jobs. He retired in 2005 as a Delta Airline Captain. He and his wife live in Peachtree City, GA.

Postcard to America

DURING MY SECOND VIETNAM tour I came home on emergency leave to deal with a "Dear John" letter from my then wife of six years. I found that she was seeing someone else and that she really wanted to end our marriage and ride off into the sunset with her new friend. Once I understood what was going on, we went to a lawyer and filed for an uncontested divorce, then I took my broken heart and headed back to Vietnam.

After a tearful farewell with my three-year old daughter at the Atlanta airport, I got on the airplane and took my window seat, in uniform. A guy in a suit sat down next to me and after we took off he started trying to get the attention of the stewardess. When she came by to see what he wanted, he asked to be moved to a different seat, making sure I heard his request. I had other things on my mind at the time and didn't think much about it, except that it was just one more insult. The stewardess moved him and we both got on with our lives.

I'd say most civilians just wanted to ignore us, and those who didn't either despised us or felt sorry for us.

Carl "Skip" Bell, Marietta, Georgia
1/4th Cav, Lai Khe, Vietnam, 1969-70
C/317th Cav, Vin Long, 18th CAC, Can Tho, HQ 1st Avn Bde, Long Binh, Vietnam 1972-3

Benediction
Claude Newby
Bountiful, Utah

BRING GOD TO MEN *and men to God* . . . the Army Chaplain Corps motto. I was a battalion chaplain in the Vietnam War, living with our grunts in the field because I knew I could understand them only by sharing the hardships they endured, the fears and uncertainties they faced. I believed that only by sharing their lot could I effectively bring God to them and them to Him, and help counter the dehumanizing effects of war on body and soul.

Some said I should wait in the rear, at the more secure battalion command center, letting combat soldiers come to me, but I knew I would have been negligent doing so because their opportunity to come to me was as rare as their need for ministry was great. The soldier heeds the sermon he sees. My presence with the grunts during their hardship and terror would speak louder than any words I could say and would reinforce the spiritual values that I tried to represent.

And so I went where the grunts went and shared what they faced and endured. I walked with them through the valley of the shadow, dug foxholes and prayed and worshipped and rejoiced and grieved quietly with them. As I ministered to these young men, I learned to love them, to admire and respect them for their determination, mutual love and courage.

GENERAL H. NORMAN SCHWARZKOPF said, "It doesn't take a hero to order men into battle. It takes a hero to be one of those men who goes into battle." He couldn't be more right. As the war years passed, those who served bore increasingly negative stereotyping and hostility. The louder and more vulgar the protests, the more the opposition seemed acceptable to that very society that sent its young men to war. Americans seemed to forget that the draft notices to its sons began: "Your friends and neighbors have selected you . . . "

In Europe where I served in 1973, while US troops were being pulled out of Vietnam, we Army chaplains were ordered to conduct special services in celebration of "Peace with honor." No! I actually felt so strongly that I wrote that I would not do that. Rather, I conducted a final memorial service for the soldiers and civilians on both sides that yet would die in Vietnam. In that special memorial service, and in numerous speaking engagements over the years, I have shared my memories of the heroes who went into battle in Vietnam. As I did so, a gnawing inside me eased but never disappeared.

This is not about how I view American involvement in Vietnam. As a chaplain, I served no war, I served soldiers, heroes, faithful souls who stepped forward when their country called while others received accolades for dodging their duty. This is about heroes, about whole life spans compressed into months, days, hours, and terrifying moments. It took a special kind of hero to answer the call in the sixties and early seventies, and I owe them for much more than memories. I owe them my life.

SEPTEMBER 13, 1966 PASSED in a blur in our home, hastened by our best efforts to wring every ounce of joy and hope from each moment. Mostly, I remember the day for its overwhelming melancholy, even before we departed for the Salt Lake City airport.

I had gathered Helga and our five children around me after dinner the night before and blessed them one by one, from the eldest to the youngest, and Helga last of all. This was a tearful occasion, both because of the pending year of separation and the unspoken awareness the next year would be full of danger.

Helga and the children, the eve of farewell.
Photo courtesy of Claude Newby, all rights reserved

At the airport Helga and I tried to reassure one another without violating our covenant of honesty; we would keep no secrets from each other and would not makes promises beyond our ability to deliver.

The moment of parting arrived, every nerve on edge. I kissed each child, gave them a few last words and then hugged and kissed Helga one last time before boarding a Boeing 727. My dear family waved until the plane taxied from their sight. My face stayed glued to the tiny, oval-shaped window long after Helga and the children were out of sight.

From Travis Air Force Base in California, after stops in Alaska and Japan, we landed in Saigon at 0230 hours on September 16. My assignment came right after breakfast. I had my wish. I was going to the First Cavalry Division (Airmobile), to serve with infantrymen. However, upon arrival at the 1st Cavalry's base camp in the Central Highlands near An Khe, I found myself assigned to the 15th Medical Battalion, and there I would remain for almost three months.

I FIRST MET DAVID Lillywhite of Snow Flake, Arizona, in a 15th Medical Battalion facility at a base camp where he was being treated. He asked me to visit his buddy, Pfc. Danny Hyde, because Hyde was going through a spiritual struggle. The next day I went forward to find Hyde, but only got to the 2nd Batallion, 8th Regiment forward support area. There I was stopped by combat action and a cautious command sergeant major. I decided to try to visit Hyde the next day, and I did, but not in the way I planned.

At 2100 hours LTC Leighton, the 15th Med Battalion CO, informed me that I was required at the bedside of a seriously wounded soldier. A chopper would land momentarily to fly me to the 95th Evacuation Hospital in Qui Nhon, about 45 miles across the mountains by the South China Sea. I arrived at the hospital about 0200 hours the next morning. The trip had taken five hours because the chopper I started the trip on crashed en route and, while we were uninjured, we had to be picked up by another aircraft.

At the hospital a female nurse led me to the bedside of the wounded soldier that I had been called to see. Before me lay Pfc. Danny Hyde, the soldier I had failed to reach the previous day. Hyde's body, what was left of it and visible, was encrusted in black, burned flesh. A leg, an arm and an eye were missing. Intense unreasonable guilt assaulted me as I ministered to Hyde. *If only I had gotten with him yesterday, maybe this wouldn't have happened to him.* Eventually the aftermath of this experience would become the catalyst that drove me to volunteer for a second tour in Vietnam.

DAVID LILLYWHITE BROUGHT ANOTHER troubled soldier to see me. The soldier sought help in dealing with a war crime. I called in the Criminal Investigators and the atrocity, known later as the *Mao Incident*, led to the court martial and conviction of four soldiers for their varying degrees of culpability in the kidnapping, rape and murder of a young, pregnant Vietnamese woman.

Daniel Lang, writer for *New Yorker Magazine*, assigned to me the alias *Gerald Kirk* in his lengthy magazine piece titled *Casualties of War*. A Hollywood movie about the Mao Incident was released about twenty years after the magazine article, also titled *Casualties of War* and starring Michael J. Fox as the young, troubled soldier that had come to me for help that wet and cold December night.

Hollywood movies and anti-warrior protesters gave the false impression that crimes of this nature by Americans were common in Vietnam, but in my experience they were not. The *Casualties of War* movie was no different, giving the false impression brutal mistreatment of civilians was routine in Vietnam. The Mao incident, heinous as it was, involved four soldiers out of more than 17,000 in the 1st Cavalry. I can tally on one hand, without using my thumb, the war crimes that I knew of during two combat tours in Vietnam. Any like-populated community in America would welcome such a low crime rate among its young, unarmed men.

AT THE 15TH MEDICAL Clearing Company, LZ Hammond, thirty-six dead Americans - sons, dads, husbands, and sweethearts - were laid out on their backs in platoon formation with their heads to the west, faces toward home. Seventy-two eyes, less the occasional empty socket or missing face, stared vacantly into the rain-laden heavens. The grunts had given their lives in a ferocious battle in the 506 Valley.

Images of contorted bodies and staring, glazed eyes burned indelibly into my mind as I moved among the bodies and prayed for the dead and for their loved ones. I also prayed for the graves registration personnel and other soldiers that surrounded the grisly scene. I even prayed for the news reporters present, though I was angry because to my mind their flashing cameras offended the memory of the dead and the sensitivities of the living.

Ten days later I would celebrate my thirtieth birthday at that same spot as the remains of twenty-five more soldiers were brought in piled in a cargo net and laid out in *dress-right-dress* formation. Many of these dead had survived the 506 Valley fight eight days before Christmas, only to die while it was still Christmas night stateside where their loved ones prayed for their safety.

Chaplains Newby (left) and Lamar Hunt (right-center) conducting a field memorial service, War Zone C, April 1969
Photo courtesy of Claude Newby, all rights reserved

I RETURNED TO DELTA Company, having been alerted that they had sustained heavy casualties and were still engaged. I was rejoining the grunts with whom I had spent the past two sleepless days and nights. The medevac chopper circled wide of the contact area and landed in a dry rice paddy north of a hamlet from which Delta Company battled the NVA and maneuvered to recover friendly troops and casualties from the rice paddy south of the hamlet. As we touched down, grunts laden with wounded buddies rushed to the chopper. I helped load the casualties onto the chopper and as it took off went to find the company commander and help where I could.

Most of the survivors of 1LT Frazer's platoon were still prone in the rice paddy. Frazer and all but four of his platoon continued to low-crawl back to our perimeter under a rain of bursting artillery shells, shrapnel and enemy machine gun and small arms fire. One man, Pfc. Neal Thomas, lay off by himself and was presumed dead. Two others, Squad Leaders Michael Dougherty and Frank LaBletta, were pinned down by an NVA machine gun close to the body of Pfc. George Sutt. The burst of machine-gun fire that started the battle had caught Sutt at almost point-blank range and hurled his body backwards, down an embankment and into the rice paddy.

Twilight was falling when Michael Dougherty, who retired some years later from the New York City Police Department, and Frank LaBletta made it obvious that they were not withdrawing without Sutt's body. All eyes were on them as they slithered to Sutt's body and grabbed his arms, bullets kicking up dirt as they raced for the perimeter. Heavy fire on the NVA from every grunt and from a *Snoopy* gunship covered them as they made it to the perimeter. LaBletta and Dougherty might easily have been killed, but they got away with Dougherty suffering a sprained ankle.

After twilight turned to the inky black of night, a yell came from the rice paddy, "Give me cover. I'm coming in!" PFC Thomas charged out of the darkness and collapsed among his buddies. Thomas had played dead after the NVA put a round through his helmet and another round laid open his buttock, top to bottom. I especially remember Thomas as he had borrowed a field hymnal to memorize uplifting hymns.

The next morning I snapped Dougherty's picture by a palm tree while he nursed his sprained ankle, the very spot I had taken cover and laid my head during the previous night. As I took the picture a grunt out in the rice paddy kicked up an electrical wire connected to an NVA-placed American *claymore* mine. Had it been triggered the previous night, it would have decapitated me.

A few minutes later, while I visited with several grunts clustered near the perimeter about 30 feet from *my* palm tree, I heard the order to prepare to move out. While they saddled up I excused myself to get my rucksack . . . WHAM! - a huge explosion ripped the area. I ran back into the dissipating smoke and raining equipment and body parts. The group I'd just visited with lay sprawled about where the explosions had slung them. Apparently, a sniper round had hit a bag of grenades on Pvt. Charles W. Kreuger's right side, setting off more than one of them just seconds after I departed. The Battalion Journal read: *0730, D/2-8 . . . received 1 round of s/a fire . . . hit grenades on 1 indiv webgear causing it to explode. 1 US KIA, 6 US WIA.* By midnight that day D/2-8 and C/1-12 sustained 5 more KIA, 20 WIA, and 2 MIA, though we recovered the MIA bodies the next day.

PERHAPS IT WAS LATENT superstition on my part, but I never kept a *short-timer's* calendar until I was thirty days short. So I was quite aware of the days remaining when I left the Bong Son Plain for what I assumed was the last time.

That morning I visited wounded troops in the hospitals in Qui Nhon and then returned to LZ English to pick up my gear and say goodbye to the battalion staff. However, my plans changed. Bravo Company had sustained several casualties and was still in contact. Chaplain Assistant Kenneth Willis and I arrived at the contact site. A few minutes later a bolt of lightning hit a PRC-25 radio, missing the radioman and me by about 20 feet. I was way too short to be taking risks like this, but I got involved doing what I could to help where lives and souls were in trouble. As twilight approached, a teenage VC nurse died on my left shoulder while I used a rope to climb up a steep ridge trying to get her to the medics.

It was time to go home to Helga and the children but it was so hard to leave. The battalion commander solved my immediate problem the next morning by gently but firmly ordering me to, "Get on my charlie-charlie and go home, Claude." He meant it and remained behind to make sure I left the field in his seat on the H-13 chopper.

I'M LYING IN A *prone shelter. Suddenly, a Chicom grenade, a potato-masher, hurtles from the darkness and lands against me. Frantically, I grab the grenade with my left hand and attempt to fling it across my body to the right. A bloodcurdling scream shatters my nightmare.*

Shocked instantly from a nightmare into reality, I see my wife above me and half out of the bed, her frantic eyes pleading to know why I have a death grip on her hair and why I seem to be trying to hurl her through our bedroom window.

In October 1967 I moved my family to Fort Bragg, North Carolina, and I really tried to adjust to the *business as usual* world and garrison-type duty. But the news about the 1968 Tet Offensive finished stirring my ghosts of Vietnam while the shrill voices of anti-soldier protesters disgusted me. My inner turmoil came to a head one day when a couple approached Helga and I as we strolled along in a city far from Fort Bragg, and while looking at my uniform the lady inquired if I might be in the Army.

"Yes ma'am, I am."

"We wonder if you can tell us how to get information about our son. He was killed in Vietnam," said the man.

"What is your son's name, his unit in Vietnam, and when did he die?"

"Our son was Pfc. Danny Hyde," she said.

It was easy to tell them about their son because I'd been intimately involved with him as he lingered between life and death a year earlier. It was hard to control a flood of emotions at the mention of Danny Hyde's name, accompanied by the belief that there could be no earthly explanation for this second extraordinary coincidence involving Danny Hyde almost to a year later. That's when I knew I had to return to Vietnam. I concealed that decision from Helga.

ON MY LAST NIGHT home before departing for Vietnam, Helga and I were finally asleep, and I dreamed . . . *The landscape is all in shades of black and*

gray in the pale moonlight. Bushes stand out from the ground in darker shadows. The damp earth presses against my chest and elbows. A patch of gray distinguishes a roughly circular clearing from the surrounding bushes.
On the other side of the clearing lies an enemy soldier who is drawing a bead on me with his AK-47. Frantically, I try to push up with my arms and fling myself to the right. But I'm too late; the AK-47 muzzle flashes and a sledgehammer-like blow slams into my forehead. Suddenly I'm thrown in the air and back. Then in an instant I'm in the air above the ground and watching my body tumble and fall, face upward, dead eyes staring blankly. I awoke wondering if I was just dreaming, or was it a vision, an omen . . . or nothing?

MARCH, 1969, I WAS back with the 1st Cav assigned to the 1st Battalion of the 5th Cavalry. I stepped from a chopper into the Michelin Rubber Plantation at the FOB of Bravo Company and shook hands and was introducing myself to Captain Bailey, the CO, when a message came over the radio.

"Gooks on the bunkers," radioed a nearby scout helicopter, "They're sunbathing!" Then, "We're taking fire!"

"Our Loach is down," radioed the pilot of the Cobra as he gave map coordinates where his pink team Loach went down. The Loach and its three-man crew had gone down less than a klick from our position just as I arrived back in the field. Half an hour later I crossed a road with the Bravo Company element that was making its way to rescue the Loach crew, or retrieve their remains. Suddenly a burst of AK-47 rounds cracked on both sides of me. I glimpsed muzzle flashes directly ahead. Several other AK-47 rifles opened up, joined a moment later by a 30-caliber machine gun and snipers in the rubber trees. Seconds later Sp4 John Bezdan, Jr., who was one row of rubber trees in front of me, was hit in the thigh and groin. Why me? That was my first impulse, as I began to low crawl toward Bezdan while wishing that instead I could crawl under my belt buckle for protection.

Early the next morning most of Bravo Company moved out from its FOB to return to an area to recon and recover rucksacks and other gear that had been left behind in a fight that took place before I arrived the previous day. We approached our objective beneath a rolling barrage of friendly artillery, brought in so close that indirect shrapnel rained down among us. Leaders and troopers alike were obviously relieved when we found the apparently undisturbed rucksacks where they had been dropped the previous morning.

While the CP set up next to a bomb crater that took up the north side of the road, I worked eastward along the road and visited the troops. I was visiting with Platoon Sergeant Gene Nuqui when Bailey ordered him to sweep his platoon northward away from the road.

Nuqui and I paused about five feet from the road when about fifteen or twenty feet to our left Sp4 Eiker pulled the pin from a grenade and leaned downward out of sight. He's fragging a bunker, I surmised. But Eiker was cut down before he could slip the grenade through the firing slit into the bunker.

A point-blank burst of 30-caliber machine gun fire erupted from that firing slit and almost took off Eiker's leg at mid-thigh. That same sweeping burst of bullets also shattered a medic's shoulder and wounded another grunt. The wounded medic survived by running wildly to the rear, past the CP which was now in the crater. Eiker, the other WIA and two unwounded grunts found themselves pinned beneath the muzzle of the enemy machine gun.

Sergeant Nuqui and I reacted instantly and reflexively when the shooting started, but slightly differently. Nuqui dropped where he stood while I dived toward the meager cover and concealment offered by a small rubber tree, a dive that halved the distance between me and an NVA soldier that was determined to kill me. My dive also left a small, machine gun-raked clearing between me and the road.

Chaplain Claude Newby in the jungle
Photo courtesy of Claude Newby, all rights reserved

From the other side of the small clearing, Nuqui called, "Here, Chaplain," and tossed a CS grenade to my left side. Then, unknown to me, Nuqui belly-crawled to rejoin the unwounded members of his platoon that had quickly pulled back across the road.

With heavy firing to my right and left and with an NVA soldier trying to get me, I very carefully raised the CS grenade to my face with my left hand, pulled the pin with my right hand, moved the grenade back down to my left side, released the lever and straight-armed the grenade to my front. Three rounds seemed to crack between my arm and left ear as I made the toss. Then, under cover of a cloud of stinging CS Smoke, I low-crawled backwards across the clearing, and then dashed across the road to rejoin Nuqui.

Enemy tubes began spitting out mortar shells about the time I reached Nuqui. At the same time Nuqui and his men jumped up and took off westward parallel to the road, running between two rows of rubber trees. I brought up the rear as fast as I could, but something much faster pursued me. Over my right shoulder I saw a line of dirt geysers overtaking me. I knew that hitting the ground probably wouldn't help because the machine-gun rounds were striking very low.

Looking ahead frantically, I spotted a small hump of earth, a termite hill between two rubber trees. I increased my speed and dived behind the eighteen-inch-high mound, knowing it had hardened like cement. Plugs of wood flew from the tree on my right, rounds hit the termite mound, cracked over me and chewed into the tree on my left.

About two hours later, after rescuing our trapped and wounded, we withdrew to the FOB under fire. I had just completed my first 24-hour day back in the field and was exhausted. Only a year to go!

AT DAWN THE DAY after Easter, a chopper picked me up from the Alpha Company 1-5 position where I had spent the night because the under-strength company had sustained sixteen wounded the previous day. The chopper delivered me to the Charlie 1-5 FOB where Captain Jim Cain had received pre-dawn orders to attack the enemy positions that had chopped up Alpha the day before.

"Move out," Cain ordered a few minutes after I arrived. I fell in with his CP group which was with the second platoon. After a moment I came onto Bill Snyder and his M-60 team. "We will have a worship service today, no matter what," I promised Snyder.

We'd maneuvered more than half the distance to the objective, advancing in three columns, when the point man of the left column ran into red ants. Everyone held in place while he did a quick striptease, which anyone familiar with Vietnamese red ants will understand. While we waited, Captain Cain crawled to me and, in violation of his own strict rule against anything other than mission-essential talking while moving in the jungle, said, "Bill Snyder is the most honest, trustworthy, cheerful and best infantryman I've ever known."

A little before noon we held up for a close air strike that we hoped would land on our objective. Freshly dug NVA fighting positions in the north-south running ditch where we waited confirmed what we already knew; the enemy knew exactly where we were, and was ready for us to attack.

As we moved out from the ditch we again passed Snyder's machine-gun team. Sergeant Hoover and Corporal James Derda of Albuquerque, New Mexico, were with Snyder. As I approached, Snyder said, "Chaplain Newby, don't look so serious. It hasn't rained, and maybe it won't rain bullets today." I chuckled as I passed, remembering things Cain had said about Snyder's effect on morale.

An NVA 30-Caliber machine gun opened up at 1435 hours, less than a minute after I passed Snyder. The opening burst killed our point man, PFC William Allen, Jr. of Cantonment, Florida, and wounded and pinned down all but a couple of men in front of the CP group.

While I hit the ground slightly to the left with my feet still on the trail, Captain Cain and two radiomen dived behind the termite mound to the right. Simultaneously, Artillery 1LT Bill Haines, his RTO, and the company medic dove to cover behind the mound on the left. Several AK-47s joined the enemy machine gun almost immediately, and we started taking fire on our left flank. Meanwhile, the platoon on our right flank moved into the tree line from the open field, unhurt, presumably because the NVA hadn't covered the field, not expecting Americans to advance out in the open.

I knew I was in the open and needed to get away from the trail, but which way? Cautiously, I raised my helmet-covered head to try to pinpoint the source of the enemy machine-gun fire. The NVA gunner answered my unasked question with a burst of machine-gun rounds that clipped off the half-inch-thick stem of a bush where it pressed against the left side of my neck. Reflexively, I dug my left cheek into the ground a split second before the second burst filled my face with stinging gravel as the bullets dug into the ground and ricocheted past my head. I decided to play dead for the moment.

In response to Cain's call for "Machine gun forward," Snyder hurried forward at a crouch, trailed by Hoover, and keeping right of the trail for the meager protection that the termite hill provided.

Snyder dropped to one knee when he came even with me and looked directly into my eyes. That look haunts me still. "Goodbye," those eyes seemed to say. And with that, he leaped forward and threw himself onto the path slightly forward of where Cain crouched behind the termite hill and directly between the enemy machine gun and me.

While Snyder dashed forward, I rolled to the left behind the relative security of the left termite hill. Just as quickly, the enemy machine gunner shifted his fire to Bill Snyder and, from ten feet away, Captain Cain watched helplessly as Snyder's head jerked backward from the impact of a bullet between and just above the eyes.

Derda arrived a moment behind Snyder and dropped to one knee behind the termite hill and fired a 90-mm recoilless rifle flechette round to our left front. While Derda was taking pressure off our left flank, Sergeant Hoover yelled from the protection of the right mound, "Gator! Snyder's hit!" Then Sergeant Hoover dived for Snyder's M-60 and died before he could pull the trigger.

"Hold this for me, Chaplain," Derda said, tossing me the strap he used to carry the heavy 90mm recoilless rifle. Then he too went around the termite hill to die trying to man the machine gun.

"Snyder is dead," Cain called to me after the sequence of death played out right before his eyes. I knew Snyder was dead and that he had intentionally died for me. More than two hours and a lot of artillery rounds later, we pulled back with our dead and wounded.

The next of kin message to Snyder's parents read that he was killed by sniper fire, not that he gave his life attacking an enemy machine-gun position, and it did not say, "Greater love hath no man than to give his life . . ."

LOOK AROUND YOU, AMERICA. Search carefully and you will find three kinds of almost invisible heroes walking in your midst.

You will be unable to see the first type of hero, except as he is reflected in the eyes and demeanor of the other two. He is the one who never returned from the wars of the past century, or who came home in a metal box or rubber bag. He is very real, but invisible to all but a few family members who grieve still and buddies who accompanied him through hell and survived the carnage. Memories of these heroes dim each time a loved one or former comrade-in-arms passes away, taking the memories with them.

The second type of heroes are the veterans still living. Every war leaves its combat veterans forever changed, forever haunted. This is true of Vietnam veterans, but not exceptionally so, though they have long been maligned and falsely stereotyped as violent, drug-crazed *dysfunctionals*. The vast majority of combat veterans of that war returned home and successfully took up life where they had left off. They had families and careers and are overwhelmingly patriotic. Most would "do it again" if called upon by their country. A few could not adjust. Such was the price they paid to serve their country.

Many Vietnam veterans share a deep longing, a dissatisfaction that is hard to locate, a gnawing reminder that something isn't right, something impossible to describe. They know the war itself was not wrong, that they were not wrong to do their duty. They know they were not war criminals, no matter what the anti-war left said, but they wonder why those who dodged or avoided the horrors of Vietnam accused others who did serve of the most heinous of crimes, and withheld sincere welcome home to those Americans who answered the call and faithfully served? These men and women are the long ignored and maligned heroes among us.

The third type of hero are the families of those who served. Their lot was hardest because while the loved one served, they wondered and worried, seldom knowing just what he endured. Even when their soldier had a chance to relax, they couldn't because they had no way of knowing. These heroes often continued to pray for the safety of a son, husband, or father for days after he fell dead or wounded because it took that long for the word to reach them. Many of these heroes on the home front endured the brunt of anti-war protests and disdain with nowhere to escape.

Barbara Conrad and her children were some of these heroes. Barbara, the widow of SP5 (Doc) Andrew Conrad, resettled her family in Flint, Michigan, following his death. None welcomed her there, as she describes in her own words:

"My neighbors resented me because my husband was killed in Vietnam. Our daughter, Cindy, the oldest of three children, was traumatized at just ten years old when taunted at school because, 'Your Daddy was stupid enough to get his

head blown off in Vietnam.' Andrew did volunteer for Vietnam. We were stationed in Japan in 1967, where he tended the steady stream of wounded guys pouring into the Army's 249th Station Hospital. Often, Andy brought guys home with him, those who were recuperating and getting ready to return to combat. Andy volunteered for combat duty in Vietnam because of these experiences, over the objections of loved ones. Then he gave his life trying to reach wounded soldiers on a terrible 'hell-top' called LZ Pat. Stupid? No! Noble and heroic? Yes."

Chaplain Claude and Helga, Newby
Tacoma, Washington, June 1987
Photo courtesy of Claude Newby, all rights reserved

I WROTE A BOOK about the men I served to pay tribute to them and their families, the ones who served so faithfully in the face of an increasingly ungrateful nation: <u>It Took Heroes: A Cavalry Chaplain's Memoir of Vietnam</u> *(Random House/Ballantine Books, 2003)*. The "friends and neighbors" that selected them to serve should know their sacrifice and devotion to their duty. Fathers and grandfathers, veterans of other wars, should know that in Vietnam their sons upheld the values that had been passed on by them.

As a former Chaplain to some of you, I want to tell you that I know there is a God in Heaven who loves you and knows what you endured, and how well you bore the mantle of responsibility that was placed upon you. I'm certain that a loving and just God smiles favorably upon you for your honorable choice to put your lives on the line for your brothers and for all those friends and neighbors that sent you to war in their stead.

And that same loving God understands the courage and sacrifice involved when a parent, spouse, or child surrenders a loved one to the sacrificial altar of service.

Fellow veterans and families, heroes, you've endured the darkest hours, and the aftermath. You fought when you had to. You returned home and helped build. Now it is time to let yourselves be healed. To this end, I offer these words of sacred hymn by Jeremiah E. Rankin (1828-1904):

"God be with you till we meet again;
By his counsels guide, uphold you;
With his sheep securely fold you.
God be with you till we meet again.
When life's perils thick confound you,
Put his arms unfailing round you,
God be with you till we meet again.
Keep love's banner floating o'er you,
Smite death's threatening wave before you.
God be with you till we meet again."

Claude and Helga Newby, 50th Anniversary
Photo courtesy of Claude Newby, all rights reserved

Claude Newby is one of only two chaplains to receive the Combat Infantry Badge while serving as a chaplain, and the only one to do so during the Vietnam War. He was wounded three times and decorated for valor four times. In 2000 the Infantry inducted him into the Order of St. Maurice (Patron Saint of Infantry) and formally designated Helga a Heroine of Infantry.
Claude and his wife Helga have seven children, six of whom are living, 30 grandchildren and 31 great-grandchildren. They reside in Bountiful, Utah.

Epilogue

I WONDER NOW AND then whether the truth about the Vietnam War will ever replace the misinformation still widely believed. I think not, but the effort to tell the true story is worthwhile.

If you are one of the many Vietnam vets who has neither told your story nor discovered the brotherhood to which you belong, I want to tell you something I have learned in recent years.

Memories of war seem to stay clear no matter how many years have passed for reasons I don't really know. Maybe those memories are more intense because the senses are never more sharp than when we are thrust into the dirty business of killing and trying to stay alive. Maybe those memories are close to the heart since we discovered to our surprise when the shooting started that we weren't fighting for our country so much as fighting for each other.

Whatever the real reasons, those clear memories are the makings of a brotherhood. I think it has always been that way for men, and women too, who fought together against an armed enemy. Many writers and poets have tried to capture the brotherhood, but by my measure only William Shakespeare succeeded.

In his play, King Henry V, Shakespeare depicts how the king inspired his men in 1415 before the battle of Agincourt on Friday, October 25, St. Crispin's Day. While King Henry V was moving his army to the English stronghold at the northern French port of Calais to winter and rearm, French forces cut him off, forcing a fight while the English were exhausted, hungry and outnumbered six to one. When the king encouraged his men, he didn't talk about the home country or the enemy, he spoke of brotherhood, including the following excerpt:

> *This story shall the good man teach his son;*
> *And Crispin Crispian shall ne'er go by,*
> *From this day to the ending of the world,*
> *But we in it shall be remembered -*
> *We few, we happy few, we band of brothers;*
> *For he to-day that sheds his blood with me*
> *Shall be my brother; be he ne'er so vile,*
> *This day shall gentle his condition;*
> *And gentlemen in England now-a-bed*
> *Shall think themselves accurs'd they were not here,*
> *And hold their manhoods cheap whiles any speaks*
> *That fought with us upon Saint Crispin's day.*

History shows the king led his men in hand-to-hand fighting that day, and the outnumbered English won that battle. What we remember most from this story, though, is the taste of combat brotherhood.

Like many of you, I put my Vietnam experience behind me to get on with life when the anti-war fervor in the country seemed a little crazy a long time ago. For over 30 years I never attended a single veteran event, I knew nothing of the brotherhood. I always declined invitations to veteran events because I was done with Vietnam and I didn't want to relive it. In 2001 I discovered how wrong I had been.

I was prompted as a late-blooming first time Dad by thoughts of the twisted version of history my children would learn about my war and the finest men I know. I finally became involved because I want my kids to know the truth. In my first meeting with others who had flown helicopters in Vietnam, I discovered the special connection with those who shared my experience.

It's hard to explain.

Now, when I enter a room full of Vietnam vets, even though they are all strangers it's almost like coming home to my family. With just a few words in our own lingo we can communicate volumes that others outside that circle will never understand. Without ever having been to the unit location where the other guy was stationed in Vietnam, I know what challenges he faced and what he required of himself to watch his brothers' backs while doing his job. I know that he feared even more than dying that he might let his buddies down, or screw up and get them killed. That's part of the brotherhood. We have an unspoken respect for each other, guys who did the same hard things at a very young age, almost like we see in them what we like about ourselves.

I've seen this discovery of brotherhood late in life with a number of other vets. When they reunite with vets they served with, they may look different now but the years evaporate and they're commiserating and swapping tales again like they had been covering each other's back the day before. With vets they meet for the first time, they are fast friends.

I'm grateful I came out of my shell to be part of this brotherhood. Now, as Joe Galloway said, I am more comfortable in their company than any other besides my own family. Among those I never knew before, many are now close friends for life, the first ones I would call for help.

If you wonder whether too much time has passed to tell your story or to make that connection to other vets, it's never too late. I hope you don't let the end of life come before you find someone who will listen and tell your story. I hope you discover your place in the brotherhood as I did, and find your life enriched, just as mine has been.

Appendix

The Good, the Bad and the Ugly

AMERICA NEVER REALLY UNDERSTOOD what happened in Vietnam. After all these years, the truth is still tangled up in knots of myths, half-truths and political agendas. Whenever I see pundits on TV comparing current events to Vietnam, I brace myself to hear something stupid, because conventional wisdom on the war is so wrong and the lessons learned from that war by pundits seem to be the wrong ones.

Schoolbooks tend to treat the war once-over-lightly with politically correct overtones, so it is little wonder we are producing generations who *believe* the war was a mistake but don't know anything about it. What really happened in Vietnam? How did we get involved and should we have been there in the first place? How did we lose the war?

Those are good questions, and we need to get the answers right for the sake of our country's history and future, never mind a little respect for Americans who fought the war. I'm going to give you my version of the answers, if you can bear to read it. I emphasize *my version* to start with some intellectual honesty. I am convinced our part in Vietnam was a noble cause that became screwed up to a fair-thee-well by politicians in Washington micro-managing our military. The generals took it from there and screwed it up some more. I also know that, even on the basic question of whether we should have been involved in the first place, there is room for informed and honest people to disagree. The trouble seems to be in the words *informed* and *honest*.

My version comes from all I have learned about the war over the years with an intense interest and a very skeptical view. I have intentionally not included references to other works because one can find references to support virtually any preconceived notion of the war; the body of written work is huge, full of deep passion and politics. I have written here what I believe to be true about the war and, no matter which side of the argument you support, the truth is not all that pretty.

As you prepare to delve into this unsavory mix, please keep one thing firmly in mind; while I tell you about the failures, the one constant is that despite all handicaps our troops did their job with honor, skill and courage. Just like today, they were the best in the world.

If you are ready to begin perhaps you should gather some comforts like a rocking chair, your favorite beverage and maybe even a fine cigar. You might also need a good grip on your patience because *my version* may turn casual reading into a bit of work since the story is too intricate to tell properly in sound bites. Maybe that is the root of our problem. Even on complex matters like the Vietnam War, we like simple answers a bit too much, turning ourselves into a herd of rats mesmerized by the media Pied Piper.

TODAY IN COMMUNIST VIETNAM Ho Chi Minh is worshiped with nearly god-like reverence, still called "Uncle Ho" even though his policies have long been abandoned. His preserved body still lies in state in Hanoi, like Lenin in Moscow and Mao in Beijing, which is fitting when you understand who he was.

Ho Chi Minh was our enemy's leader, and their icon after his death in 1969. The popular notion of Ho Chi Minh, and his well-crafted cover story, was that he was a nationalist committed to throwing foreigners out of Vietnam and uniting the country, a rebel, a freedom fighter for independence, not much different than George Washington. The truth is not quite so flattering, but historical context matters, so let's look back a couple thousand years, just in summary.

China long desired the land we know as Vietnam, likely because the river delta rice fields were attractive for a hungry population. But Vietnam is isolated from China by rugged mountains covered in thick jungle. Near the time of the birth of Christ, the Han dynasty in China overcame those obstacles and overthrew the Viet kingdom, which became part of the Chinese empire until 938 AD. There were many rebellions until the Vietnamese finally expelled their Chinese overlords after nine centuries.

In 1407 The Chinese Ming dynasty took Vietnam once again but their rule lasted only 20 years. Throughout these periods, and before and after, there have been a complex array of armed factions struggling for control, including the Vietnamese conquering their neighbors to the south to take the rich Mekong delta region where rice fields flourish.

Europeans began their foray into Vietnam in the missionary work of the 1500s, establishing a strong Catholic presence in the population that would take root and thrive. As the European countries were competing in their effort to colonize remote parts of the world, France ventured into Vietnam in the 1800s, and by the 1860s France controlled Vietnam, then Laos and Cambodia. By 1900 the French had imposed a system of taxation and state-run monopolies to make their colonization pay for itself.

While French control lasted less than a century, it had a lasting impact. The education system banned traditional Vietnamese language, writing, culture and philosophy, replaced by western methods and values. The strengthening base of Catholicism likely helped shift the educated class to western thinking, which they probably found useful in the bloated bureaucracy the French built, where they taught the Vietnamese how to turn red tape into art.

There were benefits, like a better system of roads, imported technology, improved access to medical care, etc., all contributing to prosperity. But all things Vietnamese were subordinate to anything French. Village life as the cultural center, along with the authority of the village chief, was weakened by a French centralized system of taxing individuals, something new to the Vietnamese. Traditional emperors in the city of Hue remained on the throne at the bidding of the French, mere figureheads. Over the years there were spotty local rebellions, quickly overcome by French forces, including many Vietnamese troops in the lower ranks.

As the dark specter of communism developed as a revolutionary ideology in the early 20th century, it found fertile ground in Vietnam as an antidote to exploitation by foreign colonists, adding to the many armed factions struggling for growth and control. One of those factions was the Vietnamese Nationalist Party (VNQDD), which received support from the Chinese nationalists led by Chiang Kai-shek as they waged war against Mao's communists. The VNQDD fought the French and fought the budding communist movement in Vietnam as well, assassinating collaborators with the French.

Into this complex history enters Ho Chi Minh. Born with the name Nguyen Ai Quoc in 1890, he left Vietnam at age 22 as a sailor on a French ship and wouldn't return until 30 years later in 1941 to establish and lead the communist Viet Minh. In those 30 years Ho Chi Minh spent much time perfecting his Marxist-Leninist philosophy, lived in Moscow off and on, and was a founding member of the French Communist Party in 1921. He also founded the Indochina Communist Party in Hong Kong in 1930 and became very involved in communist activities in China, Thailand and planning for communist revolution throughout Asia. Because Nguyen Ai Quoc was well known as a committed communist, he changed his name to Ho Chi Minh in an attempt to conceal the real political intentions of the Viet Minh, which he claimed was a nationalist movement. Maybe the secret was out when the Viet Minh liquidated thousands of their competitors in the VNQDD.

During WWII when Nazi Germany controlled Vichy France, their Japanese allies occupied Vietnam, leaving the French administration in place subject to Japanese authority. The Japanese helped themselves to Vietnam's rice and rubber to aid their war effort. The Viet Minh cover story was seeking independence from France and resisting the Japanese, but Ho Chi Minh's covert plan was to install a communist regime after the pattern of his Soviet mentors. And of course, since we Americans seem always to do the long-term wrong thing with short-term good intentions, our OSS, predecessor to the CIA, either bought Ho's nationalist cover story, or decided to use him anyway to our own advantage by giving aid to the Viet Minh to resist the Japanese. China's anti-communist nationalists also provided substantial arms to the Viet Minh for resisting the Japanese, but Ho Chi Minh's plans were different. The Viet Minh did negligible fighting with the Japanese, reserving their resources for their real purpose.

As the war's end drew near and Japanese fate grew evident, fearing the loss of control, the Japanese removed the French administration and arrested as many French as they could round up. With the French crippled, when the Japanese surrendered, Ho Chi Minh stepped into the power vacuum, announcing the Viet Minh's assumption of power to a puzzled crowd in Hanoi. Rather than advertising their communist agenda they wrapped themselves in Vietnamese nationalism that would appeal to the populace and declared independence for the *Democratic Republic of Vietnam,* even using words from the American Declaration of Independence and the French Declaration of the Rights of Man.

Ho Chi Minh persuaded the figurehead emperor of Vietnam, Bao Dai in his Hue palace, to abdicate and recognize Ho as the country's leader. Bao Dai was widely seen as ineffective and corrupted by the French and the Japanese, but his abdication gave Ho legitimacy in the eyes of some Vietnamese. Meanwhile, US President Harry Truman saw the urgent need to contain the spread of communism. In 1946 Winston Churchill coined the phrase *Iron Curtain* to characterize how the communists were taking eastern Europe by force and cutting off communication with the free world.

Grave concerns over the threat of communism in Vietnam, and possible spread throughout Southeast Asia, trumped any distaste for colonialism, and the US recognized French sovereignty in Vietnam. The French re-claimed their colony and with the backing of Vietnamese loyalists they pushed the Viet Minh out of Hanoi and into the Bac Viet wilderness where Ho Chi Minh planned, organized, recruited and trained for armed conflict against the French. That's when fighting between the French and the Viet Minh began in earnest and was sustained in what we call the first Indochina War.

Ho Chi Minh's fortunes improved in 1949 when Mao's communist forces took control of China and provided camps and arms to the Viet Minh. In 1950 the Soviets and Chinese diplomatically recognized the Viet Minh as Vietnam's legitimate government while the western world supported the French. That same year North Korea invaded South Korea and US forces went to war in Asia against communist forces directly aided by the Soviet Union and China.

In America, President Eisenhower took office in January 1953, the same year the Korean War Armistice was signed. He, too was keen to stop the spread of communism, and he continued Truman's *containment policy* but he was extremely wary about committing US ground forces in Vietnam, seeing geography and a situation that could consume divisions. Congress had escalated spending to support the French in Vietnam so long as it avoided direct military involvement.

In 1954 the French suffered a devastating defeat at Dien Bien Phu in northwest Vietnam. The seven-week siege of the French took its toll on both the French and the Viet Minh but neither side knew how badly they had hurt the other; while the French were pleading for US help in a hopeless situation, the Viet Minh were pleading with China for help and predicting their own imminent defeat. Operation Vulture was a proposed American plan to use massive air power, even small nuclear weapons, to bail out the French. Ironically, the president most identified with failure in Vietnam, Lyndon Johnson, was then a congressman who worked hard to make sure we did not bail out the French. President Eisenhower declined to intervene directly although he did permit some covert US air support. The French garrison at Dien Bien Phu fell in May 1954. Two months later in a long-planned Geneva Conference the French and Viet Minh agreed to a cease-fire and to partition Vietnam at the 17th parallel, dividing communist North Vietnam from pro-western South Vietnam.

While the French and the Viet Minh signed their agreement, the US and South Vietnam declined to participate. Over time the French went home, leaving a lasting French influence in the country.

While publicly miffed that he did not win the entire country, Ho Chi Minh badly needed the breathing room provided by the partition to rebuild his devastated Viet Minh for the next phase of his communist struggle. He had lost half his fighting force, his best, at Dien Bien Phu. Ho's advantage in the partition was that his provisional government in Hanoi remained somewhat intact.

In the south, some of the Viet Minh forces went underground to form the beginnings of the Viet Cong insurgency. Others forcibly took thousands of teenagers with them to the north for indoctrination, with plans to return them south to serve their needs.

At the same time in the south, with the French removed from power there was a vacuum of leadership and various factions scrambled for power amidst chaos. The French had been in charge nearly a century and even the Vietnamese professionals and government workers were disconnected from their own Vietnamese culture, having been westernized by education, training and experience. To make things worse, roads and bridges and other infrastructure in the south had been severely damaged in the war. There was much to do and America offered its help.

The partition provided a 300-day grace period during which people could move freely between North and South Vietnam before the border was sealed. About one million – mostly Catholics - fled from the communist north to the south, while about 50,000 migrated from south to north.

Free elections were recommended by the Geneva Accords for two years hence in 1956 as a means to unify the country. South Vietnam and the US were concerned the communists were certain to win a vote since the north was more densely populated and numbers of votes were weighted against the south, and Ho Chi Minh's dictatorial control of the north virtually assured a 100% rigged vote in the north for the communists. The US proposed to have the UN supervise unification elections but the communist block led by the Soviets rejected the "American plan." South Vietnam and the US refused to commit to elections without UN involvement since it seemed suicidal. Both sides violated many terms of the accords and the elections Ho Chi Minh hoped for were never held.

In 1955, Ngo Dinh Diem, a South Vietnamese Catholic who was a committed nationalist and anti-communist, unseated and succeeded Bao Dai in a referendum. Even though the voting was widely regarded as rigged, the US decided to support Diem because he seemed strong and Bao Dai had been ineffective. Besides, Diem was untainted by any French association, making him more credible to some. Diem declared the formation of the Republic of Vietnam with himself as President.

In the north, Ho Chi Minh ordered the murder of all known Trotskyites, just as Stalin in Russia had purged Trotsky and his followers in this competing form of communism that believed in democratic rule of the proletariat rather than centralized control. Trotskyism had taken deep root in Vietnam, and they were executed in sometimes creatively gruesome fashion to create a thorough, public round of terror that squashed resistance.

Terror continued when Ho let Chinese experts lead "land reforms" in North Vietnam, following Mao's pattern of executing thousands of land owners and distributing their land to peasants. In Ho's purges, public executions were planned in every village to spread the terror throughout the entire population. Many were tortured before their execution, prompted to denounce others whose names were already on the purge list, which made for perpetual rounds of torture and executions. No accurate death toll exists, but estimates range from tens to hundreds of thousands. Former land owners who survived, if the quota of executions had been filled in a local area, were the lowest caste in this new workers paradise.

While North Vietnam was under communist control, South Vietnam was a quasi-democratic and semi-corrupt regime headed by a fiercely independent Diem. Corruption had become a part of Vietnamese life under French rule. There was much dissatisfaction with Diem's autocratic style and nepotism, appointing family and friends to key positions because he trusted nobody else.

Diem recognized the need for bottom-up support and asked for US help to reach 80% of the population in outlying villages. The US was focused on military assistance and had no funds for that purpose, an opportunity lost.

Diem, whose family had been Catholic for generations, persecuted Buddhists, who protested in spectacular fashion as Buddhist monks doused themselves in gasoline and burned themselves to death in public while TV cameras sent those dramatic scenes around the world. Meanwhile, the best jobs in government and the Army went to Catholics, regardless of qualifications, adding fuel to division and strife. The fledgling Army of the Republic of Vietnam (ARVN) was inept, avoiding contact with the enemy where possible and failing to press an advantage to a victory when the enemy was engaged, frustrating US advisors. To many ordinary Vietnamese, Diem and his government were no better than the French colonists.

North Vietnam instituted severe military conscription to prepare a fighting force to invade South Vietnam, and by late 1957 terrorist attacks in the south began. By 1959 the Ho Chi Minh Trail in Laos and Cambodia, the network of jungle-covered trails the communists would use to infiltrate into all parts of South Vietnam, was operational and expanding.

Time passed. President Kennedy (JFK) took office in 1961 and escalated our involvement, increasing troop levels from just 800 to over 16,000, but still in an advisory capacity rather than ground combat. That same year in April he presided over the Bay of Pigs disaster in Cuba and in June the Soviets erected the Berlin Wall. In late 1962 JFK held fast to prevail on the Cuban Missile Crisis, but staring down the Soviets at the nuclear brink convinced him that American resolve would be tested again, that failing to stop communism in South Vietnam would fatally damage US credibility with our allies as well as the communists. Vietnam became strategically important in our Cold War with the Soviets and the Chinese communists.

In August, 1963, Diem's brother, Ngo Dinh Nhu, orchestrated raids on Buddhist pagodas across the country, killing hundreds and arresting more than a thousand monks and nuns. In response the CIA, with JFK's tacit if not overt approval, let South Vietnamese generals know that America would not oppose a coup to replace Diem. He was deposed and executed the night of November 1. JFK was quite distraught because he had intended that Diem would be safely exiled, not killed. JFK himself would be assassinated just three weeks later.

Diem was no longer a problem in South Vietnam, but his strong nationalism was gone, his firm control was gone, and a series of coups, attempted coups and government turnovers followed in rapid succession, each considered an American puppet regime.

When JFK was killed, Vice President Johnson (LBJ) was sworn in as president. His priority was the social programs of his vision of a "Great Society," not Vietnam, and our client state was deteriorating. While South Vietnam was in disarray, Ho Chi Minh's minions in the south had been strengthening their new National Liberation Front, better known as the Viet Cong, an indigenous guerrilla force organized to terrorize South Vietnam's populace. North Vietnam formed an army and built supply lines through Laos and Cambodia, the Ho Chi Minh Trail, to facilitate their invasion of South Vietnam. The NVA and the VC coordinated their efforts to impose communism on the south by force, and they were handily defeating ARVN forces.

If you are feeling a bit nauseous that America was beginning to support a corrupt and inept country, consider this. By the end of the war South Vietnam's government and army was far more stable and capable, but America has often forged alliances with unsavory regimes in order to fight a greater evil. Moreover, our interest in Vietnam was not propping up the South Vietnamese government, our interest was stopping the spread of communism.

While South Vietnam chased their tail for a while, the Viet Cong organized, recruited and trained men and women in the art of subjugating the people by terror. Before the VC entered a village they would know from the party district headquarters which villagers favored the South Vietnamese, who owned the land, which people were educated and influential. These were their execution targets to eliminate resistance, and they would gather villagers to watch as they tortured and murdered selected villagers to frighten the people into submission.

IN AUGUST, 1964, THE destroyer USS Maddox was stalked by North Vietnamese torpedo boats in international waters, and when threatened they fired on the torpedo boats. Two days later the enemy torpedo boats attacked the destroyers USS Turner Joy and USS Maddox, which prompted retaliatory air strikes. LBJ obtained from Congress the Gulf of Tonkin Resolution which gave the president powers to conduct military operations in Vietnam without declaring war. Some say the torpedo boat incident was at least partially falsified as a pretext to overcome resistance to a bombing campaign against North Vietnam. Perhaps.

Escalation began on January 31, 1965 when the Air Force 18th TAC Fighter Squadron was ordered from Okinawa to DaNang, South Vietnam along with smaller support units. LBJ disapproved a bombing campaign planned by the US military as *too provocative*, fearing it would promote escalation by the Soviets and Chinese. Instead he ordered a brief bombing campaign called Operation Flaming Dart, *proportional* reprisals for VC attacks against US bases in various locations, particularly Pleiku. On February 7, 49 F-105 Thunderchiefs flew out of DaNang to hit targets in North Vietnam, and the war was no longer confined to South Vietnam. The word *proportional* is the converse of the basic war strategy of quickly overwhelming one's enemy, but Secretary of Defense Robert McNamara devoutly believed that anything could be successfully managed with precise calculation; one can almost imagine the Joint Chiefs of Staff rolling their eyes at McNamara's doctrine of *gradual escalation*, seeking to apply just enough incremental force to prompt the desired response from the enemy. A portent of foolishness to come, the Flaming Dart targets, numbers and type of aircraft, ordinance and timing of the strikes were decided in the White House, not in the war rooms of the commanders in Vietnam.

A month later the 3rd Battalion, 3rd Marines landed to defend DaNang Air Force Base, marking the beginning of ground operations by US forces with more soon to come. In August the Marines would execute Operation Starlite, a pre-emptive strike against VC units organizing to attack our new base at Chu Lai.

In March, 1965, LBJ ordered Operation Rolling Thunder to begin, a bombing campaign intended to destroy key enemy assets, dissuade them from their operations in the south, and to interdict their flow of men and supplies. It started as an eight-week mission to *send signals* to North Vietnam and was managed by the White House with a philosophy of *gradualism*, where the US would turn up the heat to get the desired response from our enemy, and slack off or call bombing halts as a reward or incentive to our enemy to negotiate.

From the very beginning of Rolling Thunder, US military commanders were infuriated that they were prevented from complete mission planning and execution. They wanted to hit key enemy airfields and other targets, and use aerial mining to close off the Soviet supply line through Haiphong harbor, but LBJ felt that was too provocative. Haiphong harbor actually was closed that way, but not until seven years later, far too late, in 1972 as part of Operation Linebacker II. As a precaution to prevent our military from excessive aggression, Rolling Thunder targets had to be personally approved by LBJ or Secretary of Defense Robert McNamara, as did the date, time, ordinance and in some cases direction of attack. When we bombed a bridge in North Vietnam, the most effective approach path was usually oblique to the bridge line, but we often required our pilots to approach directly down the river to minimize the risk of long or short shots hitting civilians on either side of the river. Of course that rule also reduced the chances of hitting the target while maximizing the exposure of our planes and pilots to enemy anti-aircraft fire from either bank of the river. Bombing mission decisions were made by LBJ and his civilian advisors at their regular Tuesday White House luncheon meetings.

Our pilots were strictly forbidden to approach nearer than 30 nautical miles to Hanoi or 10 nautical miles to Haiphong harbor. There was also a precautionary 30-mile buffer zone to protect the Chinese border. Laos and Cambodia were off limits, where the Ho Chi Minh Trail buzzed with the flow of enemy men and supplies on their way to invade South Vietnam, to kill civilians and South Vietnamese troops and Americans.

Pilots had little discretion on targets and were prohibited from firing on an enemy Surface-to-Air Missile site under construction for fear of hitting Russian advisors. After the site was operational and firing missiles at US aircraft, then they could hit the sites. While our jet pilots risked their lives to bomb the same questionable targets again and again, approval to hit more strategic targets was denied. Separated by time, distance and an appreciation for the immediate local situation, US target decisions in Washington were often a laughingstock among our own troops, and it would have been funny if our pilots were not taking huge risks.

At the beginning, Vietnam was a jungle guerilla war fought by the south's initially lethargic ARVN with US advisors but no US ground troops. While the VC strengthened and grew, Diem had ordered the ARVN to avoid serious battles, and Diem's successors had not managed to make the ARVN strong, as they would later become. Guerilla war meant the enemy could move undetected, easily concealed until they decided to strike and melt back into the jungle after inflicting damage to military or civilian targets. VC control and influence on the populace spread.

The 173rd Airborne Brigade arrived in May, 1965, marking the beginning of major US offensive ground operations. Two months later US airmobile units arrived in Vietnam, an escalation of forces and a test of the new airmobile concept, battalion-sized units quickly transportable by helicopter to gain advantage over their foes on foot. Helicopters would play a major role in Vietnam, 12,000 of them serving in various roles, flown by 40,000 newly trained pilots over the course of the war.

In November that year, 400 of Col. Hal Moore's 1/7 Cav troops landed by helicopter at LZ X-Ray, unaware they would be surrounded by over 2,000 battle-hardened NVA for the first significant battle between the enemy and American troops. After four days of fierce fighting, the enemy withdrew, having lost nearly 700 dead compared to 80 American dead. Gen. Westmoreland congratulated Col. Moore on his victory and a favorable ratio of casualties; maybe he forgot or didn't know what Ho Chi Minh said to the French years before, "We will kill one of your men and you will kill ten of ours, yet it will be you who tires of it."

This was not a war of conventional battles on open fields with large opposing forces trying to decimate the other, at least not until the Nixon years. There were no fronts, there were large numbers of small enemy units moving into South Vietnam from the Ho Chi Minh Trail to points all over the country under the cover of jungle. US forces consumed enormous resources searching for infiltrating enemy units, and most fighting was by small units ambushing each other in the jungle, or surprise contact with brief and furious firing until the enemy broke away to hiding again. The enemy intentionally fought very close battles; they called it "grabbing the enemy by his belt" to neutralize our use of artillery and close air support. It was mostly a hit-and-run war with enemy staging areas just across the borders of Laos and Cambodia.

US Rules of Engagement required field units to call by radio for higher-up approval to fire on the enemy unless the enemy fired first and the source of fire clearly seen. The enemy often escaped while troops waited for multiple levels of approval to engage the enemy. Sometimes those approvals had to come from Vietnamese local officials, sometimes suspected to be sympathetic to the enemy or pressured by enemy threats, but even our own chain of command could be an impediment. One of my fellow Cobra pilots, Graham Stevens from Williamsburg, Virginia, spotted an enemy rocket crew in the open with their launcher during a rocket attack on Long Binh. He called for clearance to fire and was denied because of uncertainty about the location of friendly ground units in the area. He had to hold his fire, watch the enemy pack up their launcher and disappear into the trees.

The enemy would strike and when necessary retreat across the border to their sanctuary, where US forces could not follow.

US forces never seemed to be sufficient. From 1965 to 1967 American troops in Vietnam grew from around 200,000 to nearly 500,000, while ARVN numbers grew from around 500,000 to 800,000. More important than numbers, ARVN effectiveness and reliability was improving.

THUS BEGAN AMERICAN INVOLVEMENT in Vietnam, and the insanity of requiring our military to fight a war with one hand tied behind its back. LBJ called it *Limited War*. Our military was not permitted to use its primary tool of war, unrestrained and overwhelming conventional force to quickly bring the enemy to their knees. Our enemy responded only to brute force, not signals or gestures wrapped in an on-again off-again bombing campaign; they simply used bombing halts to rebuild bridges or air defenses, which were the toughest US pilots had ever encountered. The Rolling Thunder bombing campaign lasted 44 months and failed to achieve any of its stated objectives, but it did prompt the North Vietnamese to begin negotiations that would be drawn out over five years.

LBJ's concerns about containing the war were not paranoia. We had recently fought China to a stalemate in Korea and more recently had butted heads with the Soviet Union over Russian missiles in Cuba. LBJ feared the Vietnam War getting out of hand and starting WWIII. The problem with his *Limited War* policies is that they were an impediment to victory and arguably got American troops killed by the thousands.

America went on to fight a long list of offensive battles with our enemy on the ground, and a number of air offensives as well while the war plodded on year after year. Eventually, American troops in-country would number over 500,000 at the peak. It was called a *proxy war* since we were fighting the Soviets and Chinese through the North Vietnam enemy they supported. It is arguable we were winning when we withdrew in 1973.

AS IF IN COMPETITION with politicians to screw up the war, some of our own military policies worked against us. First was *individual rotation*, a policy intended to relieve the burden of entire units going to war for the duration, as had been our practice in prior wars. The goal was to limit the individual tour of duty in the war zone to one year. Each man or woman arrived to and departed from Vietnam on their own one-year schedule. At all times some number of new people in a unit were learning the ropes, making the unit less effective and possibly posing a danger to others. Sometimes those new guys held key positions, and sometimes the guy going home was vital to the unit. Just when a man was at his best, it was time to go home.

Individual rotation also affected GIs in unforeseen ways. Going home while a man's friends and unit remained in place promoted a feeling of isolation. A day after being in combat, an infantryman could be on the streets of LA wondering why everybody lived their life as if there was no war, as if nobody knew or cared his buddies were fighting and dying for their country. Returning GIs often found the only ones they could talk to, the only ones who understood them now, were other vets. Fantasies of returning home to family and friends often did not work out as expected, replaced by disappointment that resuming old relationships was not possible because they had changed. Public hostility to the war, and indifference to returning troops, made the isolation worse.

Today, as in prior wars we deploy entire units because now we can do so more quickly, and the unit benefits from cohesion, depth of training and experience with minimal disruption from new guys. When the unit's combat is over and a man is trying to decompress, to prepare to return to normal life, it helps to be surrounded by those who shared and understand the experience. There may not have been a better option for Vietnam, but individual rotation had adverse effects.

In Vietnam the way we *rotated officers* was another problem. It is important to understand the great majority of officers at all levels were outstanding in their training, capability and their performance. But the system didn't help them, especially the few weaker officers.

From lieutenants to generals, most officers were rotated through combat assignments every six months. Just when they were getting effective in their jobs, they rotated out to be replaced by a green officer who had to learn all over again.

These rotations were intended to give as many officers as possible exposure to the challenges of war that defy simulation: combat command, political difficulties dealing with the host government, logistical problems, etc. At the pentagon there was a saying, "It's a shitty little war, but it's the only war we've got!" The idea was to spread the experience, and the expectation was it would be a short war, thus the short rotations. The practice was known as *ticket punching* because officers needed the combat and command experience to be promoted.

The effect was to improve the military career of a large number of officers but at the expense of the units and men. By the time an infantry company or battalion commander learned the enemy's habits and tactics, the terrain, logistics, the strengths and weaknesses of his men and how to best use the artillery and air support available to him, he was replaced. It worked against unit cohesion, identity and effectiveness. For the men in combat units, seeing their leaders rotating in and out every six months was demoralizing and undoubtedly cost lives.

When a new CO was rotated in the field, would he be consumed with his own ego or would he listen to his junior officers and NCOs to learn that it is harder and slower to cut our way through jungle, but the easier enemy trails are mined, booby-trapped and a set-up for ambush? Would he learn before he got troops killed that it is best to avoid prior NDPs (night defensive positions) with conveniently-dug holes because if they weren't booby-trapped they were probably preset targets for enemy mortars? How fast would he learn caution sometimes called for digging in for a NDP, then after dark moving quietly to another spot and dig in again? Would he be the kind of leader to stand up to his boss to do the right thing for his men, or a small man who tried to make himself look good to superiors by risking his men on questionable ventures? The men watched their new CO and they knew; when they were put at risk by the ego of a small man unwilling to listen, they felt betrayed not just by the man, but by the system that valued the careers of officers more than their lives.

As they watched, new guys quickly learned that winning the war during their tour was not likely, and the common goal became survival, to go home alive. A good infantry CO had to learn how to balance the lives of his troops against his combat mission of finding and killing the enemy, knowing the war would go on after they went home.

EVEN THOUGH OUR TROOPS fought well, there was a monumental paradox in Vietnam, the tension between the strategic goal and the perspective of GIs, exacerbated by our *Limited War* policies.

Our GIs won every significant battle, but what did winning mean? The enemy had no infrastructure, no territory to defend, no ground for us to take and claim victory over. Permanently occupying remote ground we took from the enemy in battle was not possible or practical in a country of vast and sparsely populated jungle, mountains and deltas interlaced with a network of waterways.

Contact usually started when the concealed enemy was flushed out by our tactics, or when they chose to strike and inflict damage. Battles usually ended when the enemy withdrew and disappeared into the landscape, dragging their dead and wounded as much as they could, leaving other bodies for us to count and declare victory for that day. Body counts turned into an absurd game when some upper-level commanders used body count as the measure of success.

VC used the hit-and-run tactic, while NVA units were stronger, more disciplined, more heavily armed, more prone to stand and fight when it fit their purpose. Whether we were fighting VC or NVA depended on the location and the timeframe during the war. Even though the VC were not as well-trained or well-armed as NVA, they were ferocious and resourceful. We reduced their name in our lingo from Viet Cong to VC, Victor Charlie, Charlie, Chuck, Charles and even "Sir Charles" by some who had great respect for their fighting skill. There were other, more derogatory nicknames, as I am sure they had for us as well.

We weren't fighting to take ground from the enemy; we were fighting to stop their infiltration into South Vietnam. We actually couldn't stop them because a mountain jungle border hundreds of miles long is impossible to secure, but we certainly were slowing them down and making them pay a price. Maybe when the price became too high we would win, and maybe winning would only be evidenced by the enemy exhausting the supply of men they were willing to send down the pipeline and lose. That was the *war of attrition* idea by genius generals who thought we could wear them down by casualties just because we killed so many more of them than we lost ourselves. In the end, they lost 1.4 million while we lost over 58,000.

When a tough fight to take a hill from the enemy was over, when the bodies had been counted and exhausted, battle-weary American troops saw that we were moving out, leaving the hill empty once again, you could almost hear them ask in their minds, "What the hell did we just fight for?" If, a month later, the enemy infested that same hill and we set out for a tough fight to take it back again, those same troops knew they might die fighting to take the same hill again, and they surely wondered why. The strategic goal of stopping infiltration was lost on many troops, and the fighting they were doing appeared futile, especially when they went back to fight for that same hill the 3rd or 4th time.

Are you beginning to see what LBJ's *Limited War* cost us? Instead of bringing the enemy to their knees with our military might, we fought this war in a way that turned it into a meatgrinder that voraciously ate Asian and American men.

When President Nixon took office in 1969, he sought a way out of Vietnam. He pushed for peace talks. Nixon initiated an aggressive program of replacing the role of GIs with Vietnamese troops as they were trained and tested, a program he called "Vietnamization," a term which brought ridicule from critics. Nixon also coined the term "Peace with Honor" as the withdrawal goal of the US.

HOW ABOUT A LITTLE good news? The American military was then, just as now, the finest fighting force in the world. Our troops were well-trained and armed, and they fought with honor, skill and courage. Despite the ball and chain of *Limited War* policies, they won every significant battle against a very tough enemy, and there was every reason for their country to be proud of them, to welcome them home with gratitude for what they had done. Of course that didn't happen.

Should we have been in Vietnam at all? With the dubious background prior to American involvement, we had a decent argument to call it a civil war and just walk away. Why would the US go on to invest so much blood and treasure in Vietnam, a small third-world country on the other side of the world?

Personally, I side with Presidents Truman, Eisenhower, Kennedy, Johnson and Nixon, each of whom saw the urgency to stop the spread of communism in Vietnam. That was our noble purpose, our national interest. Informed and honest people, even other vets, can disagree with me and I would gladly buy them a cup of coffee for a friendly conversation about it. But I do get hot under the collar when that disagreement sprouts from a naïve belief that communism posed no danger to the world and that we were unrealistically fearing a boogeyman.

President Carter said we had ". . . an inordinate fear of communism . . . " which strikes me as worse than willful blindness. The Soviets had a term for people in the West who refused to see the threat from communism: useful idiots. Communism sounded good in theory but in practice brought out the worst in those in power, and its appetite was ferocious. In the aftermath of WWII the Soviets took East Germany, Estonia, Latvia, Lithuania, Poland, Hungary, Romania, Bulgaria, Czechoslovakia, Yugoslavia and others. The chokehold of communism continued to spread. China became communist in 1949, and other countries that eventually fell include Cuba, Albania, North Korea, Angola, Mozambique and, after we left the region, Vietnam, Laos and Cambodia. By the early 1980s nearly one third of the world's population lived under communism's purges, famines, show trials, comrades informing on one another and the iron fist of the party. One has to hold their eyes closed very tightly to ignore the bloodbaths of communism.

In America, I fear we are doing just that, holding our eyes closed tight, because it seems our education system is not teaching kids in school the true history of communism, that in practice communism has been the very embodiment of evil and that the human cost in the 20th century was over 100 million lives, far more if you count the misery of those who lived. We risk reliving the history we do not learn.

The free west struggled overtly and covertly with the communists for influence throughout the world in the Cold War. *Worldwide revolution, world domination* might sound sinister enough to make useful idiots snicker at conspiracy theories, but that Soviet goal was not only real and clearly stated by the reds, it was the driving force of Ho Chi Minh's life.

THE DAILY NEWS FROM Vietnam had a profound effect on the public mood about the war. Heretofore, the American public had been isolated by distance, time and censorship from the horrors of war. Now the blood and death of our own troops and the enemy, as well as suffering by non-combatants, was on TV news at the dinner hour. Seeing daily reports with the nasty ingredients of war – any war - for the first time made the Vietnam War seem . . . evil, especially when sprinkled with an occasional American misdeed or mistake that killed civilians.

Around 70 reporters and news crews lost their life doing their job in Vietnam. There were a few reporters, like correspondent Joe Galloway, who sought out the hot spots and lived through the life and death daily game of combat in Vietnam by the side of our troops, seeing up close and personal the reality of the war. Many reporters were based in secure locations like Saigon, often depending on pool information to send in reports on the same incidents that looked much like the stories other reporters filed, removed from any first-hand accounts. By not living through and seeing real action events for themselves, something was surely lost in the translation.

But those reporters on the ground in Vietnam were just one piece of the news media, which went beyond reporting as an anti-war sentiment seeped into the selection of stories, the slanting of the story tone, wording of headlines and an unmistakable undercurrent that this war should end. For example, there were significant US victories that were reported to the American public as something far different than victories.

In 1968 our enemy negotiated a cease-fire for their New Year Tet holiday, then in pre-planned treachery they attacked all over South Vietnam on the holiday, when ARVN troop strength was at its lowest, when families were gathered to celebrate. The enemy planned that the populace would rise up to support them against the South Vietnamese government and the Americans. But the people did not support the enemy attacks at all and our counterattacks decimated their forces. Notably, the ARVN performed very well in the Tet Offensive battles, their training and improved equipment and new confidence passing an extreme test. Through some very nasty battles our enemy lost an estimated 50,000 dead and three times that many wounded. They were on the brink of defeat after Tet but were re-invigorated to rebuild and continue the fight by anti-war demonstrations and the US news media, which portrayed Tet of 68 as a US military failure. Of course it didn't help that just weeks prior to the enemy's Tet attacks, our military and president had released the news that progress in Vietnam was promising and the end may be in sight. While our military performed extremely well and won a stellar victory, news reports conveyed . . . disappointment.

Our enemy's true colors were on display during Tet. When they took the city of Hue, they brought a list of over 3,000 names of doctors, nurses, teachers, business owners, elected officials and other "enemies of the people." Squads dispersed with clipboard lists of names and went from street to street, pulling people out of their homes, some executed on the spot, many gathered, tied up, executed and buried in mass graves or tied together and buried alive. The extent of the murder was not realized until in the aftermath of Tet there were so many people missing in Hue, and searches ultimately located the mass graves. Few Americans know about this event because it was virtually ignored by our media. There was no shortage of known enemy atrocities, carried out as a planned strategy to terrorize the people and crush resistance, and despite the deafening silence in the American press, that terror had been widespread long before the intensification of the Tet Offensive.

An Australian doctor in a Mekong Delta hospital, Dr. Wylie, remembers cases in which he treated people in the aftermath of visits from the VC. A pregnant woman had both legs hacked so badly with machetes they both had to be amputated. Her husband, the village chief, had been strangled as she watched, her three-year-old child shot dead. She lost the child she was carrying and the doctor said the worst that happened to her that day was that she lived. A village policeman had been held still while a VC gunman shot off his nose and then shot through his cheekbones so close to his eyes that he was blinded, but then he died. A schoolteacher knelt trying to protect herself with her arms while VC hacked her to death with a machete.

Marine Lt. Gen. Lewis Walt was known for developing the successful Combined Action Company program of sending small Marine units to live with villagers. Life magazine summed up the program, "His CAC units all had the same orders: help protect the villages, get to know the people, find the local communist infrastructure and put it out of business." Gen. Walt told of a village chief and his wife, distraught over their seven-year-old son who had been taken by the VC. Four days later he came back running and crying, both hands cut off and a sign hanging from his neck warning that if anyone in the village went to the polls in upcoming elections, their children would receive even worse treatment.

Gen. Walt also told of a hamlet near DaNang where the people were forced to watch while the village chief's tongue was cut out, his genitals cut off and stuffed in his mouth, and as he was dying his wife's womb was slashed open. Their nine-year-old son had a bamboo lance driven from one ear through his head and out the other, then they did the same to two more of their children, leaving the five-year-old daughter untouched except by what she had seen.

Our troops often arrived in villages in the aftermath of these torture and murder sessions, horrible tactics our enemy *trained their soldiers* to apply, to terrorize the Vietnamese people into submission, acts directed by the political officers embedded in every NVA and VC unit down to the platoon level. Some VC units had specially-trained terror squads to carry out atrocity duties with finely-honed skill.

The sad paradox is, when American troops became active in the war our reporters virtually ignored the common VC and NVA atrocities but leaped into scandal mode at the mere whiff of a war crime by American troops, which were rare. The American media would help anti-war activists sell the false story that our own GIs were the monsters in Vietnam, aided by the few American atrocities that were known.

VIETNAM IS KNOWN AS a war of American GIs butchering Vietnamese civilians on a widespread daily basis. If that isn't true, how could such a horrific lie get legs? To understand this we need to consider a basic principle the American public has never understood.

Historically, the American people have been sheltered from the reality of bad things that happen in war, where men are killing each other in gruesome ways with vigor and passion. We think we know what happens in battles far away but we really have no idea of the brutality, filth, fatigue and utter stupidity of it all sometimes.

Here's a little reality. War crimes, atrocities, are a part of every war, on every side. There were atrocities in our war for independence from Britain, on both sides, and in every war thereafter. America is not made immune by our honorable intentions in a war; all of us are subject to the dark side of human nature.

In Shakespeare's *Julius Caesar*, in a rage Anthony threatens to "Cry 'Havok,' and let slip the dogs of war." Anthony meant that once war gets started it is like a release of evil demons from Pandora's Box, uncontrollable murderous mayhem, savage and chaotic, blood lust feeding on the dark side of its participants.

Unspeakable things happen in combat: you get ambushed, your best friend's guts get scattered in the trees and he cries for his mom and begs you to help him while you hold him and he dies; an enemy soldier feigns bad wounds, then when an American medic takes a risk to help the enemy soldier, the enemy shoots him in the head. The pressure builds, day after day, and sooner or later even the most docile soldier lusts for murderous payback. Overwhelming passion is to be expected, predictable as a normal part of armed conflict, completely human, natural and understandable.

Because we are civilized, unlike some of our enemies we try very hard to manage our demons of war. When we go to war we raise the lid on Pandora's Box just a little, trying to release carefully selected demons that kill by the rules we set, and we try to hold back the demons of lustful revenge. We keep the worst demons in the Box by applying strict military discipline at all levels of command with officers responsible for anticipating the passions surging through their troops, keeping their troops under control, keeping the lid on the Box, keeping the demons from running amok. We cannot let the blood lust revenge happen, no matter how justified.

But however hard we try, it happens now and then anyway.

In Vietnam, the most notorious case of the demons getting loose was My Lai. On March 16, 1968, soon after the Tet Offensive, discipline evaporated in Charlie Company, 1st Battalion, 20th Infantry Regiment, 11th Brigade of the Americal Division, and our own troops murdered over 300 civilians, including children and infants, at a hamlet named My Lai in retribution for the village's support of the Viet Cong and the recent deaths of some Charlie Company GIs. These actions were in clear violation of strong military policy prohibiting indiscriminate firing on non-combatants or targets without military significance.

Amidst this egregious act of murder, some troops refused to participate, others did some shooting then had second thoughts and withdrew. An Army helicopter pilot witnessed what was happening and landed to shield a group of civilians, telling his gunner to shoot any American troop who tried to kill them.

Lt. William Calley, 1st Platoon Leader, was tried, convicted and sentenced to life in prison. With President Nixon's intervention, he spent less than five months in prison. In my opinion Calley should still be rotting in prison; he was supposed to *prevent* his men from their murder no matter how justifiable their passion, but he *led them* in the slaughter, claiming he was ordered to do so by superiors. A number of others were charged with the war crimes or the cover-up that followed, but Calley was the only one convicted.

I could argue that Calley's superiors may have ordered the murders either overtly or by suggestion, and that they should have been prosecuted. I could argue no soldier should have to fight a half-ass war with so many rules and limitations that got our own people killed. But none of those things excuses murder.

The anti-war movement seized the opportunity to exploit My Lai to prove American immorality in Vietnam. They portrayed My Lai as a routine occurrence while it was actually an anomaly, a rare and horrible war crime.

In April, 1971, John Kerry, a leader in the radical leftist group Vietnam Veterans Against the War, testified before the US Senate Foreign Relations Committee chaired by Senator Fulbright, who was desperately searching for leverage to end the Vietnam War. Kerry told the Senate Committee some fantastic stories, and his words took the nation by storm:

"... *several months ago in Detroit, we had an investigation at which over 150 honorably discharged and many very highly decorated veterans testified to war crimes committed in Southeast Asia, not isolated incidents but crimes committed on a day-to-day basis with the full awareness of officers at all levels of command. They ... had personally raped, cut off ears, cut off heads, taped wires from portable telephones to human genitals and turned up the power, cut off limbs, blown up bodies, randomly shot at civilians, razed villages in a fashion reminiscent of Genghis Khan, shot cattle and dogs for fun, poisoned food stocks, and generally ravaged the countryside of South Vietnam ...* "

Kerry went on to tell the Committee that the South Vietnamese people didn't care whether they lived under democracy or communism, that there would be no bloodbath if the communists took control, and that America was "*more guilty than any other body*" of violations of Geneva Conventions, all ridiculous claims.

Kerry said he was ashamed of his service in Vietnam and, "*We wish that a merciful God could wipe away our own memories of that service.*"

He also said, "*The country doesn't know it yet, but it has created a monster, a monster in the form of millions of men who have been taught to deal and to trade in violence, and who are given the chance to die for the biggest nothing in history ...* "

When Senator Symington asked Kerry about the rumors of drug use in Vietnam, Kerry responded, "*The problem exists for a number of reasons, not the least of which is the emptiness. It is the only way to get through it. A lot of guys, 60, 80 percent stay stoned 24 hours a day just to get through the Vietnam . . . "* Preposterous is far too kind a word here, and even Kerry must have realized that his exaggeration went too far because he hastened to revise the number he seemed to snatch out of the air.

Kerry admitted he had never personally seen an atrocity in Vietnam. That squares with the experience of reporter Peter Arnett, an outspoken critic of the war. He accompanied American troops on a number of combat missions and said he never once saw them mistreat a civilian. He said, "They didn't even think of it. Every unit I was with, [the GI's] went out of their way to be kind and decent with the people."

I wonder whether the senators might have so eagerly accepted Kerry's testimony if they could have known that, months later in a November Kansas City meeting, Kerry's VVAW radical group would take a vote on whether to assassinate US Senators John Tower, John Stennis, and Strom Thurmond for their pro-war stance. I am told that Kerry voted "no."

As with all great lies, Kerry's fabrications presented to the senate committee were propped up by kernels of truth. What about all those veterans who Kerry says "testified" in Detroit about atrocities in Vietnam?

That event, held at a Howard Johnson motel, was called "Winter Soldier," and partially funded by Jane Fonda. Kerry's testimony was of sufficient concern to Senators that military investigations were conducted, and investigative reporters conducted their own independent inquiry. They found that some of these Winter Soldier participants were imposters who had never been in the military, using the names of real veterans. Some had served in Vietnam, but were never in combat. Even from those who had actually served in combat in Vietnam, all of these investigations turned up the same result – there was no credible "testimony" of war crimes worthy of further action from John Kerry's event. In fairness I should add that maybe there were some real atrocities witnessed by real vets from Winter Soldier, and maybe those vets protected themselves by telling the investigators nothing. Maybe.

The news media was John Kerry's propaganda megaphone, but the discrediting news that investigations failed to find any war crimes warranting further action from Winter Soldier seemed to be a whisper in the media. America was sick of the war, wanted it to end and believed John Kerry. As time went on, Hollywood's version on the big screen reinforced preconceived notions. Academia had been decidedly anti-war and also had much influence on how the war was portrayed in schoolbooks, historical works and fiction. Meanwhile, the ones who fought the war tended to keep their story to themselves when they came home because nobody wanted to hear it.

And so, the great lie about American troops in Vietnam as villains or victims took deep root. Just ask the man on the street about those vets.

Millions of American troops who served well were stained by the real war crimes at My Lai, and they were further stained by the lies of John Kerry. But you would never find in news reports the important fact that American war crimes were rare violations of policy and prosecuted, while our enemy's war crimes were a deliberate part of their strategy and policy.

Contrary to the negative stereotype of American troops as *babykillers*, murderers and rapists in Vietnam, we actually helped the Vietnamese people in a number of ways. The VC wanted to discourage school until they were doing the teaching so they burned the buildings, executed teachers and mutilated children as examples. We built new schools and roads, dug wells, helped them improve farming methods, provided basic aid and took medical care to places that had never seen a doctor. As just one small example, in my off-time I worked to help a local orphanage in Bien Hoa care for the kids. And of course we fought by our Vietnamese ally's side against the communist aggressors who wanted to take their country by force.

IN MID-1970 PRESIDENT Nixon ordered a massive incursion into Cambodia against our enemy's sanctuary and their Ho Chi Minh Trail, an operation to damage the enemy's ability to strike at US troops while we were drawing down our troop level by 150,000. At last, a president untied our hands for just a little while. We won a huge military victory in Cambodia and set our enemy back an estimated two years, but the US media didn't report it as a victory, they reported the Cambodian incursion as an unfortunate expansion of the war when hopes of a US pullout were mounting. Riots exploded in protest on US college campuses across the country. While our victory against the enemy was ignored, the photo of a woman kneeling with arms outstretched near her dead friend at Kent State University flashed on TVs endlessly to portray the brutality of National Guard troops that fired live rounds. It is true the National Guardsmen should not have fired but there was little mention that the student rioting that shut down the city and the campus for four days, vandalized businesses in the small town, set fire to buildings on campus and injured fireman, policemen and guardsmen.

In March, 1972 our enemy's Easter Offensive was a massive invasion by conventional forces into South Vietnam, with 12 of the 13 NVA divisions crossing the DMZ. By then Nixon's "Vietnamization" program put most remaining US troops in advisory and flying roles with South Vietnamese troops doing the fighting on the ground. The South Vietnamese performed well and repelled the invasion. The Paris Peace talks were stalled with unreasonable demands from Hanoi, and President Nixon responded to the enemy offensive by ordering the mining of North Vietnam's harbors and initiated Operation Linebacker, a bombing campaign intended to cut North Vietnam's outside supply sources whether by land or sea. Unlike LBJ, President Nixon lifted the bombing restrictions and let the military plan and execute the bombing campaign, hitting coastal areas around Hanoi and Haiphong including SAM missile sites, anti-aircraft artillery sites, fuel storage tanks, Naval shipyards, railroad and truck stations, etc. For the first time laser-guided bombs were used and proved very effective. Noted historian Douglas Pike opined that the Linebacker bombing around-the-clock was so intense and destructive, and so disheartening to the North Vietnamese, that if it had been applied and sustained earlier in the war it could have defeated the enemy in a matter of weeks. Maybe we were so focused on stopping the 1972 Easter Offensive and forcing Hanoi back to the negotiating table to get us out of Vietnam that we couldn't see the opportunity even then for victory, but on that I confess I'm guessing. News reports covered the bombing escalation but ignored the success of the South Vietnamese on the ground. North Vietnam returned to the Paris Peace Talks.

Jane Fonda visited Hanoi in July, 1972. Her visit is best known for photographs of her cavorting with the enemy as she pretended on their anti-aircraft gun to shoot at American planes, but far worse were her multiple radio broadcasts aimed at US GIs in Vietnam, calling our president and our pilots war criminals and encouraging our troops to disobey their orders and refuse to fight. She met with some POWs, then later when they came home and told of torture, starvation and other mistreatment, Fonda called them liars. To be accurate I should point out that the POWs reported brutal systematic torture until Ho Chi Minh died in late 1969, then the new regime stopped the systematic torture and applied fewer and more selective torture sessions; apparently they expected the war to end with POW exchanges and did not want their mistreatment to be evident on the bodies of the POWs. In the aftermath of Fonda's Hanoi visit, there was a little talk of charges of treason, but public sentiment had turned against the war and a prosecutor's backbone was nowhere to be found.

There were a few images central to the anti-war theme that were never properly reported. One Pulitzer Prize-winning photo of a naked little girl named Phan Thị Kim Phúc, running down a road near Trang Bang after she tore off her burning clothes that were set afire by napalm, must have been shown on TV ten thousand times as an icon of American evil in Vietnam. It was a fine photo showing the horrors of war and the horrible effects of weapons, but it was used dishonestly to discredit American involvement because the facts omitted from the story were that the fight had been between ARVN troops and VC who had attacked a civilian village, the plane, pilot and napalm bomb were all ARVN and even the photographer, Nick Ut, was South Vietnamese. There were no Americans in the vicinity, but those facts were not reported.

Another Pulitzer Prize-winning photograph and video shown countless times was the execution of a VC man on the streets of Saigon by South Vietnam's National Police Chief, Nguyen Ngoc Loan (former general), using a revolver to shoot the man in the head. The Tet of 1968 scene is grisly to be sure, but maybe viewers would have liked to know all the facts before passing judgment, facts like the VC man had been caught with others in an assassination squad executing South Vietnamese policemen and their families, and throwing their bodies in a ditch. Those details were overlooked while the horror of the image did its work on viewers to further discredit America's involvement in the war. The prize-winning photographer, Eddie Adams, later wrote in Time magazine, "The general killed the Viet Cong; I killed the general with my camera. Still photographs are the most powerful weapon in the world. People believe them, but photographs do lie, even without manipulation. They are only half-truths. What the photograph didn't say was, 'What would you do if you were the general at that time and place on that hot day, and you caught the so-called bad guy after he blew away one, two or three American soldiers?" When Nguyen Ngoc Loan died, Adams called him a hero in a just cause, but of course that was not news.

These are just a few examples of twisted reporting that characterized the war. On the flip side our military provided daily briefings in Saigon commonly known to reporters as "the five o'clock follies" because of the rosy predictions of success and burying of bad news. Military predictions of success were notoriously wrong. Meanwhile, the White House had been lying to the American people about various aspects of the war, as evidenced by the Pentagon Papers, top secret documents released to the press and at least partially published to disclose that we had conducted bombing missions in Laos when LBJ said we had not, that we had bombed coastal North Vietnam and preplanned a troop escalation while LBJ told a different story, etc.

And so it went, as we fought in Vietnam with our hands tied, with the truth about our successes hidden from the US public while the anti-war left and our own media encouraged our enemy and demonized our own troops. What was reported and televised seemed to emphasize our enemy's accomplishments and US shortcomings, perhaps because the media and the country were weary of war. Our enemy knew they would not defeat the US in a head-on fight, and one of their slogans was, "We will win this war on the streets of New York." Perhaps they did just that.

AFTER TET OF 1968, a sea change took place in the Vietnam War. Secretary of Defense McNamara was replaced by Clark Clifford. The on-site manager of the war, General William Westmoreland, had applied a *search and destroy* policy of major sweeps to find and destroy the enemy, with success measured by body counts, ignoring the local populace and the South Vietnamese army, ARVN, since they were regarded as incompetent in the early years. Westmoreland was seeking set-piece battles with opportunities to kill large numbers of the enemy in a war of attrition, essentially fighting an unconventional war by conventional means. The enemy responded by evading in jungle cover and making quick strikes on their own terms.

Concern about Westmoreland had been building in Washington, and General Creighton Abrams, a leader of impeccable record, was appointed to replace him after being sent to serve as Westmoreland's deputy a year earlier. Abrams repaired diplomatic liaison with the US Ambassador and South Vietnam's leadership, disposed of the body count method, and began to fight a true counterinsurgency war. Abrams replaced large sweeps with smaller operations under the control of unit commanders, applied a *clear and hold* philosophy that emphasized protecting and gaining the confidence of the local populace, and partnered with the ARVN to build and strengthen their fighting ability. Abrams' methods brought promising change, but the U.S. became increasingly committed to a negotiated withdrawal, perhaps making it too late for the possibility of victory.

DID YOU EVER ASK yourself what happens when America turns against its own war? Vietnam is an example worth studying.

There has always been an extreme lunatic fringe on the left and right, and the left fringe has opposed all of our wars. The fringe is most often dismissed as irrelevant kooks, but what if they start to appear more mainstream? With Vietnam's death and misery on TV news every day, maybe the anti-war lunatic fringe didn't look quite so crazy.

As the Vietnam War ground on year after year with no end in sight, ordinary citizens who had never been activists on anything began to oppose the war, too. Young men who didn't want to serve probably found it very easy to go along with opposing the war on moral grounds, and the formerly tiny anti-war lunatic left began to seem far larger and more legitimate, especially when magnified by TV news. Even small anti-war demonstrations were emphasised in the news while counter-demonstrations supporting our troops received little media attention. Add to that some truly massive anti-war rallies and it was easy to believe opposing the war had become universal.

Opposition to the war took its place in the mainstream over time, and hostility to the war, even hostility to our own troops, became more and more acceptable behavior while supporting our own military became *passe*. Even though many citizens supported our war effort throughout on principle, supporters of the war didn't get TV coverage or ink in newspapers like opponents. The appearance was that those who opposed the war loudly and widely broadcast their feelings while those who supported our troops remained silent.

There were good people, faithful citizens, in the anti-war movement but there were also anti-American radicals who promoted violence, conducted anti-military propaganda. Our enemy claims to have had close ties with the American anti-war movement, but I don't know if that was ever proved.

When opposing the war became fashionable, so did other activities that previously would have been unthinkable, like finding both subtle and boorish ways to publicly disapprove of men in uniform. Employers actively avoided Vietnam vets, girls refused to date military guys, college professors applied their politics in grading military or vet students, military recruiting offices were ejected from college campuses and even friends and family wondered if their loved Vietnam vet had done the terrible things that people say they did because . . . *everybody knows*. Of course there were the more active types of opposition like insulting and assaulting GIs in airports as they returned from Vietnam.

Those who were loyal to their country and either joined or served when called were looked down on as losers not smart enough to escape through a college deferment, a mad dash to Canada or finding one of the doctors that specialized in false reports of health defects to keep the draft board at bay. The ones who evaded and opposed their country's military action were admired as bold and visionary.

There is a colossal and sad irony here. While the anti-war left manufactured false charges like *widespread American atrocities* in desperation for any argument against the war, they never seemed to notice the real defects embodied in LBJ's *Limited War*. Even now, people discuss bogus lessons from the Vietnam War and seem not to know what the real failings were.

SOME GIS HAD A warm welcome home from Vietnam; many did not. As America became weary of the war, some people weren't smart enough to question the war while at the same time respecting the young people we sent to fight it. The best known abuse was at California airports where protestors gathered to hurl insults and other things at returning troops, a practice that gathered steam as the war years went by. Our troops were often warned to stay out of altercations with protestors and to change into civilian clothes to avoid confrontation. It is still common for Vietnam vets to greet one another with "Welcome home!" because so many never heard those words when they returned.

Most vets I have talked to didn't experience direct hostility, but a common thread was studied indifference to their service, as if it never happened, as if it were worth nothing . . . or less. They also noticed that when protestors did turn ugly, good people seemed always to be looking the other way. It was not a proud chapter in our history.

You can't discuss the experience of Vietnam vets as if it was the same for all. Less than 20% served in combat, others served in support roles near to or far from the fighting; it takes a lot of support to put a grunt in the field, a host of logistical and command roles collectively called "beans and bullets." The terrain varied from the sweltering flat rice paddies and marshes of the southern Mekong Delta to the coastal plains, to the mountains and central highlands, most covered in jungle that varied from single to triple-canopy in thickness. A vet's experience varied with his job, his location and the enemy activity in that region. Even in the same locale, the war changed over the years with allied or enemy campaigns and with the effects of changes going on at home.

This was a time of social conflict in America and it seeped into the war. Anti-establishment hippies spread the gospel of casual sex, drugs and rock-n-roll, drop out, tune in and turn on. Cops were called pigs. People with regular jobs were sellout squares. Leftist professors encouraged anti-war activities by students who did not want to be drafted; college campuses were hotbeds of protest but while the reason was presumed to be intellectual, self-interest surely played a dominant role.

As the war years passed, shifting attitudes in the population against the war were reflected in unwilling draftees, their disdain for everything military and propensity to smoke pot and perhaps use other drugs, a common affliction of the anti-war movement. Later in the war morale in Vietnam became more of a problem as any enthusiasm for possibly winning the war evaporated and survival became each man's priority.

The civil rights movement in the US was in full swing and racial conflict sometimes erupted into riots and violence. In Vietnam, throughout the war black and white GIs who never before had close contact with the other race discovered in combat that different skin color meant nothing while brotherhood meant everything. Ed "Skip" Ragan was a new slick pilot when wounded GIs at an LZ couldn't wait for a medevac and his helicopter picked them up one day in 1968. As they flew to the hospital in Phu Bai, Skip saw a black medic doing a tracheotomy on a wounded white grunt in the back of the helicopter, but the grunt died and the medic cried like a baby, just one example of how combat changes men, erasing barriers and forging bonds. Skip never thought of race in the same way again. There was no room for racial division in combat, but late in the war, racial strife at home did find its way to the rear areas in Vietnam.

There was no single Vietnam experience.

OUR POLITICIANS NEVER DID unshackle our military to win the war. I've heard arguments that we could have won because the enemy's financial resources were limited, or that with China's help the Vietnamese would have persisted endlessly. All I know for sure is that we continued fighting the enemy in South Vietnam and never strategically took the war to the north to defeat them.

In 1972 the emerging Watergate scandal drained attention and political support away from the war, which was increasingly seen as a lost cause. The irony was that by the spring of 1972 the Viet Cong were no longer an effective fighting force, there were no NVA sanctuaries remaining and there were no major NVA units left standing in South Vietnam. By the summer of 1972, most US ground forces had been withdrawn from South Vietnam, with ARVN forces doing the fighting on the ground and US aircraft providing air support. The government of South Vietnam was stable, its Army now strong. It is arguable that US and ARVN forces had won the war militarily in the South at that point. Maybe Lewis Sorley said it best when he observed about the war's later years that the war in Vietnam "was being won on the ground even as it was being lost at the peace table and in the US Congress."

Hanoi was first driven to the Paris negotiating table by the Operation Rolling Thunder bombing in 1968, starting the on-and-off negotiations that would last for years. The aggressive bombing in Linebacker I, a response to the enemy's 1972 Easter Offensive, drove them back to the negotiating table. By late that year the enemy was once again playing games, and President Nixon turned loose the military to hit them the way they had wanted to from the very beginning. In Operation Linebacker II we were finally unshackled for 11 days in December to kick the dogshit out of our enemy.

There were many types of American aircraft destroying 18 industrial and 14 military targets in North Vietnam. Our B-52s alone flew 731 sorties in those 11 days, called the "Christmas bombing" by a disapproving media, dropping over 15,000 tons of ordinance while jet bombers dropped over 5,000 tons. The destruction was crippling, including shutting down their supply source from the Soviets in Haiphong harbor, their rail system of delivering war materiel and their rail system of receiving shipments from China. The enemy fired SAM missiles at our aircraft until their supply was depleted, and other munitions stockpiles were destroyed. They were hurt very badly and forced back to the negotiating table. Too bad we were not permitted to use such force at the beginning; many lives might have been saved in a much shorter war.

The Paris Peace Accords were signed in January, 1973, after years of halting negotiation and pushing North Vietnam back to the negotiating table with the Linebacker I & II bombing campaigns that hit enemy assets previously protected by our own longstanding prohibitions. When the agreement was finally signed, South Vietnam's President Nguyen Van Thieu accused the US of betrayal because key provisions - specifying the 17th parallel dividing line was permanent and that South Vietnam was a sovereign nation - were omitted. Instead, the agreement treated Vietnam as a single nation ". . . as recognized by the 1954 Geneva Agreements on Vietnam." That was a victory for the communists, implying that there never was a foreign aggression against South Vietnam. Sounds like a betrayal of our South Vietnam ally to me, especially since President Thieu was informed as a fait accompli, but the US seemed desperate to disengage from Vietnam at nearly any cost and claim President Nixon's promise of "Peace with honor."

US Secretary of State Henry Kissinger received the Nobel Peace Prize for his part in the long negotiations. His North Vietnamese negotiating partner, Le Duc Tho, refused to accept his Nobel Prize, saying there was no peace in his country.

Our minimal remaining military forces left South Vietnam in 1973 with North Vietnam's solemn pledge not to invade the south, and America's pledge to South Vietnam that we would intervene if they did. Prisoners of war were exchanged. The US continued to fund our South Vietnamese allies and hoped they could defend themselves. We didn't know it then but in 1973 when the few remaining US military forces went home, the North was already planning their 1975 offensive to take South Vietnam by force. Since American bombing forces were gone, North Vietnam's work on the Ho Chi Minh trail supply line could now proceed unimpeded.

In the 1974 aftermath of the Watergate scandal and President Nixon's resignation, Congress withdrew funding from our South Vietnamese ally, violating the US solemn commitment. The outcome was inevitable. North Vietnam was well-organized and well funded by the communist block while South Vietnam was unfunded, weak and chaotic since they knew what would come eventually.

On December 13, 1974, the communists crossed the DMZ in force with massive artillery and tank formations and began taking South Vietnam's northern cities. President Ford pleaded with Congress for funds to re-supply South Vietnam before it was overrun but Congress refused to keep America's pledge.

As the communist forces worked their way south, there was chaos in southern cities as South Vietnamese people, especially those who had worked with the Americans, desperately sought any means of escape because they knew their lives were in mortal danger. Civilian throngs of hundreds of thousands took to roads retreating south with ARVNs on foot and in vehicles, and were slaughtered by the communists by tens of thousands in what was called the "Convoy of Tears." There were TV scenes of crowds fighting to get aboard helicopters, mothers pleading for an overcrowded airplane to take her child so they could have a good life, rickety boats sailing for the open sea, so overloaded with people it was doubtful they would make it out of the harbor.

On April 3, With DaNang fallen and Saigon under attack, America, Australia, France and Canada started Operation Babylift to evacuate children from South Vietnam's orphanages. An estimated 2,000 infants and children were rescued.

On April 29, US Armed Forces Radio broadcast a mention that ". . . the temperature is rising. . . " followed by eight bars of the song White Christmas, the pre-arranged code announcement that Operation Frequent Wind was under way to evacuate the American citizens remaining in Saigon. Helicopters flew round trips to US ships standing by in the South China Sea, evacuating over 1,300 US citizens and nearly 6,000 Vietnamese and other foreign nationals. South Vietnamese helicopter pilots overloaded their aircraft with desperate evacuees and made unauthorized flights to ships like the aircraft carrier USS Midway; so many helicopters landed on the ship that they had to be pushed overboard to make room for others to land. With no place to land, some pilots crash-landed their helicopters into the sea close to a US ship, hoping for rescue.

The evacuation continued until the last minute on April 30 when the communist forces rolled into Saigon, and South Vietnam was conquered. The last Americans to leave Saigon were the US embassy Marine security guards, picked up by helicopter from the roof of the embassy. News reports gave the impression, often and wrongly repeated even now, that the US military was retreating in panic and disarray with the approach of the communists. American armed forces had been gone for nearly three years.

North Vietnam's guerilla war strategy never did prevail, but they did conquer South Vietnam when America refused to help, and they did it with the conventional war techniques of massive formations of infantry, armor, and artillery to ultimately overpower the South.

The victorious communists executed tens of thousands. Over one million fled to sea in unseaworthy crafts and over a quarter died by drowning, exposure, murder by pirates, thirst, starvation or disease. The US undertook Operation New Life trying to help, and while the refugee problem was far too large for us to solve, about 100,000 were given political asylum in the US.

An estimated 800,000 South Vietnamese were interred in brutal communist re-education camps, where an estimated 100,000 died over two decades of abuse, malnutrition, disease and perhaps the loss of the will to live. The communist victors bulldozed cemeteries of South Vietnamese war dead in retribution, violating deeply held reverence of ancestors as the core of the family. Still to this day, South Vietnamese veterans who fought against the north are treated badly in communist Vietnam.

Some say Eisenhower's *Domino Theory* was disproven since all of Southeast Asia did not fall to communism when Vietnam fell, but there is more to the story.

The communist Pathet Lao took Laos.

While America kept its back firmly turned, the communist Khmer Rouge took Cambodia. It seems every new communist regime launched rounds of terror in purging real and imagined enemies, denouncing and executing the educated, landowners and other influential people who might interfere with communist control, but the Khmer Rouge were particularly energetic in their bloodbaths. In a genocidal effort to resettle the population on collective farms after the fashion of Mao's disastrous Great Leap Forward in China, the Khmer Rouge tortured, executed, starved and murdered about one quarter of Cambodia's population of seven million. Executioners were very busy; in the Killing Fields of Cambodia there remains today a Chankiri tree clearly marked as the tree trunk against which infant and children's skulls were smashed as executioners swung them by the ankles, saving bullets for their parents. But America had finally extracted itself from Vietnam and refused to see.

Other countries in Southeast Asia had taken strength from America's stand in Vietnam and took early action against budding communist insurgencies in Thailand, Malaysia, Singapore, Indonesia and The Philippines. The fact they did not fall does not discredit Eisenhower's Domino Theory at all in my view; that theory was simple common sense. Indonesian leaders Suharto and Malik reported in the 1970s that they took their strength to resist communist inroads from America's commitment in Vietnam, and they successfully prevented a Chinese communist coup in October of 1965; had the communists prevailed in that coup attempt, a number of dominoes could have fallen, perhaps including the Philippines. America's stand in Vietnam appears to have helped keep a significant part of Southeast Asia free.

While we were fighting in Vietnam, the communists were expanding into the Congo, Benin, Ethiopia, Guinea Bissau, and later Madagascar, Cape Verde, Mozambique, Angola, Afghanistan, Grenada and Nicaragua. America still suffers impaired credibility for failing to keep our commitment to South Vietnam. Nevertheless, communism has dissipated from a long list of countries and there is a good argument the war we lost in Vietnam was an important part of slowing down the advance of our enemy, draining communist resources and contributing to the collapse of the Soviet Union years later. You be the judge.

We did make one monumental mistake in Vietnam. Battles are won at the point of the spear, and we won all significant battles in Vietnam. But wars are won or lost in the hearts and minds of the people back home, and the war was lost there long before the agreements were signed paving the way for America's withdrawal.

LOOKING BACK WITH THE clarity of hindsight, I see plenty of room for informed and honest people to disagree on whether we should have fought the war at all. Personally, I believe America did the right thing taking a stand against communism in Vietnam, even though politicians screwed it up terribly.

Whether we should have stayed out of Vietnam no longer matters. Whether we should have supported elections in Vietnam in 1956 no longer matters. Whether the Gulf of Tonkin incident was a sham no longer matters. Whether the government of South Vietnam was corrupt at any given point no longer matters. Whether democracy had a good chance in South Vietnam where the people were not fiercely committed to it no longer matters.

But there are some things that still do matter. What matters is whether the troops America sent to fight and die in a war received the unbridled commitment of their country to win. What matters is whether America was united behind its own troops until they were out of harm's way. What matters is whether the country delivered a debt of gratitude to its own troops when they came home. What matters is whether our media reported the truth to the American people without promoting an agenda. What matters is whether we preserved the true history of the war, and the truth about those who fought it. What matters is whether America kept its commitment to our ally.

On the things that still matter, America failed on all counts.

That's what happened in Vietnam.

In 1977 President Carter insulted the service of Vietnam veterans by issuing a blanket pardon for all of those who had broken the law by dodging the draft.

Despite the failures, speaking for every Vietnam vet I know, nobody is more proud of serving their country well, nobody loves their country more.

Now, decades later, the Vietnam vets who served their country so well have expanding waistlines and hair turning grey while America has new generations of politicians, still neither informed nor honest, and when I hear them discuss Vietnam as if we owe that communist regime an apology for any part of the war, I feel the sudden urge to puke.

Glossary

AC - Aircraft Commander, the pilot in charge who makes all decisions for the flight crew

Agent Orange – a powerful herbicide sprayed in abundance over Vietnam's jungle by US aircraft as a defoliant in an attempt to remove the enemy's cover

AK-47 - enemy 7.62mm rifle, Russian-made Kalashnikov first made in 1947, used by most cold war communist countries. Very reliable even when dirty.

AO - Area of operation

APC - Armored personnel carrier

ARA - Aerial Rocket Artillery, rockets on helicopter gunships

Arty - Artillery

ARVN - Army of the Republic of Vietnam. South Vietnamese troops, our ally. Some were highly competent when led well, many were completely undependable, some shot at US troops when they could get away with it. The ARVN became more effective later in the war.

Ash & Trash - Helicopter crew slang term for missions requiring the delivery of cargo or people from one place to another. Delivering water and ammo to a unit in the field is an example.

Autorotation – Flying a helicopter to the ground without power.

Avionics - The instrumentation in aircraft, or aviation electronics

Beehive Round –A beehive artillery round contains anti-personnel flechettes or pellets for firing point-blank at attacking enemy in desperate situations. Also used in M-79 rounds, like a shotgun shell.

Bird - Helicopter

Brass - Slang term for high-ranking officers

C-4 - Composition 4, military plastic explosive, a stable mix of ingredients that could be shaped and molded

CA - See Combat Assault

Car-15 - Shortened version of the M-16 with a telescoping stock, favored by special ops troops

CAV - The Cav was vertically integrated, meaning they had their own ground forces (Blues), slicks and Loaches (White), Cobras (Reds) and various combinations. A Loach and Cobra hunter-killer combo was called a Pink Team. The Ready Reaction Force was on standby in case any troops were in trouble and needed help.

Carabiner - D-ring device used to secure ropes for climbing safety.

C&C, Charlie-Charlie - Command and Control, typically referring to the commander's helicopter used to get him from place to place and to provide an aerial view of a battlefield below

Chalk 1, 2, 3 - Denotes the position of a helicopter in a formation, helicopters flying tightly close so they could land more or less in unison within the confines of an LZ. Chalk 1 is the first aircraft, Chalk 2 the second, etc. This was better than relying on each AC's callsign because the Chalk numbers were positional, identifying the aircraft by location.

Charlie Mike – Phonetic for CM, or *continue mission*. The general meaning is get the mission done no matter how miserable are the obstacles.

Chicken Plate - A cloth-covered ceramic chest protector plate worn by helicopter pilots to protect against small arms rounds in the chest area.

Claymore Mine - Command-detonated by a wired remote switch, called a *clacker*, sending shrapnel in the form of steel ball-bearings out to about 100 meters across a 60° arc in front of the device. It is used primarily in ambushes and as an anti-infiltration device against enemy infantry, set up around a perimeter by pressing its small stakes into the ground. It is also of some use against soft-skinned vehicles.

CO - Commanding Officer

Collective - Helicopter flight control that increased the pitch of the blades when pulled up, decreased pitch when pushed down. Increased pitch forced more air – more lift – through the blades. The collective was controlled by the pilot's left hand, including a twist grip throttle to control initial fuel feed or startup.

Concertina Wire - Large rolls of wire stretched out in loops to keep the enemy out, with razor-sharp edges instead of barbs, more effective than barbed wire.

Contact - Military term for exchanging fire with the enemy, opposing units doing their best to kill one another.

Combat Assault - Ground troops inserted by a group of helicopters in formation, landing as many at one time as will fit into the LZ. The first troops on the ground set up a perimeter to suppress enemy fire while the other troops land.

Commissary - US Army term for the on-base food store. See PX.

Corpsman – US Navy and Marines term for medic

CP - Command Post

Crew Chief - Crewman in charge of helicopter maintenance, supervised the gunner in a Huey.

CS gas - Riot control smoke grenade, incapacitates the eyes temporarily

CW2 - Chief Warrant Officer 2nd grade, promoted from Warrant Officer. See WO.

Cyclic - Helicopter control that moved the plane of the main rotor system. The cyclic was the vertical control from the floor up between the pilot's legs, with a grip containing a number of button controls, including radios and the rocket firing button.

Danger-Close - Artillery, helicopter gunship fire or ordinance dropped from airplanes or jets, intentionally closer in proximity to friendlies than normal due to exigent circumstances, like about to be overrun by a very close enemy. Heads-up term used to alert everyone to the risk of friendly fire casualties, extra care.

DEROS - date eligible for return from overseas, the end of a man or woman's tour

Di-Di - Vietnamese slang, used by our troops, for hurry, rush, run

Dear John letter - Letter from a wife or girlfriend back home telling a guy she was moving on, goodbye, adios.

DMZ - Demilitarized Zone dividing North and South Vietnam at the 17th Parallel

Dress-right-dress - A US Army term, how recruits are trained to set themselves up in formation one arm's-length in back of the line to the front, spaced one arm's-length by extending their arm to the right, fingertips touching the next soldier's shoulder. A platoon sergeant would shout the command "Dress right . . . . DRESS!" whereupon the soldiers would extend their right arm and position themselves, then the order would be given "Ready . . . FRONT!" and the troops would be in formation, at attention.

Dustoff - the radio call sign given to the first aeromedical helicopter evacuation unit in Vietnam, the 57th Medical Detachment (Hel Amb), which arrived in-country in 1962.

"El Tee" - Slang for Lt. or Lieutenant. Troops often called their platoon leader "El Tee."

ETA - Estimated time of arrival

FAC - Forward Air Controller, typically flying in a single engine Bird Dog, armed only with a few Willy Pete rockets to mark targets. The FAC pilot ran the show in the air among helicopter pilots, fixed wing and jet fighter-bombers.

Flechettes – Small nails with the ends crimped into little fins to make them fly more or less straight. Flechette rockets fired from helicopters had thousands of them, as did artillery canister rounds used for point-blank firing at the enemy. M-79 grenade beehive rounds used flechettes as the projectiles.

FNG – Fucking New Guy, a derogatory epithet referring to the inevitable mistakes new guys made, how they sometimes got themselves and others killed, the reason experienced guys stayed away from FNGs.

FOB - "Forward Observation Base," very dangerous covert duty

Free-fire zone - An area designated on maps as "free fire" meaning civilians were minimal to none, it was known to be frequented by the enemy, and approval to fire on the enemy from higher-ups was not required. The anti-war left spread the story free fire zones were an example of an immoral war, but quite the opposite was true. Unlike prior wars, when troops fired when they thought necessary, in Vietnam troops had to have clearance from higher-ups to fire, unless they were in a free fire zone, where they could use their own judgment.

FSB - Fire Support Base, a temporary or permanent base camp with a perimeter protected by concertina wire, for the purpose of providing artillery supporting fire to ground units in the area.

Grunt - An affectionate slang term for ground troop, infantry

Gunship - Either a Huey with guns and rocket pods attached to the side, or a Cobra, specifically designed as a helicopter platform for rockets and guns

Hooch - House, home, however humble

Hot refuel - Helicopters often landed at fuel points and kept their turbines running and blades turning while one of the crew pumped fuel into the aircraft, known as a hot refuel. Shutting down and restarting the aircraft consumed time and posed a small risk of not being able to crank without the aid of external power tractors.

Huey - Bell UH-1, the workhorse helicopter of the Vietnam War

I-Corps - The northernmost military division of South Vietnam for purposes of managing the war

II-Corps - Just south of I Corps, largest subdivision, highlands of South Vietnam

III-Corps - In the south half of South Vietnam, surrounding the Saigon area, including the Cu Chi area infested with enemy tunnel systems

IV-Corps - Southenmost division of South Vietnam, flat Mekong Delta, the rice-bowl of Vietnam, an area where VC were prevalent and NVA went for R&R

Indian Country - The boonies, the bush, the jungles where the enemy roamed freely, in strength. Slang from the unrealistic cowboy-Indian films that were part of our youth.

IP - instructor pilot

JP4 - The type of kerosene that fueled Huey and other turbine engine helicopters

Jungle Penetrator - A metal stretcher frame that could be dropped end-first on a cable, from a hovering helicopter, to break through jungle canopy to get to men the helicopter crew could not see.

KIA - Killed in action

Kill Zone – For helicopter pilots, any altitude between 100 feet and 1,500 feet was called the kill zone because we were within small arms range. Lower than 100 feet flew by too fast for the enemy to hit us most times, and above 1,500 feet was out of small arms range.

Kit Carson Scout - Former NVA or VC enemy soldier, now a scout for the good guys, same as a Tiger Scout

Klick - 1,000 yards, one kilometer

LOH, Loach - See scout

LP – listening post, a 1-2 man post sent outside the perimeter at night, to be an early warning "tripwire" in case of enemy attack, very dangerous.

LRP, LRRP - Long Range Recon Patrol, six man teams in extreme camouflage, lightly armed, inserted by helicopter at dusk or sunrise, deep in enemy territory for intel gathering. If compromised, emergency extraction required.

LZ - Landing Zone, an area cleared of trees so one or more helicopters can land to insert troops

M-16 - American 5.56mm rifle used in Vietnam

M-79 - American grenade launcher (hand-carried)

MACV - Military Assistance Command, Vietnam. MACV controlled all of the US Armed Forces in Vietnam. MACV had two basic missions: seeking to engage the enemy in combat on the ground and territorial waters of the Republic of Vietnam, and to provide assistance to the constitutional government of Vietnam in building a free society capable of defending itself.

McGuire rig - a 15' x 3" nylon strap attached to heavy duty rope dropped from a helicopter to extract soldiers from the jungle where no suitable LZ existed. Typically used for LRRP teams needing quick extraction with the enemy hot on their trail.

Medevac - Same as Dustoff except in some units the Medevac callsign indicated they were armed with M-60 machine guns

Minigun - 6-barrell electrically-driven gun, 7.62mm ammo in belts, fires 4,000 rounds per minute on full speed, over 66 rounds per second, every 5th round a tracer. Firing a minigun sounds like a very loud burp, aim like a garden hose. The firepower is devastating, cuts down small trees. Miniguns were turret-mounted in Cobra helicopter gunships, mounted on the sides of C-model Huey gunships, and sometimes hand-aimed from the back of helicopters. See Snoopy.

MOS – military occupational specialty

Movement - Term for seeing or hearing something suspected to be the elusive, hiding enemy, but not yet in contact. When someone on the radio said "We have movement" all would stand by expecting contact with the enemy.

Number One! – in Vietnamese slang, means "the best" or "very good" while **Number Ten!** meant "very bad." The idea was a sliding scale between one and ten but we never heard anything but one and ten. Number Ten Thousand meant the very worst.

NVA - North Vietnamese Army, well trained, uniformed, organized, well armed and tough enemy that infiltrated the south through the Ho Chi Minh Trail they built in Laos and Cambodia just across the South Vietnam border. Typically stronger, larger, more organized fighting force than the Viet Cong, which tended to strike by surprise then disappear into the jungle. NVA were also known as PAVN, People's Army of Vietnam.

OH-6 - See scout

OH-13- See scout

OH-58 - See scout

OinC - Navy term, Officer in Charge

OJT - On-the-job training

P-38 - Small, simple folded metal can opener found in a box of C-rations, known for its' reliability to consistently do the job, often worn on dog tag chains.

Pax - Slang for "passenger" used by some helicopter crews

Peter Pilot - Slang for co-pilot, a learning stage before promotion to AC or Aircraft Commander

Point - The lead man in a ground unit, usually moving in single file, preventing grouping to make it difficult for the enemy to hit more than one man at a time. The point man was most vulnerable, most likely to be first exposed to enemy fire, most likely to be hit when the shooting started. Point was typically rotated to spread the risk, but some crazy guys actually preferred walking point, preferred relying on themselves alone on point.

POL - Petroleum, oil and lubricants, a helicopter refuel point

PRC -25 - The hand-held radio used by ground troops, called "Prick-25"

Probe - An enemy testing our defenses by provoking a response

PSP - Perforated Steel Plating, steel interlocking forms used to lay over the ground to form a runway

PTSD - Post Traumatic Stress Disorder, an umbrella definition for suppressed bad memories which manifest themselves in unexpected ways as one gets older, like in their 50s

Pucker Factor - Term used by helicopter pilots to measure the degree of danger or fear, referring to their butt-cheeks clenching the seat.

PX - Post Exchange, US Army term for on-base store selling non-food items. See Commissary.

PZ - Pickup Zone, same as LZ except slang for extraction instead of insertion of troops

Radios - Helicopter pilots monitored 3 different radios at once: UHF for air-to-air contact with other aircraft, VHF for air-to-ground such as the control tower at an airfield, and FM for ground troop contact. In hot situations all three radios might be chattering and at the same time the aircraft crew yelling on the intercom, and a pilot had to be quick to keep it all straight.

Ready Reaction Force - Forces on standby to reinforce troops in the field in contact with the enemy. See Cav

Revetment - An L-shaped, chest-high area to park a helicopter, made of sand bags or other material to absorb shrapnel from explosives. It protected at least one side of the helicopter.

ROKs - Republic of Korea ground troops, very tough and ruthless with the enemy

RPG - Rocket propelled grenade (small rocket launcher)

R&R - Rest & Recuperation. US troops in Vietnam had two weeks of R&R they could use in places like Hawaii, Australia, Hong Kong or Thailand. Our enemy had their own R&R spots in Cambodia near the Ho Chi Minh Trail and in Vietnam, notably IV Corps south of Saigon.

RTO - Radio Telephone Operator who carried the PRC-25 radio, always by the CO's side to keep him in contact with higher-ups, artillery, air support, other ground units

Ruck, rucksack - A soldier's backpack containing food, clothes, gear etc.

Ruff-Puffs - RF/PF: Regional Forces and Popular Forces of South Vietnam, not unlike our National Guard, armed and operating as security forces close to their homes. Not very effective against main-line enemy units.

Satchel charge - Explosives in canvas case

Sampan - Small boat used in rivers and canals

Scout helicopter - Scout pilots had a dangerous job of flying low and slow to search for trails, campfires, food caches and other signs of the enemy, and to draw enemy fire to expose their position. Scouts were OH-13 Bell helicopters, like in the TV show MASH early in the war, OH-58 Bell Jet Rangers late in the war, but mostly Scouts were known to fly the OH-6 Cayuse built by Hughes, known as the "Loach," a slang derivative from LOH for Light Observation Helicopter.

Sit-rep - Situation report, summarize what the hell is going on

Six – directly behind a man, since his 12 o'clock is straight ahead. Watching a man's six is watching his back, covering him. Also, any radio call-sign ending in "six" denoted the commanding officer of that unit.

SKS - Russian semi-automatic carbine

Slack – 2nd man in a patrol, behind the point man and watching the point's six and every other direction, looking out front as point looked down for enemy signs. Slack gave point adjustments to keep on right heading.

Slick - UH-1 Huey helicopter used to move troops and materiel, called "slick" because they were not outfitted as gunships, but were "slick" on the sides.

Sniffer - On a "sniffer" mission a Huey helicopter would fly a grid pattern low over the treetops trailing a device on a wire, lowered to just over the trees. The device was connected by wire to a machine in the helicopter that detected certain chemicals, like ammonia, the dominate component of urine. Positive readings might mean a concentration of humans, or it might mean a concentration of frogs or monkeys. Sniffer missions were used to detect enemy forces hidden by jungle.

Snoopy, Spooky, Puff - AC-47 camouflage-painted cargo plane, with miniguns lined up on one side, each one able to fire 4,000 rounds per minute, 66 per second, every 5th round a tracer, aimed by careful banking of the aircraft in a precise turn that kept fire on target. Called Snoopy, Spooky and Puff the Magic Dragon because the tracer fire from aircraft to ground curved with flight and looked like a dragon's fiery breath. Snoopy was called in extreme circumstances. Succeeded by AC-130.

SOG - Studies and Observations Group (SOG) was a special projects cover name to conceal their real purpose, which was unconventional warfare through clandestine missions to unmentionable places. They were small teams of very bold, highly trained special ops men, controlled directly out of MACV.

Tiger Scout - Former NVA or VC enemy soldier, now a scout for the good guys, same as a Kit Carson Scout

TOC - Tactical Operations Center

UHF guard - 243.0, a UHF known to and monitored by all pilots because it was used for emergency broadcasts for things like a Mayday call when shot down to announce the location to aid rescue, and to announce warnings to pilots of a B-52 strike.

VC - Viet Cong, the enemy locals in South Vietnam working to help the NVA overthrow the south and install a communist government

WIA - Wounded in action

WO, WO1 – Warrant Officer, typically helicopter pilot during Vietnam War. These were non-commissioned officers given officer status as warranted by their high level of responsibility and need for limited command authority to get things done. The WO was the first grade for a new Warrant Officer, referred to in slang as *Wobbly One*. See CW2.

XO - Executive Officer, 2nd in command under CO

CPSIA information can be obtained at www.ICGtesting.com
Printed in the USA
LVOW08*0810020416

481893LV00002B/12/P